IET MATERIALS, CIRCUITS AND DEVICES SERIES 53

VLSI Architectures for Future Video Coding

Other volumes in this series:

Volume 2 **Analogue IC Design: The current-mode approach** C. Toumazou, F.J. Lidgey and D.G. Haigh (Editors)
Volume 3 **Analogue–Digital ASICs: Circuit techniques, design tools and applications** R.S. Soin, F. Maloberti and J. France (Editors)
Volume 4 **Algorithmic and Knowledge-Based CAD for VLSI** G.E. Taylor and G. Russell (Editors)
Volume 5 **Switched Currents: An analogue technique for digital technology** C. Toumazou, J.B.C. Hughes and N.C. Battersby (Editors)
Volume 6 **High-Frequency Circuit Engineering** F. Nibler *et al.*
Volume 8 **Low-Power High-frequency Microelectronics: A unified approach** G. Machado (Editor)
Volume 9 **VLSI Testing: Digital and mixed analogue/digital techniques** S.L. Hurst
Volume 10 **Distributed Feedback Semiconductor Lasers** J.E. Carroll, J.E.A. Whiteaway and R.G.S. Plumb
Volume 11 **Selected Topics in Advanced Solid State and Fibre Optic Sensors** S.M. Vaezi-Nejad (Editor)
Volume 12 **Strained Silicon Heterostructures: Materials and devices** C.K. Maiti, N.B. Chakrabarti and S.K. Ray
Volume 13 **RFIC and MMIC Design and Technology** I.D. Robertson and S. Lucyzyn (Editors)
Volume 14 **Design of High Frequency Integrated Analogue Filters** Y. Sun (Editor)
Volume 15 **Foundations of Digital Signal Processing: Theory, algorithms and hardware design** P. Gaydecki
Volume 16 **Wireless Communications Circuits and Systems** Y. Sun (Editor)
Volume 17 **The Switching Function: Analysis of power electronic circuits** C. Marouchos
Volume 18 **System on Chip: Next generation electronics** B. Al-Hashimi (Editor)
Volume 19 **Test and Diagnosis of Analogue, Mixed-Signal and RF Integrated Circuits: The system on chip approach** Y. Sun (Editor)
Volume 20 **Low Power and Low Voltage Circuit Design with the FGMOS Transistor** E. Rodriguez-Villegas
Volume 21 **Technology Computer Aided Design for Si, SiGe and GaAs Integrated Circuits** C.K. Maiti and G.A. Armstrong
Volume 22 **Nanotechnologies** M. Wautelet *et al.*
Volume 23 **Understandable Electric Circuits** M. Wang
Volume 24 **Fundamentals of Electromagnetic Levitation: Engineering sustainability through efficiency** A.J. Sangster
Volume 25 **Optical MEMS for Chemical Analysis and Biomedicine** H. Jiang (Editor)
Volume 26 **High Speed Data Converters** A.M.A. Ali
Volume 27 **Nano-Scaled Semiconductor Devices** E.A. Gutiérrez-D (Editor)
Volume 29 **Nano-CMOS and Post-CMOS Electronics: Devices and modelling** Saraju P. Mohanty and Ashok Srivastava
Volume 30 **Nano-CMOS and Post-CMOS Electronics: Circuits and Design** Saraju P. Mohanty and Ashok Srivastava
Volume 32 **Oscillator Circuits: Frontiers in design, analysis and applications** Y. Nishio (Editor)
Volume 33 **High Frequency MOSFET Gate Drivers** Z. Zhang and Y. Liu
Volume 34 **RF and Microwave Module Level Design and Integration** M. Almalkawi
Volume 35 **Design of Terahertz CMOS Integrated Circuits for High-Speed Wireless Communication** M. Fujishima and S. Amakawa
Volume 38 **System Design With Memristor Technologies** L. Guckert and E.E. Swartzlander Jr.
Volume 39 **Functionality-Enhanced Devices: An alternative to Moore's law** P.-E. Gaillardon (Editor)
Volume 40 **Digitally Enhanced Mixed Signal Systems** C. Jabbour, P. Desgreys and D. Dallett (Editors)
Volume 43 **Negative Group Delay Devices: From concepts to applications** B. Ravelo (Editor)
Volume 47 **Understandable Electric Circuits: Key concepts, 2nd Edition** M. Wang

Volume 58 **Magnetorheological Materials and Their Applications** S. Choi and W. Li (Editors)
Volume 60 **IP Core Protection and Hardware-Assisted Security for Consumer Electronics** A. Sengupta and S. Mohanty
Volume 62 **VLSI and Post-CMOS Devices, Circuits and Modelling** R. Dhiman and R. Chandel (Editors)
Volume 68 **High Quality Liquid Crystal Displays and Smart Devices, vol. 1 and vol. 2** S. Ishihara, S. Kobayashi and Y. Ukai (Editors)
Volume 69 **Fibre Bragg Gratings in Harsh and Space Environments: Principles and applications** B. Aïssa, E.I. Haddad, R.V. Kruzelecky, W.R. Jamroz
Volume 70 **Self-Healing Materials: From fundamental concepts to advanced space and electronics applications, 2nd Edition** B. Aïssa, E.I. Haddad, R.V. Kruzelecky, W.R. Jamroz
Volume 71 **Radio Frequency and Microwave Power Amplifiers, vol. 1 and vol. 2** A. Grebennikov (Editor)

VLSI Architectures for Future Video Coding

Edited by
Maurizio Martina

The Institution of Engineering and Technology

Published by The Institution of Engineering and Technology, London, United Kingdom

The Institution of Engineering and Technology is registered as a Charity in England & Wales (no. 211014) and Scotland (no. SC038698).

First published 2019

The Institution of Engineering and Technology
Michael Faraday House
Six Hills Way, Stevenage
Herts, SG1 2AY, United Kingdom

www.theiet.org

British Library Cataloguing in Publication Data
A catalogue record for this product is available from the British Library

ISBN 978-1-78561-710-2 (hardback)
ISBN 978-1-78561-711-9 (PDF)

Typeset in India by MPS Limited

Contents

Preface

Recently, video coding has become pervasive due to the availability of devices able to act as video co-decoders. Stemming from the MPEG2 standard, several new algorithms and techniques have been proposed to improve the coding efficiency of modern video-coding systems. These techniques include different parts of the compression/decompression chain, such as transform stages, entropy coding, motion estimation, intra-prediction and filtering. For these reasons, in the last 20 years, video compression started facing new challenges. High-quality/High-resolution sequences, virtual reality and immersive gaming have pushed video-compression standards toward new directions. In 2013, the high-efficiency video-coding (HEVC) standard, jointly developed by Video Coding Experts Group (VCEG) and Moving Picture Experts Group (MPEG) in a Joint Collaborative Team on Video Coding (JCT-VC), has been published. HEVC is able to achieve twice better compression efficiency compared to the Advanced Video Coding (AVC) standard. Namely, the same subjective video quality is achieved at twice lower bitrate compared to the high profile of AVC. Thus, with the HEVC standard, the computational complexity has become really impressive. Moreover, one of the key objectives of HEVC was to focus on high-resolution video coding. All these aspects brought to new challenges from the implementation point of view, fostering researchers to find new approaches and paradigms to reach efficient memory handling, low-power consumption and high parallelization. As a consequence, new highly optimized Very Large Scale Integration (VLSI) architectures were proposed in the literature showing that these results can be obtained by combining deep knowledge of implementation aspects with a good background in video-coding fundamentals. One of the effects of this cross-layer approach has been to bring the approximate computing paradigm as a promising solution for future video coding VLSI architectures. Concurrently, the JCT-VC started working on the next video-compression standard: versatile video coding (VVC) which is expected to be finalized in 2020, thus opening a new scenario for hardware implementation.

This book tries to cover the main aspects related to the implementation of VLSI architectures tailored to future video coding. For this reason, the book addresses several hardware-implementation-related topics, including scalable transform architectures, joint algorithm-architecture design, high throughput architectures and low-power techniques for high-resolution video coding, 3D video coding and frame memory compression. Moreover, the books also include important examples and cases of study, by showing how the approximate computing paradigm can be exploited to reduce the complexity of the transform stage and to design reconfigurable and multifrequency architectures. Finally, the book provides some future perspectives by presenting new tools and algorithms, which are likely to be part of the forthcoming VVC standard.

Maurizio Martina

Chapter 1

Scalable transform architectures for video coding

Sonda Ben Jdidia[1], Maher Jridi[2], Pramod Kumar Meher[3], Nouri Masmoudi[1] and Ayman Al Falou[2]

In spite of the recent advances in telecommunication standards, communication networks still have limited bandwidths and storage capacity. Therefore, video compression has drawn increasing importance since high-resolution video contents have become more and more used in various fields. These requirements raise the need for high-performance video-compression technologies able to reduce the amount of data to be transmitted or stored by compressing the input video signal into a bitstream file. Improving the coding efficiency was always one of the crucial issues of various compression standards that aim to get the most compact representation of the reconstructed video, with a high subjective quality. The high-efficiency video coding (HEVC) comes to respond to these requirements. However, the increased consumption of high-quality multimedia content has pushed the international communication companies to put much effort to better enhance video-coding techniques. In this perspective, an upcoming video-coding standard to be known as versatile video coding (VVC) has emerged aiming to improve the coding efficiency of the current HEVC codec. Improvements on rate distortion (RD) performance that came with both HEVC and VVC have brought an increased complexity in the majority of the coding modules, which makes it difficult to implement on hardware systems with real-time encoding. This chapter focuses on the transform coding stage as one of the most computationally demanding modules. For HEVC, efficient approximation algorithms in addition to reconfigurable and scalable architecture have been developed in order to decrease the computational complexity of the transform module. The main objectives are to meet low power and real-time processing constraints while maintaining a compression gain and a satisfying video quality. Field programmable gate array (FPGA) implementation results and comparisons with existing works confirm the efficiency of the proposed approximations since they contribute in reducing time and power consumption, optimizing the hardware resources and bringing peak signal-to noise ratio (PSNR)

[1]Université de Sfax, Ecole nationale d'Ingénieurs de Sfax (ENIS), Laboratoire d'Electronique et de Technologie de l'Information (LETI), Sfax, Tunisie
[2]ISEN Yncrea Ouest, L@bISEN, Brest, France
[3]C V Raman College of Engineering, Bhubaneswar, India

improvement as well. Similarly, for the adaptive multiple transform (AMT) introduced in the transform module of the VVC, approximations were done for discrete cosine transform (DCT)-II and discrete sine transform (DST)-VII transforms since they are statistically the most used ones among the five predefined types. Bitrate (BR) reduction with a slightly degradation of video quality and a less use of hardware resources are the main contributions of the proposed approximations.

1.1 Introduction

Video-coding efficiency is strongly related to the ability to provide the maximum of compression ratio while conserving good-quality decoded videos, as well as the capability to get real-time video broadcasting and minimum of energy consumption [1]. Nowadays, various video services are increasingly handling HD (high-definition) and ultra HD resolutions such as television broadcasting, video surveillance, telemedicine, streaming video, Internet Protocol Television. Therefore, broadcasting, displaying, transmitting or storing video contents with high quality would be impossible without video compression.

The latest video compression standard H.265/HEVC approved in April 2013 [2] was the outcome of common efforts between both the international organization, namely, the International Telecommunication Union-Telecommunication (ITU-T) video-coding experts group and the Standardization/International Electrotechnical Commission (ISO/IEC) moving picture experts group (MPEG) working jointly under a Joint Collaboration Team on Video Coding (JCT-VC), established in January 2010 and dedicated to the exploration of coding technologies [3]. Many users are extensively considering this codec since it brings 50% of BR gain while keeping the same video quality of its predecessor H.264/AVC (advanced video coding) [4]. The saving brought by this codec with respect to its predecessor H.264/AVC is mainly related to the exploitation of new features and tools such as additional intra-prediction and inter-prediction modes, larger transform unit sizes and new structures for picture partitioning [5]. The increased complexity that comes with this standard is up to 40%–70% compared to AVC [6,7]. Therefore, the challenge to design lower power and hardware-efficient architecture becomes more crucial especially for video services that are limited by a storage capacity and a battery lifetime such as mobile devices. As the transform stage brings huge complexity because of the amount of hardware resources used in computation and because it consumes most of the power and computation time [8], the optimization of this module will be highly recommended to get low-power video applications. Transform optimization is more supported since profiling results of different classes of the HM (HEVC test model) reference software have shown that the time required for the transformation and quantization modules represents about 25% of the total time [9]. To carry on investigating the increased demand for permanent improvements in video compression and to serve the need for recent video applications with higher resolutions (spatial resolution up to ($4K \times 2K$, $8K \times 4K$)) [10], efforts have been carried out targeting to ameliorate the subjective video quality, get less bandwidth requirement and minimize the computational complexity of HEVC codec [11]. In this perspective, a new group called Joint Video Exploration Team (JVET) was created at the 13th MPEG meeting in Geneva in October 2015 by ITU-T and ISO/IEC organizations supporting the exploration of beyond HEVC video codec [11]. This group was renamed to Joint Video Experts Team investigating on the new standard proposal to be known as VVC from April 2018 [10]. It is planned to be complete before the close of 2020 [12]. The new standard proposal is expected to reduce the BR by half while keeping the same visual quality and maintaining the computational complexity when compared to HEVC [11]. This group has developed a new reference software called Joint Exploration test Model (JEM) based on HEVC reference HM16.6 [11] dedicated to the generation of new technical contributions. Recently, the reference software HM-16.6-JEM-5.0[1] provides about 29%–35% better coding gain than HEVC to the detriment of higher complexity (up to $\times 12$ and $\times 10$), respectively, at the encoder and the decoder sides [13]. This complexity is caused by the improvements cognized per almost all of the coding stages particularly the transform module where additional transform types and larger transform unit sizes have improved this module [14]. Transform stage optimization is more needed in this case.

This chapter focuses on reducing the complexity of the transform module by means of the approximation approach. This type of computation relies on the aptitude of different applications to tolerate some quality loss of the resulted videos [15]. In this perspective, various DCT optimization algorithms using approximate computing are proposed in the literature. In this chapter, a comparison will be done with the works below. A signed DCT (SDCT), proposed in [16], is obtained by the application of the signum function to the elements of the DCT. The proposed transform involves only addition and subtraction operations since the elements of the SDCT can be either 0 or ± 1. When applied to 8×8 blocks, results show that the SDCT maintains the decorrelation and energy compaction of the DCT and performs closely to the DCT in image compression. Bouguezel–Ahmad–Swamy (BAS) [17] have proposed an orthogonal multiplication-free algorithm for transform of order $N > 4$ with low computational complexity and good performance when applied to image compression. Bayer *et al.* [18] have suggested an approximate orthogonal 16×16 transform matrix composed only of 0 and ± 1 coefficients. The proposed transform shows good results when implemented in hardware in addition to an improvement of PSNR when applied to image compression. Authors in [19] have developed a method that generates binary transforms exploiting the Walsh–Hadamard transform (WHT). Binary DCT (BDCT) and binary discrete Hartley transform (DHT) (BDHT) are obtained by applying the proposed method, respectively, to the DCT and the DHT. Results show that the BDCT represents a good approximation of the DCT and has the same advantages as the WHT, while the BDHT is able to approximate the DHT very well. Four different modes have been defined in [20] which are MODE0, MODE1, MODE2 and MODE3 indicating, respectively, the exact DCT computation, the first approximation, the second approximation and the WHT computation.

[1]More recent reference test models are developed but we have used the HM-16.6-JEM-5.0 for simulation experiments.

The contributions of this chapter are depicted as follows:

1. First, we propose a review of recent state-of-the-art approximate transforms used in HEVC as well as improvements for the transform module in VVC.
2. Second, we recall the method stated in [21] to get an approximation of DCT II of the VVC. Hardware architecture are derived accordingly.
3. Third, a new algorithm to approximate the DST-VII transform is proposed. The coding efficiency of the modified transform matrices of the VVC is evaluated by means of BR and PSNR criteria.

This chapter unfolds as follows: a review of scalable transforms in HEVC is given in Section 1.2. An overview of video-coding concept in VVC standard is presented in Section 1.3 where approximate algorithms for DCT-II and DST-VII are described along with experimental results. Finally, Section 1.4 concludes this chapter.

1.2 Review of scalable transforms in HEVC

1.2.1 Transform coding in HEVC

Transform coding allows removing the correlation from the residual signal considered as the difference between the input and the predicted signal. This stage is responsible for compacting the energy by converting the residual samples from the spatial to spectral domain [22]. This operation is performed by the DCT which is widely used due to its orthogonal structure and its close performance to the Karhunen–Loève Transform [23]. The majority of video-coding standards schemes exploits the DCT such as JPEG, MPEG-1, MPEG-2, H.261, H.263 and AVC/H.264 [18]. In HEVC, for square transform blocks of sizes 4×4, 8×8, 16×16 and 32×32, an integer form of DCT is used to encode inter-predicted blocks while the DST is used to encode the 4×4 luma intra-predicted residual blocks [1].

A two-dimensional (2D) transform is obtained by applying the one-dimensional (1D) transform, respectively, on the rows and the columns of the transform block. The coefficients of the transform matrix are obtained by calculating the transform basis functions. They are multiplied by a scaling factor and rounded to get the transform coefficients [5]. Basis function of N-point DCT is computed in order to obtain the N elements $c(i,j)$ DCT matrix C_N. These transform coefficients are defined according to the following equation:

$$C(i,j) = \varepsilon_i \times \sqrt{\frac{2}{N}} \times \cos\left(\frac{(2j+1)i\pi}{2N}\right) \quad \forall\ 0 \le i,\ j \le N-1$$
$$\text{with } \varepsilon_i = \begin{cases} \dfrac{1}{\sqrt{2}}, & \text{if } i = 0 \\ 1, & \text{if } i > 0 \end{cases} \tag{1.1}$$

where i and j correspond, respectively, to the row and column indexes of the transform matrix and ε_i represents the scale factor serving to normalize the transform matrix.

The general equation responsible for computing the N-point 1D transform is

$$X = C_N \cdot x \tag{1.2}$$

where $x[x_0, \ldots, x_N]$ represents the residual input vector, C_N is the N-point transform matrix and $X[X_0, \ldots, X_N]$ is the corresponding output vector.

As for the coefficients of an N-point DST-VII transform matrix, they are computed through the following equation:

$$S(i,j) = \sqrt{\frac{4}{2N+1}} \times \sin\left(\frac{\pi(2i+1)(j+1)}{2N+1}\right) \quad \forall \quad 0 \le i,j \le N-1 \tag{1.3}$$

where i and j are, respectively, the row and column indexes of the transform matrix. The use of the 4×4 DST-VII does not bring much more computational complexity; therefore, more interest is devoted in reducing the complexity that comes with the DCT-II transform matrices of the HEVC standard.

1.2.2 Approximate DCT algorithms and their hardware architecture for HEVC

The high arithmetic complexity brought by exact DCT is related to the substantial amount of addition and multiplication operations that consume most of the power and computation time. Therefore, to get fast algorithms, matrix approximations are required especially for higher length DCTs. Three features are considered while approximating DCTs. The first one concerns obtaining low computational complexity. The second one is about achieving a compression performance comparable to the exact DCT by getting an approximation scheme with low error energy and the last one is related to its ability to work for DCTs of higher length.

1.2.2.1 Recursive sparse matrix decomposition

Algorithm description

Taking advantage of the even/odd symmetry property of the rows vectors of the DCT matrix, C_N can be decomposed into a matrix product according to the following equation:

$$C_N = \frac{1}{\sqrt{2}} \times M_N^{per} \times T_N \times M_N^{add} \tag{1.4}$$

where T_N represents a block sparse matrix expressed as follows:

$$T_N = \begin{bmatrix} C_{N/2} & 0_{N/2} \\ 0_{N/2} & S_{N/2} \end{bmatrix} \tag{1.5}$$

where $0_{N/2}$ is the $((N/2) \times (N/2))$ zero matrix and the block submatrix $S_{N/2}$ constitutes the odd rows of the first $N/2$ columns of $\sqrt{2}C_N$. The permutation matrix M_N^{per} is defined as

$$M_N^{per} = \begin{bmatrix} P_{N-1,N/2} & 0_{1,N/2} \\ 0_{1,N/2} & S_{N-1,N/2} \end{bmatrix} \tag{1.6}$$

where $0_{1,N/2}$ is a row of $N/2$ zeros and $P_{N-1,N/2}$ is a $((N-1) \times (N/2))$ matrix defined as follows:

$$P^{(i)}_{N-1,N/2} = \begin{cases} 0_{1,N/2}, & \text{if } i = 1,3,5,\ldots,N-1 \\ I_{N/2(i/2)}, & \text{if } i = 0,2,4,\ldots,N-2 \end{cases} \quad (1.7)$$

where $I_{N/2(i/2)}$ corresponds to the $(i/2)$th row vector of the $((N/2) \times (N/2))$ identity matrix. As for M_N^{add}, it is represented as follows:

$$M_N^{add} = \begin{bmatrix} I_{N/2} & J_{N/2} \\ I_{N/2} & -J_{N/2} \end{bmatrix} \quad (1.8)$$

where $J_{N/2}$ is an $((N/2) \times (N/2))$ anti-diagonal with all ones values matrix and zeros elsewhere.

As the DCT matrix is decomposed as the product of three matrices as mentioned in (1.4), then the computational complexity reduction of C_N is related to reducing the computational cost of these three matrices. M_N^{per} serves as a permutation matrix which does not require any arithmetic operation. M_N^{add} does not involve much computational complexity since it requires only N signed additions. Hence, reducing the arithmetic cost of N-point DCT with the use of the approximation approach depends on simplifying the T_N matrix by approximating its submatrices.

To find $\hat{C}_{N/2}$ and $\hat{S}_{N/2}$ which designate, respectively, the approximation matrices of $C_{N/2}$ and $S_{N/2}$, the smallest size of DCT is considered because it allows the computation of DCT of higher lengths. Since the implementation of size 2 and 4 DCT requires only adders, the approximation procedure is applied for transforms of size N where $N \geq 8$. In our case, the 8-point approximate transform is considered as the basis block responsible for the computation of higher length DCTs. The approximation of C_8 is done according to the method used in [24] since it provides the best trade-off between the arithmetic complexity and the quality of the reconstructed image. This method approximate C_8 by $\hat{C}_8 = \lfloor 2C_8 \rceil$ where $\lfloor . \rceil$ represents the rounding-off operation. The signal flow graph (SFG) of $\hat{C}_8$ is shown in Figure 1.1. The output sequence $\{X_n\}, n \in [0,7]$ is obtained by $X = \hat{C}_8 \times x^t$, where $x = \{x_0, \ldots, x_7\}$ represents the input sequence (residuals).

To get $\hat{S}_8$, the idea consists in approximating the submatrix S_8 by $\hat{C}_8$. Such approximation benefits from the good compression performance that come with $\hat{C}_8$. It allows getting orthogonal transform matrix, since $\hat{C}_8$ is orthogonalizable, which can be represented by a regular signal-flow graph. It also contributes in an efficient hardware implementation since it exploits scalability and reconfigurable properties. Therefore, when generalizing this approximation approach for N-order transforms, the equation of the approximate N-point DCT-II transform is presented as

$$\hat{C}_N = \frac{1}{\sqrt{2}} \times M_N^{per} \begin{bmatrix} \hat{C}_{N/2} & 0_{N/2} \\ 0_{N/2} & \hat{C}_{N/2} \end{bmatrix} M_N^{add} \quad (1.9)$$

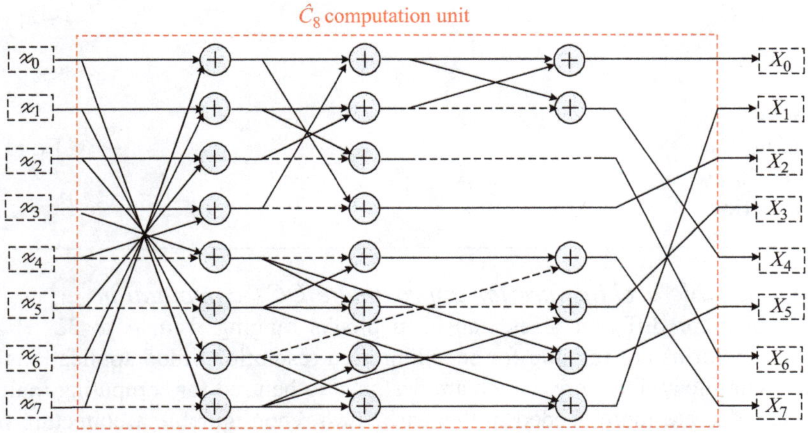

Figure 1.1 Signal flow graph (SFG) of $\hat{C}_8$. Dashed arrows represent multiplications by -1

The following equation of D_N can be calculated according to (1.10) since the matrix $\hat{C}_N$ is orthogonalizable:

$$D_N = \sqrt{(\hat{C}_N \times (\hat{C}_N)^t)^{-1}} \tag{1.10}$$

where $(.)^t$ indicates matrix transposition. D_N serves to compute $C_N^{orth} = D_N \times \hat{C}_N$ which can be used in data compression instead of $\hat{C}_N$ since $(C_N^{orth})^{-1} = (C_N^{orth})^t$. C_N^{orth} brings the same computational cost as $\hat{C}_N$ because the diagonal matrix D_N can be incorporated into the scaling in the quantization stage. The term $1/\sqrt{2}$ is integrated into the quantization step in such a way that we get a multiplierless architecture. The computation of the approximate DCT matrices for sizes $N \geq 8$ is done according to Algorithm 1.

Algorithm 1 Generation of proposed DCT matrix [25]

---

function PROPOSED DCT(N) ▷ N power of 2, $N \geq 8$

$N_0 \leftarrow \log_2 \frac{N}{8}$ ▷ N_0 is the number of 8-sample blocks

$\hat{C}_{\frac{N}{2^{N_0}}} \leftarrow \lfloor 2C_8 \rfloor$

while $N_0 > 0$ **do**

$\tilde{N} \leftarrow \dfrac{N}{2^{N_0-1}}$

Calculate $M_{\tilde{N}}^{per}$, $M_{\tilde{N}}^{add}$ ▷ (1.6), (1.8)

Calculate $\hat{C}_{\tilde{N}}$ ▷ Eq (1.9)

$N_0 \leftarrow N_0 - 1$

end while
Calculate D_N ▷ Eq (1.10)
return $\hat{C}_N$,D_N
end function

Reconfigurable architecture for approximate DCT computation
HEVC opts for different sizes transform blocks ranging from 4 to 32. Higher length transforms are suitable for encoding high-resolution video applications [5]. Reconfigurable architecture, which are designed to be used for computing multiple DCT lengths, are more needed in this case. Two reconfigurable architecture have been designed to compute the proposed approximate DCTs. The first one regroups sizes 8 and 16 while the second architecture allows computing 8-, 16- and 32-point DCTs. The reconfigurable architecture, shown in Figure 1.2, is designed to compute a 16-point approximate DCT or two 8-point approximate DCT in parallel. It includes five essential units that consist of a 16-point input adder unit (IAU) whose function is depicted in (1.8), a 16/8-point computation selection unit, two 8-point approximate DCT units and an output permutation unit (OPU) whose function is depicted in (1.6). The input adder generates $a(i)$ and $b(i), i \in [1:7]$ which correspond, respectively, to the sum and the difference of the input signals $x_i, i \in [1:7]$. These signals are fed to a 16/8-point computation selection unit composed of 16 MUXes. The first 8 MUXes choose between $[a(0), a(1), \ldots, a(7)]$ or $[x_0, x_1, \ldots, x_7]$ to be the inputs of the first 8-point DCT unit according to whether it is used for computing a 16-point DCT or 8-point DCT. The remaining MUXes select between $[b(0), b(1), \ldots, b(7)]$ or $[x_8, x_9, \ldots, x_{15}]$ to be fed to the second 8-point DCT unit according to whether it is a 16-point DCT or 8-point DCT computation. The 14 MUXes of the OPU serve to select and reorder the transformed coefficients based on the size of the selected DCT. The control signal $Sel16$ considered as input of the MUXes selects the inputs and performs the permutation of the outputs depending on the choice of the transform size. When $Sel16 = 1$, the computation of 16-point DCT is allowed and, when $Sel16 = 0$, the computation of a pair of 8-point DCT in parallel is enabled.

The area efficient architecture is designed as depicted in Figure 1.3. The proposed reconfigurable architecture is supplied for 8-, 16- and 32-point DCT computation. It is realized using 32-point IAU, two 16-point IAUs, four 8-point DCT units along with three control blocks containing 64 2:1 MUXes and 30 3:1 MUXes. The first control block indicates whether 32-point DCT or lower length will be computed. The decision is made through the selection input signal $Sel32$. When $Sel32 = 1$, the input data correspond to those of the 32-point DCT, otherwise, inputs of lower length DCT are considered. The second control block designates whether the DCT size is higher than 8 through the selection input signal $Sel16$. When $Sel16 = 1$, DCT of higher than 8 (size 16 and 32) is allowed; otherwise,

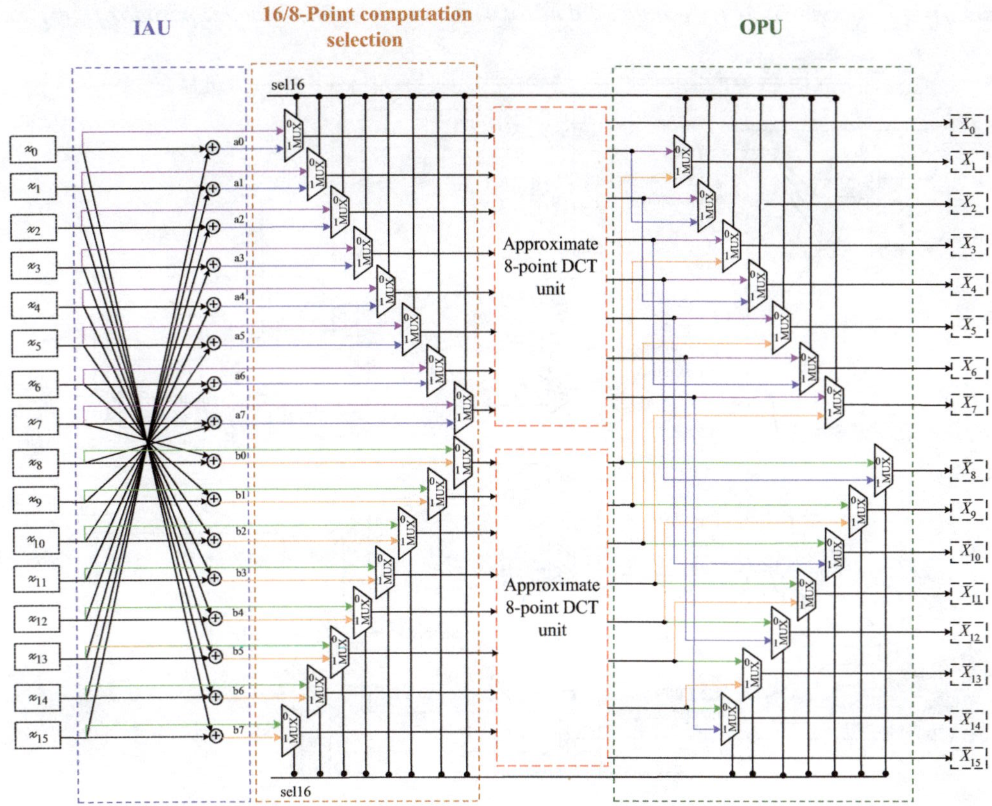

Figure 1.2 Reconfigurable architecture for approximate computing of 8- and 16-point DCT

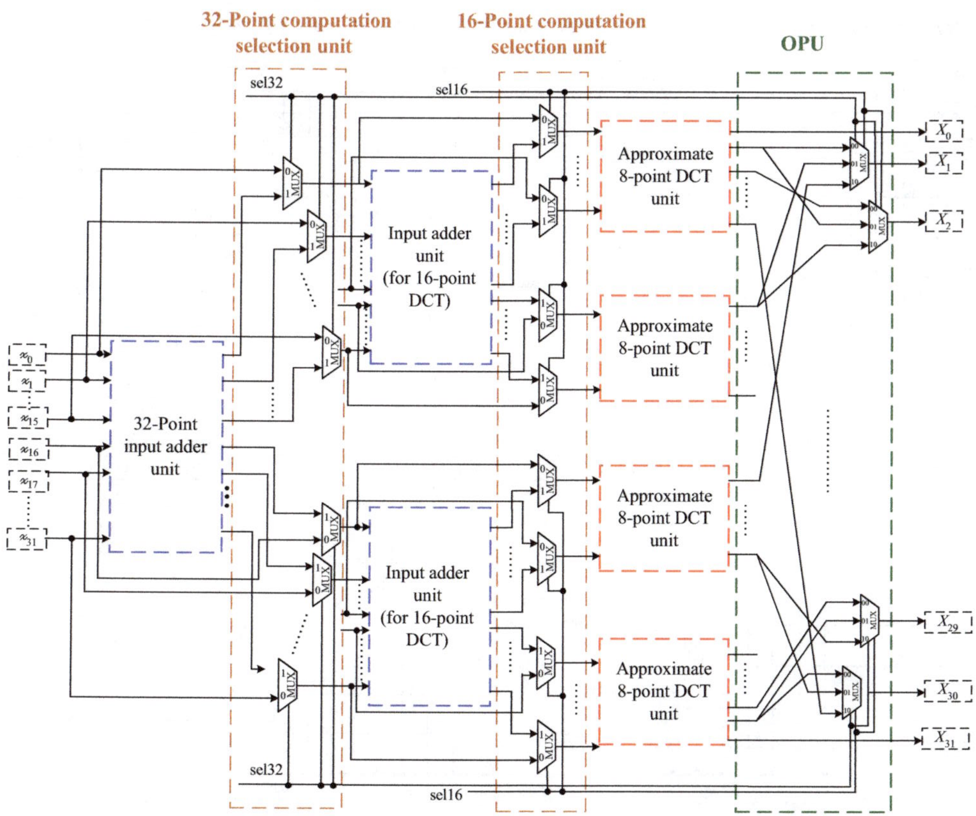

Figure 1.3 Proposed reconfigurable architecture for approximate DCT of lengths N = 8, 16 and 32

the DCT length is 8. The third control block ensures output reordering according to the choice of the size of the DCT. The control signals of the 3:1 MUXes, which are *Sel*16 and *Sel*32, indicate whether the 32 outputs are furnished by four 8-point parallel DCTs (when $\{Sel32, Sel16\}_2$ are equal to $\{00\}$), two parallel 16-point DCTs (when $\{Sel32, Sel16\}_2$ are equal to $\{01\}$) or 32-point DCT (when $\{Sel32, Sel16\}_2$ are equal to $(\{11\})$.

1.2.2.2 Sparse matrix decomposition based on a new 4-point DCT approximation

Algorithm description

The recursive sparse matrix factorization derived by (1.9) is recalled in order to get approximate transforms. This decomposition is applied to any power-of-2 length DCTs higher than 4, but this study is restricted to size 8, 16 and 32 transforms. For the smallest DCT kernel, which is size 4 transform, the transform matrix is given as

$$C_4 = \begin{bmatrix} 64 & 64 & 64 & 64 \\ 83 & 36 & -36 & -83 \\ 64 & -64 & -64 & 64 \\ 36 & -83 & 83 & -36 \end{bmatrix} \tag{1.11}$$

We apply the following equation to the elements of the original DCT matrix to get the approximate matrix:

$$\hat{C}_4(i,j) = 32 \times R\left(\frac{C_4(i,j) + 32 \times \mathrm{sgn}(C_4(i,j))}{64}\right) \quad \forall\ i,j \tag{1.12}$$

where $R(.)$ denotes the rounding operation and sgn(.) refers to the sign function. When applying this equation to the original 4-point DCT, the obtained approximate matrix is given by

$$\hat{C}_4 = \begin{bmatrix} 64 & 64 & 64 & 64 \\ 64 & 32 & -32 & -64 \\ 64 & -64 & -64 & 64 \\ 32 & -64 & 64 & -32 \end{bmatrix} \tag{1.13}$$

Merits of considering $\hat{C}_4$ as an approximation of C_4 consist in the orthogonalization property of the approximate matrix in addition to its closeness to the original one. It is composed of coefficients that are power-of-two integers, which are replaced only by shift operations when realizing multiplications in hardware.

Signal flow graph of 4-point approximate DCT

Size 4 approximate transform is considered as the basis computational block since it will be reused to get higher length transforms. Therefore, it seems important to have an SFG with a reduced number of addition and shift operations.

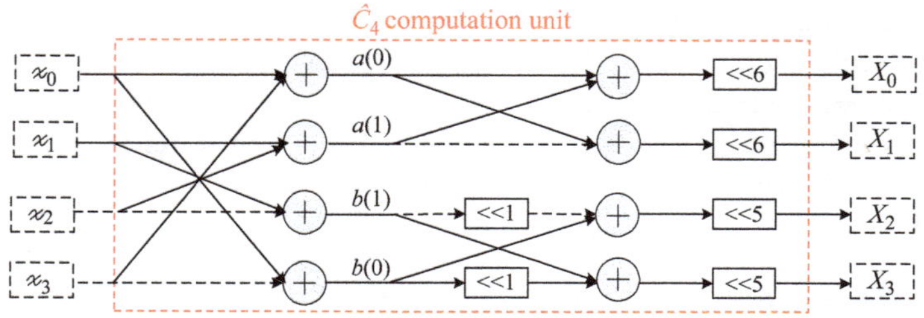

Figure 1.4 Architecture for 4-point integer 1D DCT approximate computing

For an input sequence $\{x_0, x_1, x_2, x_3\}$, the corresponding outputs $\{X_0, X_1, X_2, X_3\}$ can be written as

$$\begin{bmatrix} X_0 \\ X_2 \end{bmatrix} = 64 \times \begin{bmatrix} 1 & 1 \\ 1 & -1 \end{bmatrix} \times \begin{bmatrix} x_0 + x_3 \\ x_1 + x_2 \end{bmatrix} \tag{1.14}$$

$$\begin{bmatrix} X_1 \\ X_3 \end{bmatrix} = 32 \times \begin{bmatrix} 2 & 1 \\ 1 & -2 \end{bmatrix} \times \begin{bmatrix} x_0 - x_3 \\ x_1 - x_2 \end{bmatrix} \tag{1.15}$$

The proposed architecture is shown in Figure 1.4 where the dashed arrows indicates multiplications par (-1), $a(0)$ and $b(0)$ represent, respectively, the sum and the difference between x_0 and x_3, while $a(1)$ and $b(1)$ correspond, respectively, to the sum and the difference between x_1 and x_2.

Higher lengths approximate DCT architecture: case of 8-point DCT computation

Two possible architecture to implement 8-point DCT are depicted in Figure 1.5. First, the parallel architecture, represented in Figure 1.5(a), involves a pair of 4-point DCT in addition to an IAU and an OPU. The IAU serves to generate the inputs calculated as sums of the pairs of pixel values for the first 4-point transform block while assigning the differences of those pairs of pixel values to the second transform block. As for the OPU, it is used to reorder the outputs. 16- and 32-point transforms can be designed, respectively, by a pair of 8-point DCT and a pair of 16-point DCT in addition to an IAU and an OPU for both.

The second architecture is shown in Figure 1.5(b) where the reconfigurable "input Add/Sub" unit allows computing either addition or subtraction of the pairs of pixel values according to the control-bit *EO* of the four reconfigurable adder/subtractor circuits. When $EO = 0$, the summation of the pairs of pixel values is realized to get the outputs with even indices $\{X_0, X_2, X_4, X_6\}$, and when $EO = 1$, we get outputs with odd indices $\{X_1, X_3, X_5, X_7\}$ after subtracting the pairs of pixel values.

1.2.3 Complexity analysis

The computational complexity brought by an *N*-point approximate DCT is evaluated through the number of arithmetic operations required while computing (generally, it

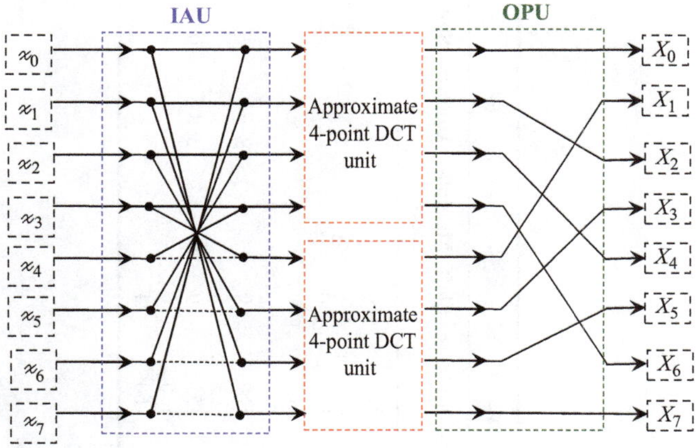

(a)

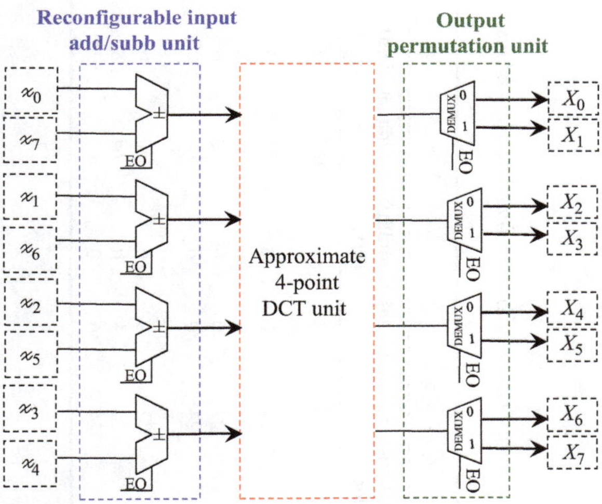

(b)

Figure 1.5 Approximate 8-point 1D DCT architecture: (a) parallel architecture for approximate 8-point 1D DCT and (b) area-constrained architecture for approximate 8-point 1D DCT

is equal to N^2 multiplications and $N \times (N-1)$ additions). The implementation of the transform by matrix multiplication results in high complexity due to the amount of additions and multiplications brought by the generated products. Several works opt for multiplierless transform-coding structure. Multipliers are replaced by adders and shifters since shift operation does not require any combinational components and uses only rewiring while implementing it in hardware [25]. However, shift-add operations increase the bit-width which increases the hardware complexity of the arithmetic units that follow the shift-add operation. Table 1.1 summarizes the number of

Table 1.1 Computational complexity of several 1D DCT approximation algorithms

Architecture	$N = 4$			$N = 8$			$N = 16$			$N = 32$			$N = 64$		
	Multiplication	**Addition**	**Shift**	**Multiplication**	**Addition**	**Shift**	**Multiplication**	**Addition**	**Shift**	**Multiplication**	**Addition**	**Shift**	**Multiplication**	**Addition**	**Shift**
Ref algorithm, [20] MODE0	4	8	2	22	28	4	86	100	8	342	372	16	–	–	–
SDCT [16]	–	–	–	0	24	0	0	64	0	0	160	0	–	–	–
BAS [17]	–	–	–	0	24	4	0	64	8	0	160	16	–	–	–
DCT of [18]	–	–	–	0	22	0	0	72	0	–	–	–	–	–	–
BDCT [19,20] MODE3	–	–	–	0	24	0	0	64	0	0	160	0	–	–	–
DCT of [25]	–	–	–	0	22	0	0	60	0	0	152	0	0	368	0
DCT of [26]	0	8	2	0	24	4	0	64	8	0	160	16	–	–	–
DCT of [21]	0	8	2	0	24	6	0	64	12	0	160	24	–	–	–

arithmetic operations brought by different approximation algorithms. Our proposed approximate algorithm of [25] has the lowest hardware cost since it involves the lowest number of additions and does not involve any shift operations when compared to all works stated in Table 1.1. Concerning our work of [26], no multiplication operations are used. It shares the same number of adders as that of [16,17,19] while it is slightly higher than that of [25]. As for our work proposed in [21], it has the highest algorithmic complexity among the works stated in Table 1.1. When compared to the reference algorithm that uses the DCT of the HM software (reference HEVC software) [27], which is based on matrices of [28], our algorithm provides less algorithmic complexity. This work has no additional addition operations with a slightly higher number of shifts for $N = 8, 16, 32$ when compared to those of [17,19,26].

1.2.4 Synthesis results

For a future hardware implementation on FPGA device, our proposed designs of [21,25,26] have been implemented in RTL using VHDL language and then synthesized with XST tool using the Digilent evaluation board of the Spartan 6 LX45T. Synthesis results of approximate and reference designs of 1D DCT for different lengths are depicted in Table 1.2. The minimal consumption of LUTs corresponds to our proposed approximate design of [25], which has also the lowest area-delay product (ADP). However, the number of LUTs increases slightly when using our proposed structure of [21] because it requires a higher number of shift

Table 1.2 Synthesis results of approximate and reference designs of 1D DCT for different lengths

Design		**Area (LUT)**	**Time (ns)**	**Power (mW)**	**ADP**	**EoC (pJ)**
Architecture	*N*					
BDCT [19]	8	240	2.31	2.93	554	2.53
	16	672	2.34	6.29	1,572	3.67
	32	1,760	2.37	16.71	4,171	6.18
Algorithm of [20]	4	304	4.33	3.02	1,316	9.8
	8	1,392	4.98	13.26	6,932	41.27
	16	5,002	5.65	55.68	28,261	137.63
	32	18,772	6.21	209.13	115,552	365.25
DCT of [25]	8	218	2.31	2.92	503	2.52
	16	624	2.34	6.26	1,460	3.66
	32	1,656	2.37	16.64	3,925	6.16
DCT of [26]	4	76	2.28	1.41	173	1.6
	8	240	2.31	2.93	554	2.53
	16	672	2.34	6.29	1,572	3.67
	32	1,760	2.37	16.71	4,171	6.18
DCT of [21]	4	76	2.28	1.41	173	1.6
	8	244	2.31	2.93	564	2.53
	16	680	2.34	6.29	1,591	3.67
	32	1,776	2.37	16.71	4,209	6.18

operations when compared to methods in [25,26]. The same minimum clock cycle duration is obtained by the existing approximation methods since no extra slices or delay is used by shift operators. Finally, the energy per output coefficient (EoC) parameter was evaluated by multiplying the power consumption by the clock cycle duration and then by the number of cycles required to get the first output coefficient, divided by N to get the average. Approximate designs of [19,26] share the same value of EoC; however, it is slightly higher for the case of [21,25]. Note that power consumption estimate is given by Xpower tool.

1.3 Video-coding concept in VVC standard

1.3.1 VVC encoder scheme

The VVC block diagram is similar to previous standards such as HEVC. This block-based video coding is used to compress an input video signal to get a bitstream. It is based on the same hybrid encoder structure of temporal prediction (inter-prediction mode), spatial prediction (intra-prediction mode) and the 2D transform coding [6]. The main difference between VVC and HEVC encoder structure is the introduction of the adaptive loop filter (ALF) block and the different improvements applied to almost all of the coding modules. The coding process unfolds as shown in Figure 1.6.

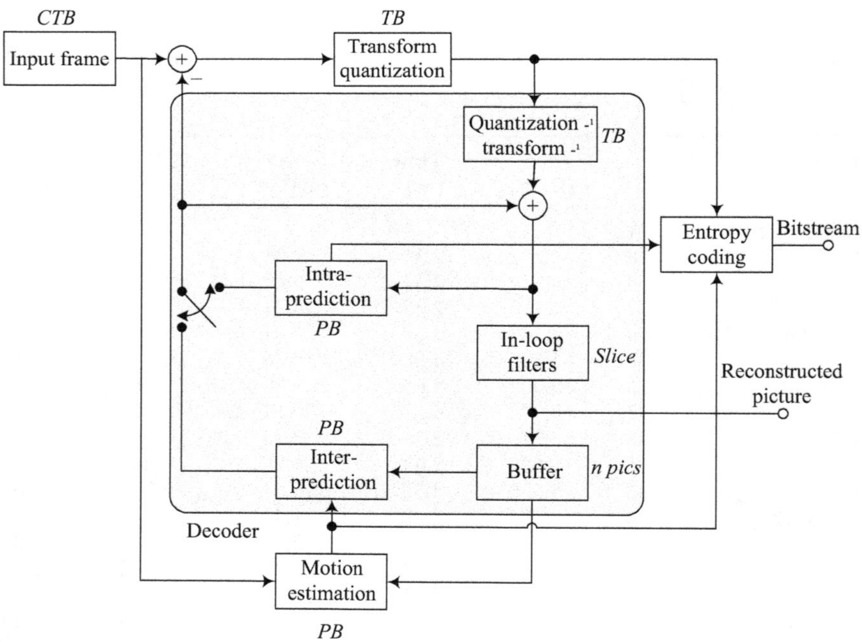

Figure 1.6 A simplified schematic block of VVC standard [30] (CTB, coding tree block; PB, prediction block; TB, transform block)

Each frame of the input video, which is partitioned into square blocks called CTB (coding tree block), is fed to the encoder. The first frame is encoded using only intra-prediction that exploits the spatial redundancy to remove the data correlation present in the coded frame. For the remaining frames of the video sequence, either intra-prediction or inter-prediction is used. The latter uses time redundancy to select the motion vector of each block relative to a reference image. Inter-prediction is also called motion-compensated prediction since it compensates the motion of objects between successive frames of the input video [29]. Once the block is predicted, the residual signal, which is the difference between the reference block and the predicted block, will be decorrelated and compacted into few coefficients in the frequency domain due to the transform coding [30]. The obtained transformed coefficients are quantified in order to eliminate the least significant ones. The quantized coefficients, the intra-prediction information, the motion vectors as well as the reference indexes for the inter-prediction that are useful for reproducing the prediction signal at the decoder side are coded by entropy coding CABAC (Context-Adaptive Binary Arithmetic Coding) to provide a bitstream representing the output of the coding scheme [31]. A decoder loop is built-in into the HEVC encoder acting on the quantized transform coefficients by applying inverse quantization and inverse transform. They are added to the prediction signal to be filtered by the in-loop filters, namely, the deblocking filter, the simple adaptive offset filter and the ALF. They serve to assess and transmit the optimal filter parameters to the decoder in order to improve the quality and remove the prediction defects of the reconstructed image [32]. The generated frame is saved in the buffer in order to be used as a reference image for inter-coded images.

1.3.2 Transform coding for VVC standard

One key feature of the transform module in VVC standard is the introduction of an AMT with five different transform types for both rows and columns, involving DCT-II, DCT-V and DCT-VIII in addition to DST-I and DST-VII [14]. These transforms are computed on square or rectangular blocks made of $N \times M$ samples where N and M may be 4, 8, 16, 32, 64 or 128. They are given by the basis functions shown in Table 1.3. A mode-dependent transform candidate selection is applied to the residual of the intra-coded block. Three transform subsets are defined as shown in Table 1.4. According to the intra selected mode, two transform sets are selected: one for the horizontal and one for the vertical processing. As for inter-coded blocks, only DST-VII and DCT-VII transform types are used for all inter modes and applied for horizontal and vertical transform.

1.3.3 Statistical analysis

In order to obtain an optimized transform stage in terms of computational complexity, a statistical study on the transform types and sizes was conducted to perform the approximation approach on the most used ones. Different experiments with videos from class A1 and A2 are carried out using HM-16.6-JEM5.0 [14] reference encoder. The encoder configuration is specified as random access (RA)

Table 1.3 Transform basis functions of DCT-II/V/VIII and DST-I/VII

Transform type	Basis function $T_i(j), i,j = 0,1,\ldots,N-1$
DCT-II	$T_i(j) = \omega_0 \cdot \sqrt{2/N}\cos((\pi \cdot i \cdot (2j+1))/2N)$, where $\omega_0 = \begin{cases} \sqrt{\frac{2}{N}}, & i = 0 \\ 1, & i \neq 0 \end{cases}$
DCT-V	$T_i(j) = \omega_0 \cdot \omega_1 \sqrt{2/(2N-1)}\cos((2\pi \cdot i \cdot j)/(2N-1))$, where $\omega_0 = \begin{cases} \sqrt{\frac{2}{N}}, & i = 0 \\ 1, & i \neq 0 \end{cases}, \omega_1 = \begin{cases} \sqrt{\frac{2}{N}}, & j = 0 \\ 1, & j \neq 0 \end{cases}$
DCT-VIII	$T_i(j) = \sqrt{4/(2N+1)}\cos((\pi \cdot (2i+1) \cdot (2j+1))/(4N+2))$
DST-I	$T_i(j) = \sqrt{2/(N+1)}\sin((\pi \cdot (i+1) \cdot (j+1))/(N+1))$
DST-VII	$T_i(j) = \sqrt{4/(2N+1)}\sin((\pi \cdot (2i+1) \cdot (j+1))/(2N+1))$

Table 1.4 Predefined transform candidate subsets

Transform set	Transform candidates
0	DST-VII, DCT-VIII
1	DST-VII, DST-I
2	DST-VII, DCT-V

configuration and the quantization parameter (QP) is set to 22 and 37 values for each test sequence:

- The statistical results on transform types are depicted in Table 1.5. Overall, DCT-II and DST-VII are used more frequently than the other types, respectively, with an average of 47% and 33%. DCT-VIII takes the third rank with roughly 11%. DCT-V and DST-I are the least used types with an average of 5% and 4%. Only one exception is presented in the case of the ToddlerFountain video sequence from class A1. In fact, DST-VII exceeds DCT-II for all QP values and the DCT-V becomes more used than the DCT-VIII for QP = 37.
- The statistical results on the transform sizes are summarized in Table 1.6. First of all, it is noted that size 2 and 128 transforms are used only when the DCT-II transform is used. It can be noticed that the choice of the transform size depends on the texture of the video sequence. In fact, the frequency of using small units increases in the case of textured videos where many details are represented. This choice is favored when using small QP value where many spatial details are kept. For textured videos like CampfireParty and CatRobot, size 8 transform takes the first rank and size 4 transform takes the second rank

Table 1.5 Statistics on transform types (all given results are in percentage)

Class	Resolution	Sequence	QP	DCT-II	DST-VII	DCT-VIII	DCT-V	DST-I
A1	4,096 × 2,160	Tango	22	39.45	36.35	12.14	6.87	5.18
			37	54.45	34.53	7.21	2.65	1.15
	3,840 × 2,160	Drumbs100	22	34.77	33.63	28.37	1.59	1.64
			37	48.57	35.36	8.86	4.63	2.58
	3,840 × 2,160	CampfireParty	22	50.69	29.49	10.67	4.99	4.17
			37	68.61	23.02	2.95	3.41	2.01
	4,096 × 2,160	ToddlerFountain	22	28.62	42.32	11.15	9.90	8.00
			37	13.02	51.80	12.66	13.72	8.80
A2	3,840 × 2,160	CatRobot	22	52.36	27.55	14.06	3.50	2.53
			37	63.60	27.78	4.96	2.31	1.35
	3,840 × 2,160	TrafficFlow	22	39.02	33.84	13.16	6.43	7.56
			37	59.32	30.87	4.64	2.72	2.45
	3,840 × 2,160	DaylightRoad	22	39.05	32.38	15.62	7.09	5.87
			37	48.36	36.43	7.87	5.31	2.03
	4,096 × 2,160	Rollercoaster	22	54.79	29.47	13.02	1.38	1.34
			37	60.92	29.28	6.04	2.26	1.49
Average				47.22	33.38	10.84	4.92	3.63

when the QP value is equal to 22. However, when QP is equal to 37, size 16 transform takes the first rank and size 8 transform takes the second rank. This choice of small transform sizes is justified by the presence of more details compared to the other videos. For less-textured videos, which contain both smooth and textured areas such as ToddlerFountain, TrafficFlow and DaylightRoad, small transform sizes are used for QP value equal to 22. However, there is a trend towards larger transform blocks. In fact, when using QP = 37, the total use of size 16 and 32 transforms occupies almost 70% for the case of ToddlerFountain and TrafficFlow and it is nearly 60% for the case of DaylightRoad video. The third case regroups videos with less spatial complexity compared with the two cases stated before such as Tango and RollerCoaster. In this case, large units are used when QP is equal to 22 and 37. For QP = 22, the transform size 16 takes the first rank and the transform size 8 takes the second rank, while for QP = 37, the transform size 32 takes the first rank and the transform size 16 takes the second rank. The last case is about the video sequence Drumbs100 with large smooth areas and few textured ones. In this case, the use of large transform sizes is clear. In fact, when QP is equal to 22, the use of size 16, 8 and 32 transforms reaches, respectively, 34%, 28% and 20% of the total use of the transform sizes. When increasing QP, size 16 and 32 become the most used ones.

Based on these statistical results, optimized algorithms for DCT-II and DST-VII were designed. The approximation approach will be carried out first on small transform sizes, and it will be generalized for larger ones.

Table 1.6 Statistics on transform sizes (all given results are in percentage)

Class	Resolution	Sequence	QP	Size 2	Size 4	Size 8	Size 16	Size 32	Size 64	Size 128
A1	4,096 × 2,160	Tango	22	0.24	14.43	30.62	32.64	17.37	4.42	0.29
			37	0.01	0.62	7.71	29.23	40.42	19.25	2.76
	3,840 × 2,160	Drumbs100	22	1.66	11.97	28.42	33.91	20.06	3.98	0.01
			37	0.46	5.93	21.98	36.78	26.53	7.47	0.85
	3,840 × 2,160	CampfireParty	22	4.98	30.62	38.68	19.75	5.33	0.62	0.02
			37	0.93	8.95	24.57	33.00	23.72	7.85	0.98
	4,096 × 2,160	ToddlerFountain	22	1.87	24.54	42.03	26.40	5.07	0.10	0
			37	0.03	2.88	16.36	36.51	33.20	10.85	0.18
A2	3,840 × 2,160	CatRobot	22	3.46	28.02	37.43	20.55	8.01	2.14	0.39
			37	0.43	7.75	24.33	33.29	24.31	8.64	1.24
	3,840 × 2,160	TrafficFlow	22	0.52	29.46	45.81	17.01	5.58	1.47	0.16
			37	0.20	3.76	21.50	45.46	23.35	4.72	1.00
	3,840 × 2,160	DaylightRoad	22	1.71	28.07	39.33	21.08	7.91	1.78	0.11
			37	0.12	7.04	22.54	33.01	25.78	9.43	2.08
	4,096 × 2,160	Rollercoaster	22	0.51	7.90	26.95	35.35	21.61	6.91	0.76
			37	0.03	1.40	11.95	35.04	34.09	12.84	4.64
Average				1.07	13.33	27.51	30.56	20.15	6.41	0.97

1.4 Approximation for DCT-II transform and its hardware architecture

1.4.1 Algorithm description

The different sizes of the DCT-II matrices of VVC are calculated from the equation related to DCT-II basis vectors given in Table 1.3, to which we applied addition and shift operations as stated in [14]. To alleviate the presentation, we have displayed only 4- and 8-point DCT-II as they serve to compute higher length DCTs. They are represented, respectively, by (1.16) and (1.17):

$$C_4 = \begin{bmatrix} 256 & 256 & 256 & 256 \\ 334 & 139 & -139 & -334 \\ 256 & -256 & -256 & 256 \\ 139 & -334 & 334 & -139 \end{bmatrix} \tag{1.16}$$

$$C_8 = \begin{bmatrix} 256 & 256 & 256 & 256 & 256 & 256 & 256 & 256 \\ 355 & 301 & 201 & 71 & -71 & -201 & -301 & -355 \\ 334 & 139 & -139 & -334 & -334 & -139 & 139 & 334 \\ 301 & -71 & -355 & -201 & 201 & 355 & 71 & -301 \\ 256 & -256 & -256 & 256 & 256 & -256 & -256 & 256 \\ 201 & -355 & 71 & 301 & -301 & -71 & 355 & -201 \\ 139 & -334 & 334 & -139 & -139 & 334 & -334 & 139 \\ 71 & -201 & 301 & -355 & 355 & -301 & 201 & -71 \end{bmatrix} \tag{1.17}$$

C_8 involves a large number of arithmetic operations required for future hardware implementation. To get the approximate matrix for 8-point DCT-II, we have applied the algorithm detailed in [21] as it brings a good performance whether in hardware or in software. The proposed 8 × 8 DCT approximation allows generating approximate DCT for either lower length, which is size 4 transform, or higher lengths, which consists of 16- and 32-point transform. To do that, we proceed first by the odd/even decomposition detailed in section. C_8 is decomposed into a product of three matrices:

$$C_8 = M_8^{per} \times T_8 \times M_8^{add} \tag{1.18}$$

where

$$M_8^{per} = \begin{bmatrix} 1 & 0 & 0 & 0 & 0 & 0 & 0 & 0 \\ 0 & 0 & 0 & 0 & 1 & 0 & 0 & 0 \\ 0 & 1 & 0 & 0 & 0 & 0 & 0 & 0 \\ 0 & 0 & 0 & 0 & 0 & 1 & 0 & 0 \\ 0 & 0 & 1 & 0 & 0 & 0 & 0 & 0 \\ 0 & 0 & 0 & 0 & 0 & 0 & 1 & 0 \\ 0 & 0 & 0 & 1 & 0 & 0 & 0 & 0 \\ 0 & 0 & 0 & 0 & 0 & 0 & 0 & 1 \end{bmatrix} \tag{1.19}$$

$$T_8 = \begin{bmatrix} 256 & 256 & 256 & 256 & 0 & 0 & 0 & 0 \\ 334 & 139 & -139 & -334 & 0 & 0 & 0 & 0 \\ 256 & -256 & -256 & 256 & 0 & 0 & 0 & 0 \\ 139 & -334 & 334 & -139 & 0 & 0 & 0 & 0 \\ 0 & 0 & 0 & 0 & 355 & 301 & 201 & 71 \\ 0 & 0 & 0 & 0 & 301 & -71 & -335 & -201 \\ 0 & 0 & 0 & 0 & 201 & -355 & 71 & 301 \\ 0 & 0 & 0 & 0 & 71 & -201 & 301 & -355 \end{bmatrix} \quad (1.20)$$

and

$$M_8^{add} = \begin{bmatrix} 1 & 0 & 0 & 0 & 0 & 0 & 0 & 1 \\ 0 & 1 & 0 & 0 & 0 & 0 & 1 & 0 \\ 0 & 0 & 1 & 0 & 0 & 1 & 0 & 0 \\ 0 & 0 & 0 & 1 & 1 & 0 & 0 & 0 \\ 1 & 0 & 0 & 0 & 0 & 0 & 0 & -1 \\ 0 & 1 & 0 & 0 & 0 & 0 & -1 & 0 \\ 0 & 0 & 1 & 0 & 0 & -1 & 0 & 0 \\ 0 & 0 & 0 & 1 & -1 & 0 & 0 & 0 \end{bmatrix} \quad (1.21)$$

As M_8^{per} and M_8^{add} do not involve any additional computational complexity, the approximation of C_8 is done by separately approximating the even-indexed and odd-indexed points, which constitute T_8. One of the key properties of the DCT is that the transform coefficients of a $2M \times 2M$ DCT matrix are equal to the first half of the even basis vectors of a $2M + 1 \times 2M + 1$ DCT matrix. Therefore, the approximation of the even part of the T_8 corresponds to approximating the C_4 basis vectors. The approximation is done as stated in [33] to obtain the following matrix:

$$\hat{E}_4 = \hat{C}_4 = 128 \times \begin{bmatrix} 2 & 2 & 2 & 2 \\ 2 & 1 & -1 & -2 \\ 2 & -2 & -2 & 2 \\ 1 & -2 & 2 & -1 \end{bmatrix} \quad (1.22)$$

To approximate the first half of the odd basis vectors of C_8, we processed by the same approach stated in [21]. Two types of elements are considered: extremum values and the rest. The original extremum values are rounded in order to insure decorrelated DCT output coefficients, and the remaining elements are approximated by their positions instead of their values. This idea is motivated by the high correlation existing between neighbor pixels of natural images and especially those with high resolutions. Every input block of these images will serve to multiply the basis vectors of the DCT matrix. Extremum values have higher importance than the other coefficients which are approximated by their distance to the extremum. By doing this, the elements of the approximate matrix can have the following values $\{0, \pm1, \pm2\}$. Minimum and maximum values are approximated, respectively, by 0 and ±2 and the remaining ones are approximated to ±2 or ±1 according to their positions to the maximum and minimum values. Applying this procedure to $\{\pm 355, \pm 301, \pm 201, \pm 71\}$, which constitute the elements of the first half of the odd basis vectors of the original 8-point DCT-II matrix, maximum and minimum values (i.e., ± 355 and ± 71) are approximated, respectively, to ± 2 and 0 and the rest elements are approximated depending on their signs and positions to the extremum values. For example, the coefficient 301 of the second row, second column is approximated by 2 since its position is closer to 355 than 71. However, the same coefficient situated in the fourth row, first column is approximated to 1 because its position is closer to -71 than -355. By doing this, the approximate matrix is given by

$$\hat{O}_4 = 128 \times \begin{bmatrix} 2 & 2 & 1 & 0 \\ 1 & 0 & -2 & -2 \\ 2 & -2 & 0 & 1 \\ 0 & -1 & 2 & -2 \end{bmatrix} \tag{1.23}$$

By replacing the even and odd parts by their approximate matrices, we obtain the following approximation for DCT-II size 8 which requires few material resources:

$$\hat{C}_8 = 128 \times \begin{bmatrix} 2 & 2 & 2 & 2 & 2 & 2 & 2 & 2 \\ 2 & 2 & 1 & 0 & 0 & -1 & -2 & -2 \\ 2 & 1 & -1 & -2 & -2 & -1 & 1 & 2 \\ 1 & 0 & -2 & -2 & 2 & 2 & 0 & -1 \\ 2 & -2 & -2 & 2 & 2 & -2 & -2 & 2 \\ 2 & -2 & 0 & 1 & -1 & 0 & 2 & -2 \\ 1 & -2 & 2 & -1 & -1 & 2 & -2 & 1 \\ 0 & -1 & 2 & -2 & 2 & -2 & 1 & 0 \end{bmatrix} \tag{1.24}$$

Multiplication by 128 ensures conformity with VVC requirements. It will be incorporated in the scaling stage. To get lower length DCT, which is size 4 transform, the matrix $\hat{C}_8$ is down-sampled while higher length DCTs are generated according to the decomposition approach proposed in "Algorithm description" section.

1.4.2 Approximate 8-point transform architecture

The 8-point 1D approximate DCT is computed according to the SFG depicted in Figure 1.7. For a given input $x = \{x_0, \ldots, x_7\}$, the output $X = \{X_0, \ldots, X_7\}$ is obtained by means of combination between addition and shift operations.

1.4.3 Reconfigurable designs for 1D/2D DCT computing

1.4.3.1 Reconfigurable design for 4- and 8-point 1D DCT computing

A reconfigurable architecture that allows computing 8-point or a pair of 4-point approximate DCT is presented in Figure 1.8. It includes two $\hat{E}_4$ computation units ($\hat{E}_4$ CU), one $\hat{O}_4$ CU, a reusable 8-point IAU as well as a permutation and size selection unit. The reusable 8-point IAU contains the 8-point IAU from Figure 1.7 and 12 AND gates. When this architecture is used to compute 8-point DCT, the

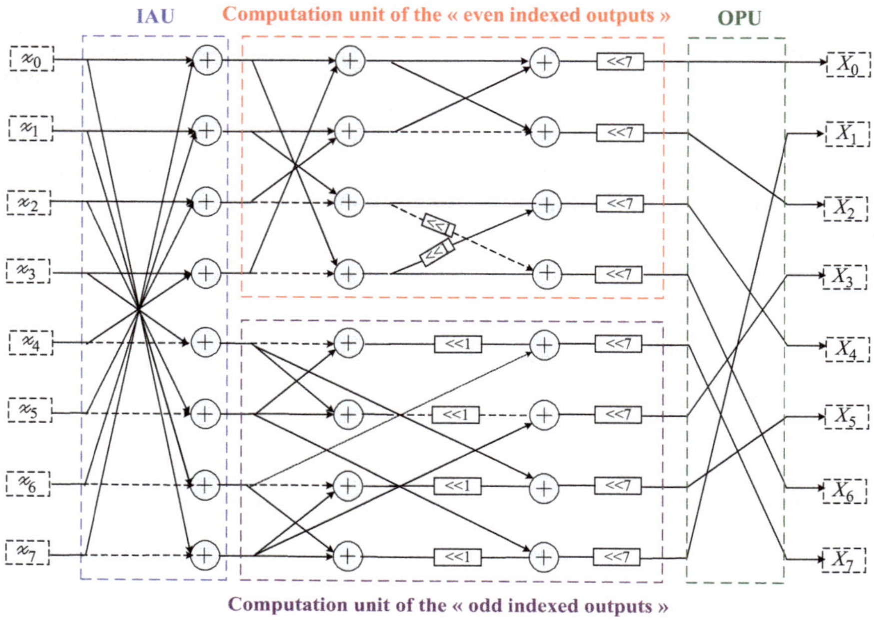

Figure 1.7 Signal flow graph for the computation of 8-point 1D DCT (dashed arrows represent multiplication by −1)

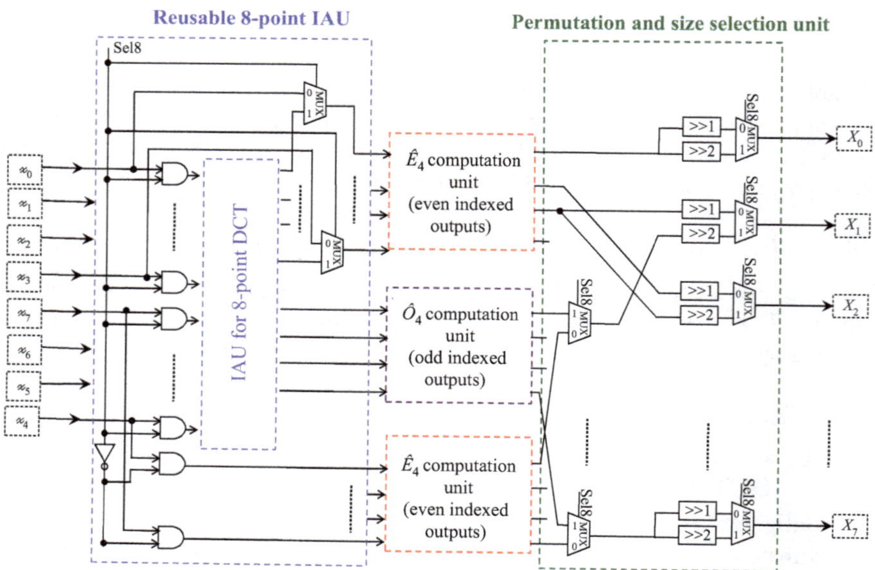

Figure 1.8 Reconfigurable design for the computation of 8-point or a pair of 4-point approximate DCT

control bit *Sel*8 goes high enabling the first $\hat{E}_4$ CU and the $\hat{O}_4$ CU as well as disabling the second $\hat{E}_4$ CU. The first $\hat{E}_4$ CU is fed by the first half of the 8-point IAU to compute the even 8-point DCT coefficients, while the $\hat{O}_4$ CU generates the odd ones. One the other hand, when this architecture is used to compute a pair of 4-point DCT, the control bit $Sel8 = 0$ to deselect the 8-point IAU. The first $\hat{E}_4$ CU computes the DCT of the inputs $\{x_0, \ldots, x_3\}$, while the second $\hat{E}_4$ CU is responsible for computing the DCT of the inputs $\{x_4, \ldots, x_7\}$. The permutation and size selection unit ensures 1 or 2 shift operations, respectively, for the computation of 4-, or 8-point DCT as stated in [14].

1.4.3.2 Reconfigurable design for 32-point 1D DCT computing

A 32-point approximate DCT, shown in Figure 1.9, is derived according to the decomposition method detailed in "Algorithm description" section. The proposed design allows computing one 32-point DCT, two 16-point DCTs, four 8-point DCTs or eight 4-point approximate DCTs. This choice is done according to the control bits *Sel*32, *Sel*16 and *Sel*8 that enable, respectively, the computation of 32-, 16-point or 8-point approximate DCT. For the case of 32-point DCT when $Sel32 = 1$, the reconfigurable 32-point IAU serves to calculate the addition and the difference of input pixels $(v0 = x_0 + x_{31}, \ldots, \quad v31 = x_0 - x_{31})$. However, when $Sel32 = 0$, the IAU is not activated, then $v0 = x_0, \ldots, \quad v31 = x_{31}, \ldots$. The structure of the reconfigurable IAU for both 16- or 32-point DCT is different from the one proposed in Figure 1.8. In fact, they require only 8-point DCT units; however, 8-point DCT necessitates 4-point DCT and a $\hat{O}_4$ CU.

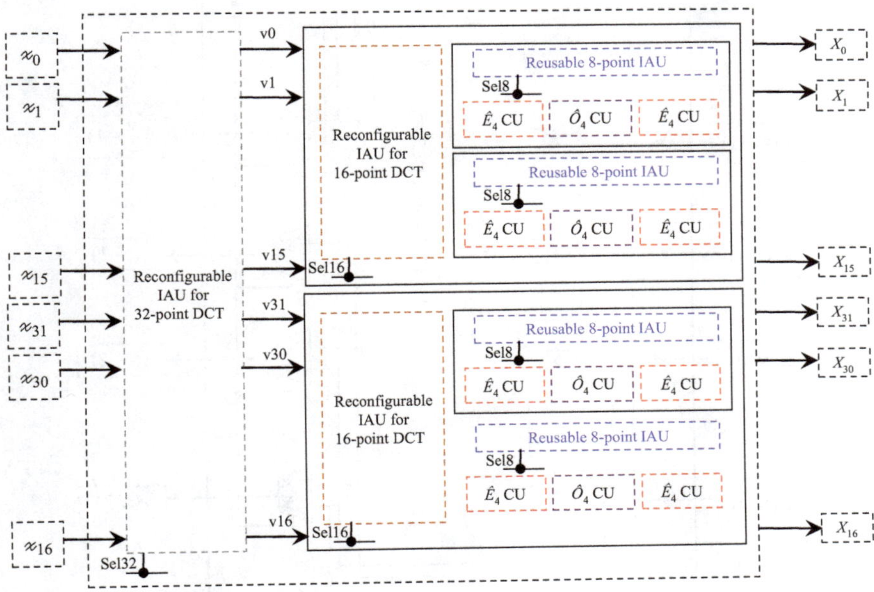

Figure 1.9 Reconfigurable design for a 32-point approximate DCT computing

The reconfigurable IAU is generalized for N-point DCT as illustrated in Figure 1.10. The control bit Sel_N decides whether the computation of N-point DCT or a pair of N/2-point DCTs is allowed. For the first case, $Sel_N = 1$ enables the computation of the IAU for N-point DCT. For the second case, the computation of the N-point IAU is bypassed.

1.4.3.3 Reconfigurable architecture for 2D approximate DCT computing

The N-point 2D DCT is obtained by performing two N-point 1D DCT, first row-wise and then column-wise [34]. The first stage of 1D transform computation corresponds to applying an N-point transform to each row of an $N \times N$ residual block. The intermediate matrix, which is stored in a transposition buffer, is considered as input for the second stage of 1D transform computation applied to each column of the intermediate matrix to get the final $N \times N$ transformed matrix. The proposed reconfigurable 1D architecture is designed for the computation of 4-point and 8-point DCT as shown in Figure 1.11. It is reused to get 2D DCT, and it can be

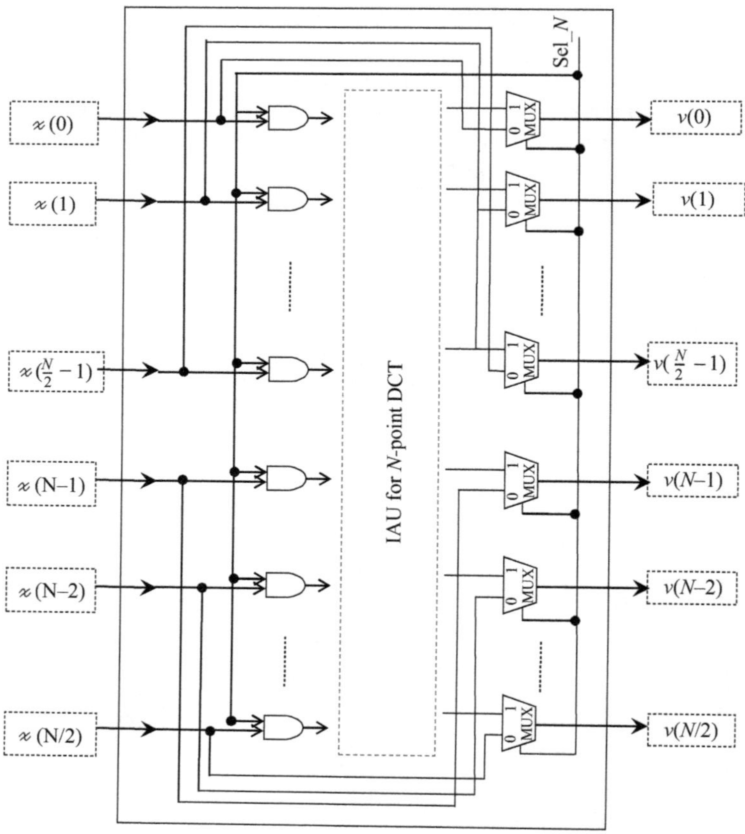

Figure 1.10 Reconfigurable IAU for the computation of N-point DCT

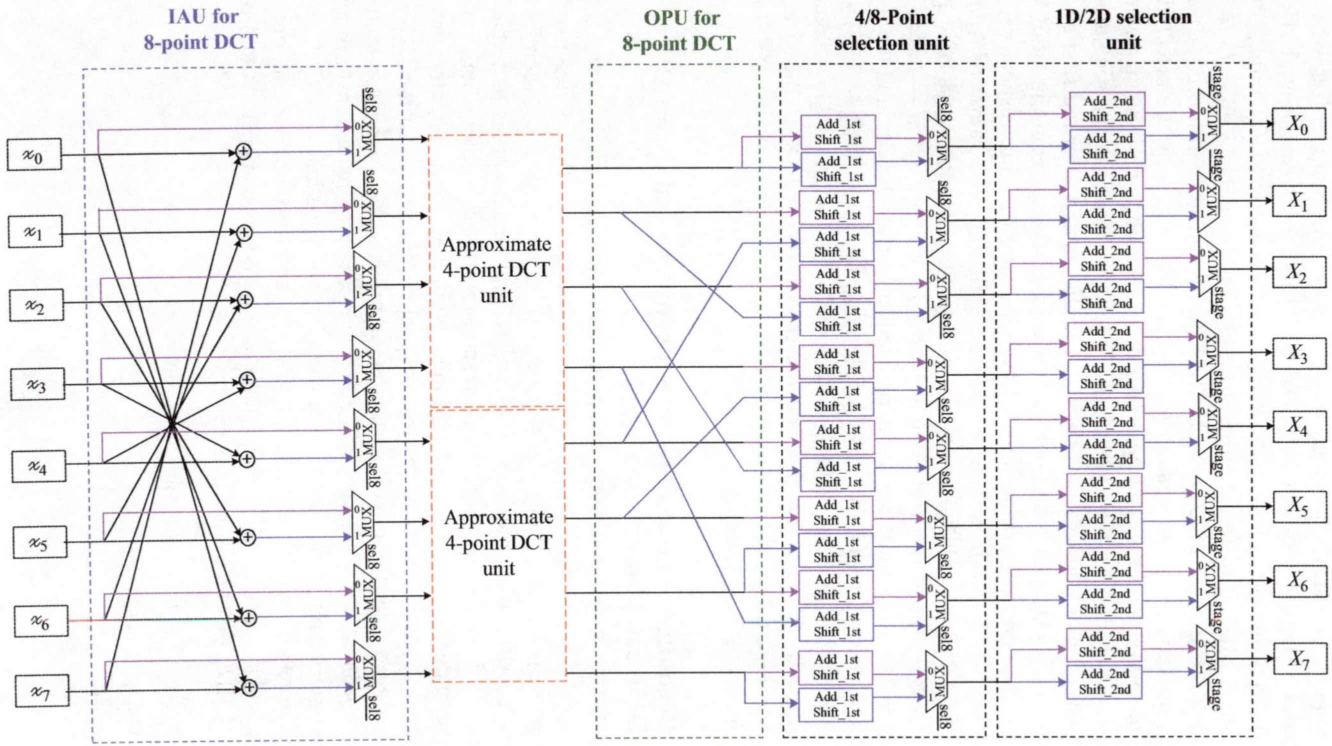

Figure 1.11 Reconfigurable architecture for 4- and 8-point 1D DCTs computing

extended to compute higher length DCTs. A folded structure of 2D DCT is obtained by using one 1D DCT module along with a transposition buffer, whereas a fully parallel structure requires two 1D DCT and a transposition buffer. It includes an IAU that provides inputs data to the two 4-point DCT units. Such a unit contains 8 MUXes to select either size 4 (when the control bit $Sel8 = 0$) or size 8 (when the control bit $Sel8 = 1$) DCT computation is allowed. After the stage of computation, the outputs are fed to the OPU in order to be selected and reordered according to the DCT size. The 1D/2D selection unit indicates whether the architecture is used for the computation of the first stage or the second stage of the 2D DCT according to a bit control. A scaling stage after the first and the second stage of 2D DCT is done in order to meet the scaling requirement of VVC and to keep the norm of residual block [28]. They are defined, respectively, by the two blocks: Add-1st/Shift-1st and Add-2nd/Shift-2nd. The values of Add-1st, Shift-1st, Add-2nd and Shift-2nd are defined in [14]. For the case of 4-point 2D DCT ($sel8 = 0$), $\{Add - 1st = 16, shift - 1st = 5\}$ and $\{Add - 2nd = 512, shift - 2nd = 10\}$ operations are done, respectively, after the first and the second forward DCT. However, for the case of 8-point 2D DCT ($sel8 = 1$), $\{Add- 1st = 32, shift - 1st = 6\}$ and $\{Add - 2nd = 1024, shift - 2nd = 11\}$ operations are done, respectively, after the first and the second forward DCT. Based on the architecture of Figure 1.11, the scaling stage is done by the combination of 4/8-point selection unit and 1D/2D selection unit.

1.4.3.4 Reconfigurable architecture for inverse and forward DCT computing

Hardware sharing between forward and inverse transforms is an important feature since the same device should support video recording as well as video playback such as mobile applications [35]. The symmetry property existing between forward and inverse transforms of the VVC standard is exploited in order to reduce the surface occupied by the hardware block responsible for computing both forward and inverse transforms. In the other hand, exploiting the symmetry property existing between the elements of the inverse transform matrices of the DCT-II of the VVC, an even–odd decomposition is done in order to factorize these transform matrices into submatrices [36]. For an N-point inverse approximate transform matrix C_N^T, inverse transform is done according to the following equation, $V = C_N^T U$, where the input vector is $U = [u(0), u(1), \ldots, u(N-1)]^T$ and the corresponding output is $V = [v(0),\ v(1), \ldots, v(N-1)]^T$. Applying the even–odd decomposition to 4-point inverse DCT (IDCT), outputs are obtained by

$$\begin{bmatrix} v(0) \\ v(1) \\ v(2) \\ v(3) \end{bmatrix} = \begin{bmatrix} w(0) + w(2) \\ w(1) + w(3) \\ w(1) - w(3) \\ w(0) - w(2) \end{bmatrix} \tag{1.25}$$

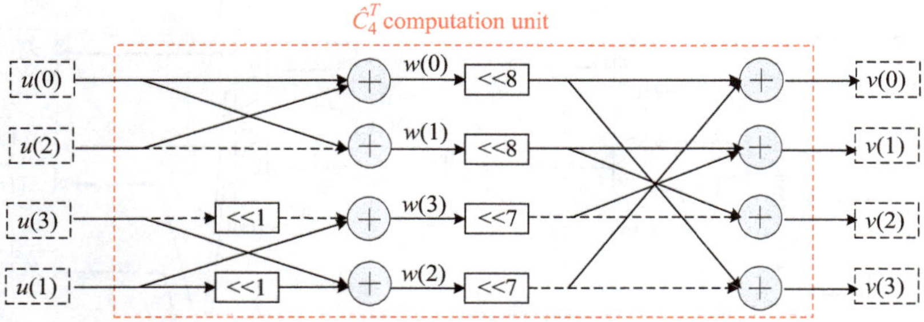

Figure 1.12 Proposed architecture for approximate 4-point integer 1D IDCT

where

$$\begin{bmatrix} w(0) \\ w(1) \end{bmatrix} = 256 \times \begin{bmatrix} 1 & 1 \\ 1 & -1 \end{bmatrix} \times \begin{bmatrix} u(0) \\ u(2) \end{bmatrix} \tag{1.26}$$

$$\begin{bmatrix} w(2) \\ w(3) \end{bmatrix} = 128 \times \begin{bmatrix} 2 & 1 \\ 1 & -2 \end{bmatrix} \times \begin{bmatrix} u(1) \\ u(3) \end{bmatrix} \tag{1.27}$$

The proposed SFG of the approximate 4-point 1D inverse DCT is shown in Figure 1.12.

The fact that forward and inverse transforms can share the hardware blocks that implement the kernel matrices supports the idea of the unified design that regroup forward and inverse of transforms. This design is not limited to size 4 transform. It can be extended to higher lengths such as 32-point transform as shown in Figure 1.13. Forward or inverse transform computation is selected due to the control signal Inv/fwd of the 64 MUX/DEMUX used by the design.

1.4.4 Video-coding performance

1.4.4.1 Simulation conditions

Simulations were done for 8 video sequences from classes A1 and A2 selected from the 28 sequences stated in the common test conditions document from JVET [37]. The 10-bit test sequences have different resolutions and frame rate. They are encoded covering a set of frames, which constitute one GOP (group of pictures) as stated in Table 1.7. All tests were performed on the recent standard VVC using the reference software model HM-16.6-JEM-5.0 [14], compiled using Microsoft Visual Studio C++ Compiler and carried on a Windows 7 OS platform with Intel core TM i7-3770 @ 3.4 GHz CPU and 12 GB of RAM. All videos were evaluated using RA configuration. They are encoded using four different quantization parameters QP: 22, 27, 32 and 37.

1.4.4.2 Evaluation criteria

In order to evaluate the coding efficiency of the proposed approximation algorithms, the modified transform matrices are integrated into the reference software model

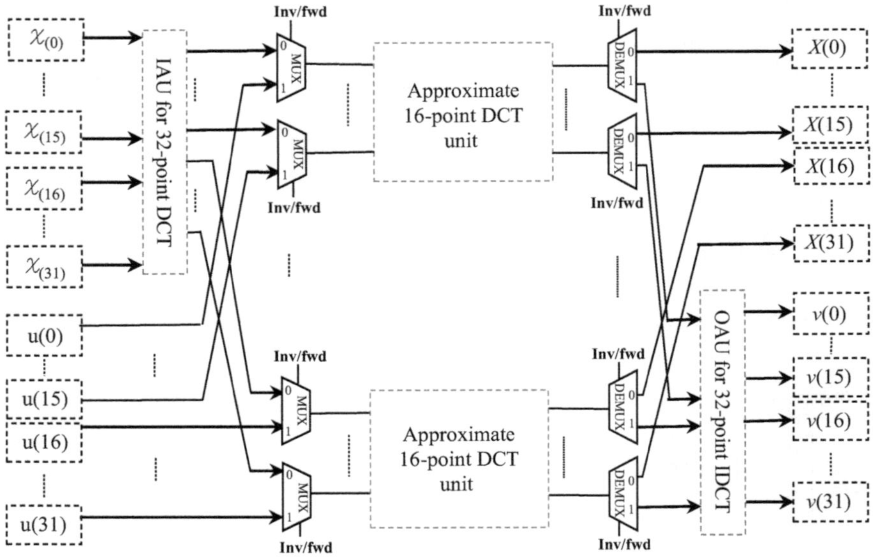

Figure 1.13 Unified design for approximate 32-point 1D transform (without scaling)

Table 1.7 Test sequences characteristics

Class	Sequence	Resolution	Frame rate (fps)	Bit depth	Intra period	Frames to be encoded
A1	CampfireParty	3,840 × 2,160	30	10	32	33 = 1 GOP
	Drumbs100	3,840 × 2,160	100	10	96	97 = 1 GOP
	Tango	4,096 × 2,160	60	10	64	65 = 1 GOP
	ToddlerFountain	4,096 × 2,160	60	10	64	65 = 1 GOP
A2	CatRobot	3,840 × 2,160	60	10	64	65 = 1 GOP
	TrafficFlow	3,840 × 2,160	30	10	32	33 = 1 GOP
	DaylightRoad	3,840 × 2,160	60	10	64	65 = 1 GOP
	Rollercoaster	4,096 × 2,160	60	10	64	65 = 1 GOP

HM-16.6-JEM-5.0 [14]. A first modification is done only for size 4, 8 and 16 DCT-II matrices as they exceed 71% of use compared to higher sizes. The second modification concerns the different transform sizes of the DCT-II. In these two cases, the proposed modification was compared to the original algorithm, and the performance was measured by means of criteria given in Table 1.8. $\Delta YUV - PSNR$ (dB) and $\Delta BR(\%)$ keep up with the formula summarized in the table where $PSNR_p$ and $PSNR_o$ correspond, respectively, to the PSNR index of the proposed and the original algorithm. PSNR values refer to $YUV - PSNR$ which is a weighted sum of the

Table 1.8 Evaluation criteria

Criteria	Description	Formula
$\Delta YUV - PSNR$ (dB)	PSNR loss	$PSNR_p - PSNR_o$
ΔBR (%)	Bitrate increase	$(Bitrate_p/(Bitrate_o - 1)) \times 100\%$
$BD - PSNR$ (dB)	Bjøntegaard PSNR	Defined in [39]
$BD - rate$ (%)	Bjøntegaard Bitrate	Defined in [39]

Table 1.9 Experimental results of the two versions of modified DCT-II matrices compared with the state-of-the-art algorithm

Class	Sequence	First modification		Second modification	
		ΔBR (%)	**$\Delta YUV - PSNR$ (dB)**	**ΔBR (%)**	**$\Delta YUV - PSNR$ (dB)**
A1	CamFireParty	−1.75	−0.16	−1.81	−0.16
	Drumbs100	−0.54	−0.09	−0.72	−0.12
	Tango	−0.64	0.03	0.42	−0.05
	ToddlerFountain	−0.24	−0.12	2.35	−0.09
A2	CatRobot	0.16	−0.06	−0.11	−0.06
	TrafficFlow	−0.09	0.01	0.51	−0.06
	DaylightRoad	−0.92	−0.04	−0.92	−0.05
	RollerCoaster	−0.05	0.04	0.11	−0.12
Average		−0.55	−0.05	−0.02	−0.09

PSNR values for the luma (Y) and chroma (U, V) components [38]: $PSNR_{YUV} = (6 \times PSNR_Y + PSNR_U + PSNR_V)/8$. $Bitrate_p$ and $Bitrate_o$ represent, respectively, the BR of the proposed algorithm and the original one. Bjøntegaard (BD) metrics [39], as BD-PSNR and BD-BR, are used to validate the efficiency of our proposed algorithms.

1.4.4.3 Results and discussions

Table 1.9 evaluates the performance of our proposed approximation algorithm applied first to size 4, 8 and 16 DCT-II transform matrices and secondly to all transform sizes. For the first modification, the BR saving is averaging 0.55% while maintaining a good quality for the majority of videos. In fact, the best case corresponds to the video CamFireParty where the BR saving reaches 1.75%, while the worst case presents a BR degradation of 0.16% for CatRobot video. As for quality evaluation, for the best case, the proposed modification ameliorates the quality by 0.04 dB, while in the worst case, quality degradation is only 0.16 dB which is invisible by the human eye. Concerning the second modification, the performance is lower than the first one. The average of BR saving is about 0.02% with a good quality for the majority of videos. The best case corresponds to the video

Table 1.10 Bjøntegaard results of the two versions of modified DCT-II matrices versus the state-of-the-art algorithm

Class	Sequence	First modification		Second modification	
		BR-rate (%)	**BD-PSNR (dB)**	**BR-rate (%)**	**BD-PSNR (dB)**
A1	CamFireParty	5.55	−0.12	5.06	−0.11
	Drumbs100	4.77	−0.08	6.27	−0.11
	Tango	−2.38	0.03	4.78	−0.05
	ToddlerFountain	2.68	−0.1	4.89	−0.18
A2	CatRobot	4.86	−0.07	4.32	−0.06
	TrafficFlow	−0.65	0.01	5.65	−0.06
	DaylightRoad	3.47	−0.03	4.02	−0.03
	RollerCoaster	−1.73	0.04	5.23	−0.12
Average		2.07	−0.04	5.03	−0.09

CamFireParty where the BR saving is 1.81%, while the worst case presents a BR degradation of 2.35% for ToddlerFountain video. As for quality evaluation, for the best case, the second modification degrades the quality by 0.05 dB, while in the worst case, quality degradation is 0.16 dB.

The comparison was done also by means of BD metrics whose method consists in evaluating numerical averages between different RD curves [39]. Table 1.10 depicts the results of BD-PSNR and BD-BR-rate. For the first modification, results show that the best case corresponds to a BR saving of 2.38% in BD-rate for Drumbs100 video, whereas the worst case represents a BR loss of 5.55%. As for quality evaluation, an improvement is achieved by Tango, TrafficFlow and RollerCoaster videos, while a slight degradation, which is not perceptible, is obtained by the other ones. As for the second modification, the BR loss varies between 4.02% and 6.27%, and the BD-PSNR loss is between 0.03 and 0.18 dB.

1.4.4.4 RD curves

Figure 1.14 depicts the RD curves of the original HM-16.6-JEM-5.0 algorithm, the first approximation of the DCT-II transform (only size 4, 8 and 16 are modified) and the second approximation of the DCT-II transform (all sizes are modified) for the test sequences Drumbs100 from class A1 and DaylightRoad from class A2, respectively, depicted in Figure 1.14(a) and (b) taken at QPs of 22, 27, 32 and 37. The vertical and the horizontal axes represent, respectively, the PSNR (dB) and the BR (kbit/s) and the four represented points of the curves correspond to the four QP parameters. The curves indicate that our proposed algorithms have almost the same coding efficiency as the original one. They are slightly shifted which mean that the modified transform matrices have an effect on PSNR and BR values. They result in BR increases with a slight degradation in PSNR.

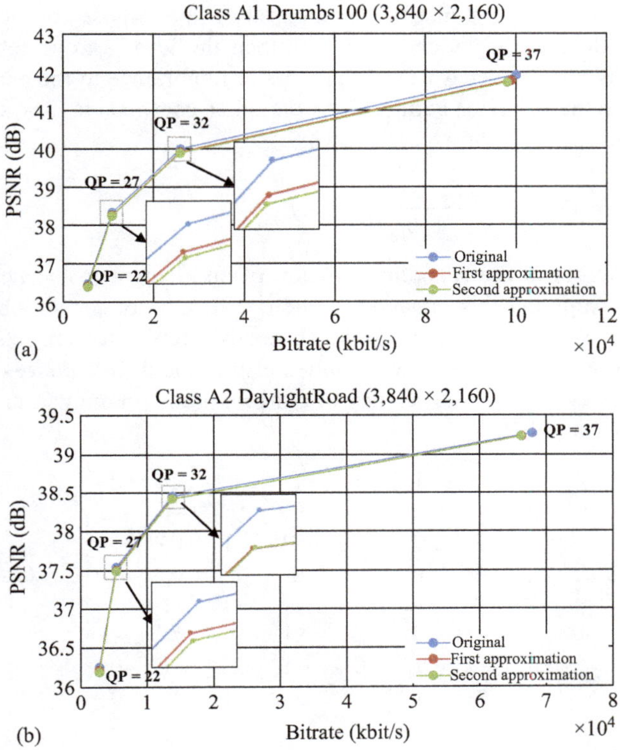

Figure 1.14 Rate-distortion curves for two test videos with QPs 22, 27, 32 and 37

1.5 New approximation for DST-VII transform

1.5.1 Algorithm description

The computation of *N*-point DST-VII of VVC where $N \in \{4, 8, 16, 32, 64, 128\}$ is done according to the equation related to DST-VII basis vectors given in Table 1.3, to which we applied addition and shift operations as stated in [14]. The 8-point DST-VII is presented as follows:

$$S_8 = \begin{bmatrix} 65 & 127 & 185 & 237 & 280 & 314 & 338 & 350 \\ 185 & 314 & 350 & 280 & 127 & -65 & -237 & -338 \\ 280 & 338 & 127 & -185 & -350 & -237 & 65 & 314 \\ 338 & 185 & -237 & -314 & 65 & 350 & 127 & -280 \\ 350 & -65 & -338 & 127 & 314 & -185 & -280 & 237 \\ 314 & -280 & -65 & 338 & -237 & -127 & 350 & -185 \\ 237 & -350 & 280 & -65 & -185 & 338 & -314 & 127 \\ 127 & -237 & 314 & -350 & 338 & -280 & 185 & -65 \end{bmatrix} \quad (1.28)$$

Since the 8-point DST-VII matrix is composed of lines whose coefficients are the same having different signs and positions, then the idea is to search an approximation of this line in such a way we give more importance to the extremums and we keep the same behavior compared to the other coefficients. The equation that ensures this property is shown as follows:

$$\hat{S}_8(i,j) = 128 \times \text{fix}\left(\frac{S_8(i,j)}{160}\right) \quad \forall\ i,j \tag{1.29}$$

where *fix*(.) refers to the rounding function of an element to the nearest integer toward zero. Applying this equation to the coefficients of S_8, we obtain the following matrix $\hat{S}_8$ depicted in (1.30) which requires few hardware resources when implementing it in hardware. It was applied also to the different sizes of the DST-VII in order to evaluate the efficiency of our proposed approximation measured in terms of PSNR and BR:

$$\hat{S}_8 = 128 \times \begin{bmatrix} 0 & 0 & 1 & 1 & 1 & 1 & 2 & 2 \\ 1 & 1 & 2 & 1 & 0 & 0 & -1 & -2 \\ 1 & 2 & 0 & -1 & -2 & -1 & 0 & 1 \\ 2 & 1 & -1 & -1 & 0 & 2 & 0 & -1 \\ 2 & 0 & -2 & 0 & 1 & -1 & -1 & 1 \\ 1 & -1 & 0 & 2 & -1 & 0 & 2 & -1 \\ 1 & -2 & 1 & 0 & -1 & 2 & -1 & 0 \\ 0 & -1 & 1 & -2 & 2 & -1 & 1 & 0 \end{bmatrix} \tag{1.30}$$

1.5.2 Video-coding performance

To evaluate the proposed approximate algorithm of DST-VII, the modified matrices of different sizes are integrated into the reference software model HM-16.6-JEM-5.0 [14]. The performance of the proposed approach is measured by the criteria mentioned in Table 1.8. Table 1.11 depicts the results of our proposed

Table 1.11 Performance of the proposed algorithm (with modified DST-VII transform matrices) versus the state-of-the-art algorithm

Class	**Sequence**	ΔBR **(%)**	$\Delta YUV - PSNR$ **(dB)**
A1	CamFireParty	−0.10	−0.04
	Drumbs100	−0.42	−0.03
	Tango	−0.60	−0.02
	ToddlerFountain	0.78	−0.04
A2	CatRobot	−0.63	−0.02
	TrafficFlow	−0.79	−0.03
	DaylightRoad	−0.97	−0.02
	RollerCoaster	0.07	−0.03
Average		−0.33	−0.03

Table 1.12 Bjøntegaard results of the approximate DST-VII transform for VVC

Class	Sequence	BD-rate (%)	BD-PSNR (dB)
A1	CampfireParty	1.60	−0.04
	Drumbs100	1.08	−0.02
	Tango	1.11	−0.01
	ToddlerFountain	1.86	−0.07
A2	CatRobot	1.02	−0.01
	TrafficFlow	1.33	−0.01
	DaylightRoad	1.03	−0.01
	Rollercoaster	1.45	−0.04
Average		1.33	−0.03

approximation algorithm versus the original one. It can be seen that the average of BR saving reaches 0.33% at the expense of a slight quality degradation of 0.03 dB. The best case corresponds to DaylightRoad video where the BR saving is 0.97%, while the worst case represents a BR loss of 0.78% for ToddlerFountain video. As for the video quality, the degradation varies from 0.02 to 0.04 dB.

Table 1.12 denotes the encoding comparison results of our proposed approximation of DST-VII with respect to original algorithm using BD metrics. The results denote a BR (BD-rate) increase of 1.33% in average and a quality degradation (BD-PSNR) of 0.03 dB in average. This proves that the coding performance loss is tolerable. The best and the worst case correspond, respectively, to the BD-rate loss values of 1.02% and 1.86% for CatRobot and ToddlerFountain videos. Video quality degradation is negligible with 0.01–0.07 dB BD-PSNR loss.

Figure 1.15(a) depicts the RD curves of the original HM-16.6-JEM-5.0 algorithm and the algorithm based on the approximation of DST-VII done for Drumbs100 and DaylightRoad videos. The evaluation of our proposed algorithm shows a BR saving with a slight degradation is video quality which is visually indistinguishable by human eye.

1.6 Conclusion

Video services are already the large majority of data traffic. They increasingly handling high-resolution contents which has led to significant issues whether in storage or transmission. Therefore, the increased demand for better video compression has intensified two essential issues which are the coding efficiency and the complexity of video compression systems. After the success of HEVC in providing a better coding efficiency with respect to previous standard, recently, the VVC which is the next generation of video-coding standard is aiming to get a higher coding efficiency.

This chapter has targeted to decrease the computational complexity brought by the transform module of HEVC and VVC by means of approximate computing. It has addressed the complexity evaluation as well as the design of efficient hardware

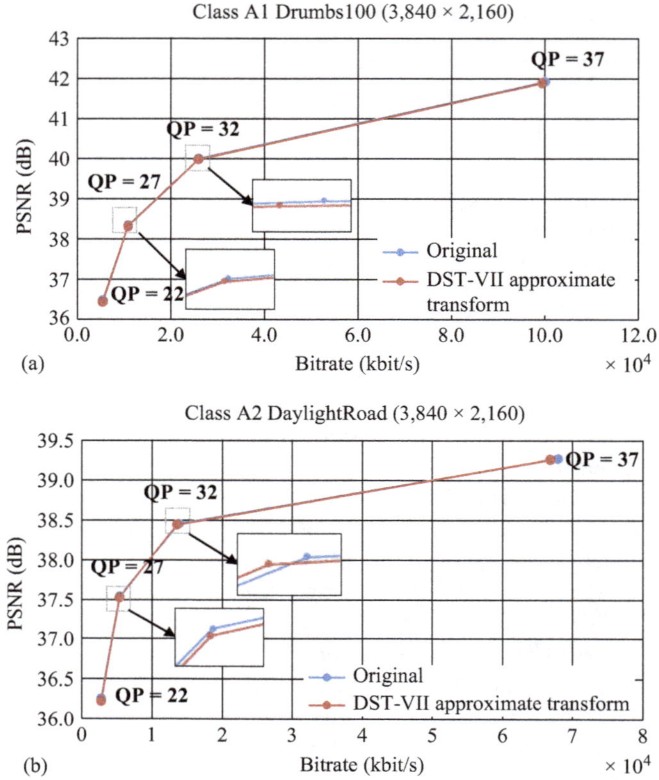

Figure 1.15 Rate-distortion curves for two test videos with QPs 22, 27, 32 and 37

architecture. First, this chapter has reviewed recent state-of-the-art approximate transforms used in HEVC. Second, an overview of video-coding concept in VVC standard was provided focusing on the transform module. Third, approximations of the DCT-II of VVC in addition to new approximation for the DST-VII have been conducted. The coding efficiency of the modified transform matrices of the VVC was evaluated by means of BR and PSNR criteria.

References

[1] Ohm JR, Sullivan GJ, Schwarz H, *et al.* Comparison of the coding efficiency of video coding standards—including high efficiency video coding (HEVC). IEEE Transactions on Circuits and Systems for Video Technology. 2012;22 (12):1669–1684.

[2] Kim IK, McCann K, Sugimoto K, *et al.* Hm10: High efficiency video coding (HEVC) test model 10 encoder description. In: Proc. Joint Collab. Team Video Coding; 2013.

[3] Sullivan GJ, and Ohm JR. Recent developments in standardization of high efficiency video coding (HEVC). In: Applications of Digital Image Processing XXXIII. vol. 7798. International Society for Optics and Photonics; 2010. p. 77980V.

[4] Hanhart P, Rerabek M, De Simone F, *et al.* Subjective quality evaluation of the upcoming HEVC video compression standard. In: Applications of Digital Image Processing XXXV. vol. 8499. International Society for Optics and Photonics; 2012. p. 84990V.

[5] Sullivan GJ, Ohm JR, Han WJ, *et al.* Overview of the high efficiency video coding (HEVC) standard. IEEE Transactions on Circuits and Systems for Video Technology. 2012;22(12):1649–1668.

[6] Correa G, Assuncao P, Agostini L, *et al.* Performance and computational complexity assessment of high-efficiency video encoders. IEEE Transactions on Circuits and Systems for Video Technology. 2012;22(12):1899–1909.

[7] Xu M, Li T, Wang Z, *et al.* Reducing complexity of HEVC: a deep learning approach. IEEE Transactions on Image Processing. 2018.

[8] Raffin E, Nogues E, Hamidouche W, *et al.* Low power HEVC software decoder or mobile devices. Journal of Real-Time Image Processing. 2016; 12(2):495–507.

[9] Bossen F, Bross B, Suhring K, *et al.* HEVC complexity and implementation analysis. IEEE Transactions on Circuits and Systems for Video Technology. 2012;22(12):1685–1696.

[10] Jens Rainer O, and Gary JS. Versatile Video Coding—towards the next generation of video compression. In: Picture Coding Symposium 2018: "Bridging the Gap" Invited Talk; 2018. Available from: http://www.pcs2018.com/uploads/1/1/0/0/110091239/pcs2018_vvc_overview_06_26.pdf.

[11] Marzuki I, and Sim D. Overview of potential technologies for future video coding standard (FVC) in JEM software. IEEE Transactions on Smart Processing & Computing. 2018;7(1):22–35.

[12] Multimedia CA. Beyond HEVC: Versatile Video Coding project starts strongly in Joint Video Experts Team; 2018. Available from: https://news.itu.int/versatile-video-coding-project-starts-strongly/.

[13] Ohm J, and Sullivan G. Meeting report, the 119th SC 29/WG 11 meeting, 2017-07-17/21, Torino, Italy [SC 29/WG 11 N 16922]. In: ISO/IEC JTC 1/SC 29 Coding of Audio, Picture, Multimedia and Hypermedia Information; 2017.

[14] Chen J, Alshina E, Sullivan G, *et al.* JVETE1001: Algorithm Description of Joint Exploration Test Model 5. Joint Video Exploration Team (JVET), Geneva, Tech Rep.; 2017.

[15] Mercat A, Bonnot J, Pelcat M, *et al.* Smart search space reduction for approximate computing: a low energy HEVC encoder case study. Journal of Systems Architecture. 2017;80.

[16] Haweel TI. A new square wave transform based on the DCT. Signal processing. 2001;81(11):2309–2319.

[17] Bouguezel S, Ahmad MO, and Swamy M. A novel transform for image compression. In: Circuits and Systems (MWSCAS), 2010 53rd IEEE International Midwest Symposium on. IEEE; 2010. p. 509–512.

[18] Bayer FM, Cintra RJ, Edirisuriya A, *et al.* A digital hardware fast algorithm and FPGA-based prototype for a novel 16-point approximate DCT for image compression applications. Measurement Science and Technology. 2012; 23(11):114010.

[19] Bouguezel S, Ahmad MO, and Swamy M. Binary discrete cosine and Hartley transforms. IEEE Transactions on Circuits and Systems I: Regular Papers. 2013;60(4):989–1002.

[20] Masera M, Martina M, and Masera G. Adaptive approximated DCT architectures for HEVC. IEEE Transactions on Circuits and Systems for Video Technology. 2017;27(12):2714–2725.

[21] Jridi M, Alfalou A, and Meher PK. Efficient approximate core transform and its reconfigurable architectures for HEVC. Journal of Real-Time Image Processing. 2018;16:1–11.

[22] Masera M. Transform Algorithms and Architectures for Video Coding; 2018.

[23] Clarke R. Relation between the Karhunen Loève and cosine transforms. In: IEEE Proceedings F-Communications, Radar and Signal Processing. vol. 128. IET; 1981. p. 359–360.

[24] Cintra RJ, and Bayer FM. A DCT approximation for image compression. IEEE Signal Processing Letters. 2011;18(10):579–582.

[25] Jridi M, Alfalou A, and Meher PK. A generalized algorithm and reconfigurable architecture for efficient and scalable orthogonal approximation of DCT. IEEE Transactions on Circuits and Systems I: Regular Papers. 2015;62 (2):449–457.

[26] Jridi M, and Meher PK. Scalable approximate DCT architectures for efficient HEVC-compliant video coding. IEEE Transactions on Circuits and Systems for Video Technology. 2017;27(8):1815–1825.

[27] Bossen F. Common test conditions and software reference configurations. Document JCTVC-L1100. 2013.

[28] Budagavi M, Fuldseth A, Bjøntegaard G, *et al.* Core transform design in the high efficiency video coding (HEVC) standard. IEEE Journal of Selected Topics in Signal Processing. 2013;7(6):1029–1041.

[29] Vivienne S, Madhukar B, and Gary J. High Efficiency Video Coding (HEVC): Algorithms and Architectures. Integrated Circuits and Systems. Springer International Publishing Switzerland; 2014.

[30] Wien M. High efficiency video coding. Coding Tools and Specification. 2015:133–160.

[31] Ohm JR, and Sullivan GJ. High efficiency video coding: the next frontier in video compression [standards in a nutshell]. IEEE Signal Processing Magazine. 2013;30(1):152–158.

[32] de la Rocha Gomez-Arevalillo A. Investigating the Adaptive Loop Filter in Next Generation Video Coding; 2017.

[33] Ben Jdidia S, Jridi M, Belghith F, *et al.* Low-complexity algorithm using DCT approximation for POST-HEVC standard. In: Pattern Recognition and Tracking XXIX. vol. 10649. International Society for Optics and Photonics; 2018. p. 106490Y.

[34] Sjövall P, Viitamäki V, Vanne J, *et al.* High-level synthesis implementation of HEVC 2-D DCT/DST on FPGA. In: Acoustics, Speech and Signal Processing (ICASSP), 2017 IEEE International Conference on. IEEE; 2017. p. 1547–1551.

[35] Budagavi M, and Sze V. Unified forward+ inverse transform architecture for HEVC. In: Image Processing (ICIP), 2012 19th IEEE International Conference on. IEEE; 2012. p. 209–212.

[36] Chang CW, Hsu HF, Fan CP, *et al.* A fast algorithm-based cost-effective and hardware-efficient unified architecture design of 4 × 4, 8 × 8, 16 × 16, and 32 × 32 inverse core transforms for HEVC. Journal of Signal Processing Systems. 2016;82(1):69–89.

[37] Suehring K, and Li X. JVET Common Test Conditions and Software Reference Configurations, document JVET-B1010. In: 2nd Meeting, San Diego, CA, USA; 2016.

[38] Grois D, Nguyen T, and Marpe D. Performance comparison of AV1, JEM, VP9, and HEVC encoders. In: Applications of Digital Image Processing XL. vol. 10396. International Society for Optics and Photonics; 2018. p. 103960L.

[39] Bjøntegaard G. Calculation of average PSNR differences between RD-curves. VCEG-M33; 2001.

Chapter 2

Joint algorithm-architecture design of video coding modules

Cláudio M. Diniz[1], Brunno Abreu[2], Mateus Grellert[1], Felipe Martin Sampaio[3], Daniel Palomino[4], Fábio Luís Livi Ramos[5], Bruno Zatt[4] and Sergio Bampi[2]

2.1 Introduction

Along the last three decades, video coding algorithms have been evolving driven by the popularization of digital videos and by the non-stopping demand for higher quality, larger resolutions, larger frame rates and immersion. Starting from a niche market in the early 1990s, digital videos became omnipresent with applications ranging from digital cinema to personal recording including video surveillance, telemedicine, video conferencing, video streaming, etc. Video coding experts have been enabling such evolution through the release of generations of video coding standards. H.261, published in 1988, and MPEG-1, published in 1993, represent the first video coding standards of a long-standing series that include MPEG-2, H.263, MPEG-4 Part 2, H.264/AVC (Advanced Video Coding) and the state-of-the-art High Efficiency Video Coding (HEVC), released in 2013.

The evolution of video coding standards brought impressive coding-efficiency improvements over the last years. However, as one can expect, such efficiency comes at the cost of increased computational complexity. For instance, the encoding complexity increase observed from H.264/AVC to its successor, the HEVC, ranges from 1.2× to 3.2× [1]. In turn, the current implementation of the upcoming Versatile Video Coding (VVC) standard can be 18.8× more complex than the reference implementation of HEVC (as shown in Section 2.3). Therefore, there is a need for high-throughput video coding solutions capable of employing optimized/simplified algorithms with no/reduced coding-efficiency losses.

[1]Graduate Program on Electronic Engineering and Computer Science, Catholic University of Pelotas, Pelotas, Brazil
[2]Informatics Institute – PGMICRO & PPGC, Federal University of Rio Grande do Sul, Porto Alegre, Brazil
[3]Campus Farroupilha, Federal Institute of Rio Grande do Sul, Farroupilha, Brazil
[4]Graduate Program on Computer Science, Federal University of Pelotas, Pelotas, Brazil
[5]Bachelor Program in Computer Engineering, Federal University of Bampa, Bagé, Brazil

According to CISCO [2], 71% of the worldwide internet video traffic will be generated/consumed using mobile devices by 2022. Given the battery/energy restrictions of such mobile devices, energy-efficient video (de)coding becomes mandatory leading to the need for dedicated hardware architecture. However, some algorithms employed by coding standards are not suitable for hardware implementation demanding hardware-friendly optimizations. As a result, it has been demonstrated [3–5] that high-throughput energy-efficient video coding systems can be achieved by jointly designing algorithms and hardware architectures.

Although every new video coding generation brings new tools, features and particularities, they follow a common hybrid video coding template. In this sense, it is possible to learn from challenges related to current video coding generation in order to anticipate the characteristics of future video coding standards and design-efficient custom architecture.

This chapter presents a variety of custom hardware modules conceived through joint algorithm-architecture design for the HEVC standard. Based on that and considering experiments performed over VVC software, we map such solutions to the context of future video coding and highlight the major challenges to be faced in the near future. A discussion regarding the video coding evolution and its state of the art, including a qualitative comparison between HEVC and VVC, is presented in Section 2.2. Section 2.3 brings an experimental comparison between HEVC and VVC reference software implementations highlighting coding efficiency and computational complexity perspectives. In order to bring a comprehensive discussion, this chapter covers most of the major modules of the video encoder: rate–distortion optimization (RDO) (Section 2.4), inter-frame prediction (Section 2.5), intra-frame prediction (Section 2.6), transforms (Section 2.7), in-loop filter (Section 2.8) and entropy coding (Section 2.9). Conclusions and upcoming challenges are presented in Section 2.10.

2.2 Video coding evolution and state of the art

2.2.1 Evolution of video coding standards

The most popular international video coding standards were developed along three decades by International Telecommunication Union (ITU-T) Video Coding Experts Group (VCEG) and by International Organization for Standardization (ISO) and International Electrotechnical Commission (IEC) Motion Picture Experts Group (MPEG). H.261 [6] is the first video coding standard developed by the VCEG. Other popular standards developed by ISO/IEC and/or ITU-T are MPEG-1 [7], H.262/MPEG-2 [8], H.263 [9], MPEG-4 Part 2 [10] and H.264/AVC [11].

The state-of-the-art video coding standard is HEVC [12] (ITU-T H.265 and ISO/IEC MPEG-H part 2). HEVC was developed by the Joint Collaborative Team on Video Coding (JCT-VC), formed by experts of VCEG and MPEG, to support ultra-high resolution video (beyond 1,920 × 1,080 pixel) with better compression than H.264/AVC. HEVC standard achieves 40% bit rate reduction (on average) compared to H.264/AVC for a similar objective video quality [1]. The support of larger block sizes (to deal with increased spatial resolution), the introduction of new flexible quadtree block partitioning structure and several advanced coding tools contribute to achieve such result [13]. However, HEVC encoder software (HEVC Test Model—HM) [14] is up to 3.2× more complex than H.264/AVC's Joint Test Model encoder [1].

Other video codecs exist, e.g., VP8, VP9, Daala, and Thor. Recently, Alliance for Open Media is developing the royalty-free AV1 codec [15]. Recent analysis shows that AV1 encoder is 6.8× more complex than HM encoder to achieve similar compression [16]. Another work found similar result and concludes that AV1 cannot transform its increased complexity into compression efficiency [17]. This chapter will focus on future standardization efforts of ITU-T and ISO/IEC.

The increasing demand for increased resolution and different video formats, such as high dynamic range and 360° videos, led ISO/IEC and ITU-T to establish the Joint Video Exploration Team (JVET) in October 2015 to study a new standard to succeed HEVC. After evidences collected from experiments on the Joint Exploration Model, JVET issued in October 2017 a call for proposals [18] with the goal to reduce bit rate of approximately 50% at the same subjective quality compared to HEVC [19]. Twenty-two groups responded to the call for proposals by April 2018. After that, JVET was renamed as Joint Video Experts Team and named the new standard as VVC. The first VVC draft [20] and reference software called VVC Test Model 1 (VTM 1) [21,22] started with a minimal set of common coding tools from the received proposals, while other tools are still under test. The publication of the standard is planned to 2020. The next section overviews and compares HEVC and VVC codecs.

2.2.2 Overview of HEVC and VVC codecs

Upcoming VVC video codec keeps the same hybrid video-compression concept that is used in previous video coding standards, e.g., H.264/AVC and HEVC, a block-based codec with prediction, transforms and entropy coding. Figure 2.1 shows the main components of a video encoder that follows this concept. First, the captured input video frames, represented in YCbCr color format, are partitioned into blocks of square size. Each block is processed by the encoder modules shown in Figure 2.1.

Prediction, transforms, quantization, and entropy-coding compose the forward path of video encoder. Intra-frame prediction generates a predicted block of samples based on reconstructed neighbor samples of the current frame. Inter-frame prediction is formed by motion estimation (ME) and motion compensation (MC) that generate a prediction based on samples from reference frames. ME searches in the reference frames for a block that is the most similar with the current block to be encoded. A motion vector that indicates the displacement between the most similar block (best block) and current block positions is sent in the bitstream. MC reconstructs the block samples indicated by the motion vector. A residue, i.e., the differences between predicted block and original block, is calculated and delivered to transforms and quantization module. Transforms convert the samples to frequency domain to decorrelate and concentrate the energy in a few coefficients. Quantization

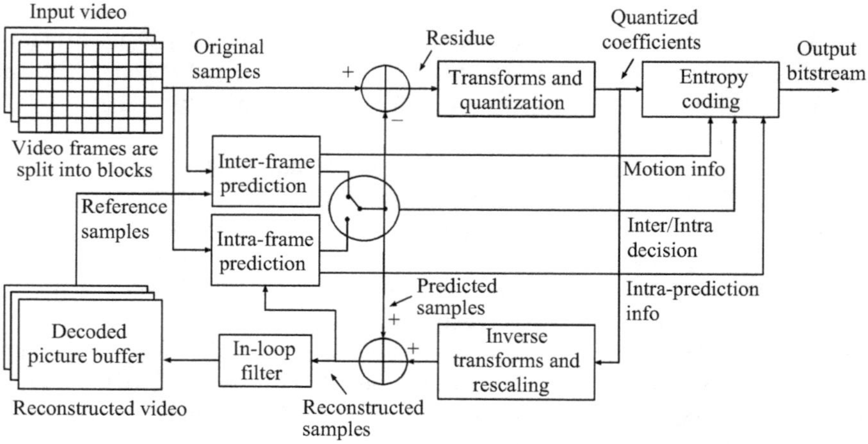

Figure 2.1 Generic video encoder diagram

eliminates some small coefficients that are not relevant to the human eye, incurring into coding losses. The quantization is controlled by a quantization parameter (QP) that is directly proportional to the strength of coding loss and controls the quality of reconstructed video. Every coding information (motion vectors, intra-prediction modes, intra/inter mode decision and residue) is delivered to the entropy coding module that generates the final bitstream of coded video, which will then be transmitted at a certain bitrate (BR).

The inverse path, which avoids mismatch of the reconstructed frames between the encoder and the decoder, is formed by inverse transforms and rescaling followed by in-loop filter. Inverse transforms convert quantized coefficients back to spatial domain that are further rescaled to the magnitude level of predicted samples. Intra-frame prediction is applied to unfiltered reconstructed samples. Inter-frame prediction uses input from in-loop filter that is applied to remove blocking artifacts caused by a strong quantization step. The output samples of in-loop filter module are stored into a decoded picture buffer (DPB), which are used as reference frames by inter-frame prediction. Since reconstructed frames are different from original frames (because of quantization), the quality loss must be evaluated with quality metrics, such as the peak signal-to-noise ratio (PSNR).

Figure 2.2 shows the main components of a video decoder. It looks like the inverse path of video encoder. The coded video bitstream input is delivered to entropy decoding module. The quantized coefficients are processed by inverse transforms and rescaling module, while other information, e.g., motion vectors, intra-frame prediction modes, are delivered to the prediction modules. Inter-frame prediction contains only the MC module. Predicted samples are added to the output of inverse transforms and rescaling to reconstruct blocks that are filtered by in-loop filter to produce the decoded video to be displayed and stored into DPB to be used as reference frames for inter-frame prediction.

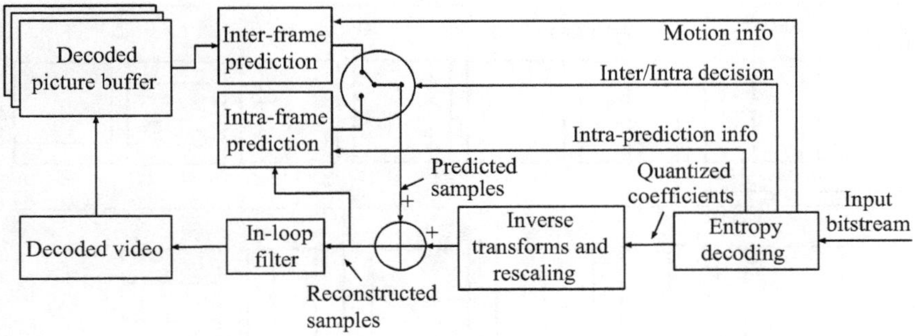

Figure 2.2 Generic video decoder diagram

Table 2.1 Tools currently supported by the HEVC and VVC reference encoders

	HEVC (HM) [12]	**VVC (VTM) [20,22]**
Block sizes (luma samples)	64 × 64–4 × 4	128 × 128–4 × 4
Partitioning structures	CU, PU, TU in a CTU quadtree structure	CTU quadtree with nested multi-type binary/ternary tree
Intra-frame prediction	33 Directional modes, DC and planar modes	65 Directional modes, DC and planar modes
Inter-frame prediction	1/4-pixel luma MV precision, 1/8-pixel chroma MV precision, merge mode	Affine motion compensation, up to 1/16 MV precision (in affine mode), adaptive motion vector resolution (AMVR)
Transforms	32 × 32–4 × 4 DCT, 4 × 4 DST	64 × 64–2 × 2 DCT, DST
Quantization	QP = [0, 51]	QP = [0, 63]
Entropy	CABAC	CABAC
In-loop filter	SAO, deblocking filter	SAO, deblocking filter, ALF

The main differences between HEVC and VVC standard is observed within the main codec modules, as shown in Table 2.1. It is important to note that HEVC standard has been concluded back in 2013, so the coding tools are defined in [12] and do not change. On the other hand, VVC is still under development, at the time of this book printing; thus some coding tools may be included or removed until the conclusion of the standard. Hence, we consider the coding tools included in the VVC Draft 1 and in VTM 1 [20,22].

HEVC and VVC divide video frames into coding tree units (CTUs), composed by one luma coding tree block (CTB) and two chroma CTBs. The maximum size of luma CTB in HEVC is 64 × 64 samples. Chroma CTBs have the maximum size of 32 × 32 samples when 4:2:0 color subsampling format is used. VVC increased the maximum CTB sizes to 128 × 128 luma samples and 64 × 64 chroma samples to support higher spatial resolutions.

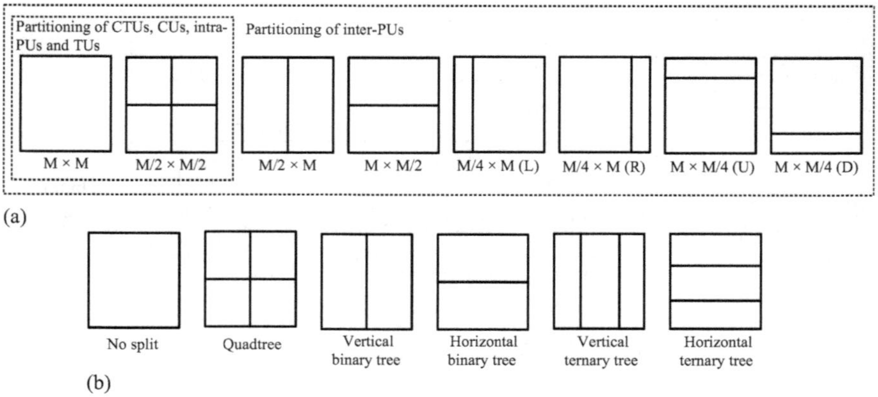

Figure 2.3 Block partitioning in (a) HEVC and (b) VVC standards

In HEVC, CTUs can be recursively partitioned in a quadtree manner into one or four coding units (CUs) with a restriction that the minimum size of luma CTB is 8 × 8 samples. Leaf CUs must be further partitioned into prediction units (PUs) and transform units (TUs) (as independent quadtrees) with the partitioning modes shown in Figure 2.3 with a minimum size of luma CTBs of 4 × 4 samples. Some partitioning modes are restricted to inter-PUs. Transform sizes define the size of TUs ranging from 32 × 32 down to 4 × 4 luma samples.

In VVC, a CTU is partitioned into CUs using a quadtree with nested multi-type tree structure, i.e., the quadtree leaf CUs nodes are partitioned into binary or ternary partitions, as shown in Figure 2.3. There are no PUs and TUs in VVC, so the leaf CUs are used for prediction and transforms without any further partitioning, unless the leaf CU is larger than the maximum transform size of 64 × 64 samples. The minimum size of luma CTB is 4 × 4 samples. VVC increased the number of partitioning ways because the binary and ternary partitions are not restricted to inter-PUs but can be used for intra-prediction and transforms (restricted to the maximum transform size is 64 × 64 luma samples and the minimum block size is of 4 × 4 luma samples).

Other coding tools included in HEVC and VVC are detailed in the following sections. Before them, we conduct an application analysis and comparison between HEVC and VVC reference software in Section 2.3.

2.3 Video coding application analysis

As previously stated, the VVC encoder aims at achieving a 50% compression gain for the same image quality compared with its predecessor, the HEVC. In order to achieve this, new tools are constantly being tested and compared with the current

Table 2.2 Setup for experiments

HEVC encoder	HM 16.9
VVC encoder	VTM 1.1 with SIMD optimizations
QP	22, 27, 32, 37
Frame count	150
Temporal structure	Random access
Computing platform	Intel Core i5-7400 CPU @ 3.00 GHz, 4 GB RAM

implementation in terms of coding complexity and compression gain. The tools presented in Table 2.1 are the ones currently implemented in the reference software implementation of each encoder, namely, HM [14] and VTM [21].

If we compare both implementations based solely on the supported tools, it is natural to expect a higher complexity in VVC encoders, as more partitioning possibilities, intra-directions, inter-prediction modes and transforms are supported. To mitigate this issue, the VTM software supports parallel processing as well as single instruction/multiple data (SIMD) optimizations. The SIMD kernels include interpolation filters, buffer allocation and distortion metrics, and none of them has any impact on the coding efficiency. The parallelism tools, however, break dependencies in the CABAC (context-adaptive binary arithmetic encoder (BAE)) context tables, which affects the encoding performance to some extent.

This section will discuss how the HM and VTM encoders compare in terms of encoding complexity and compression efficiency. Table 2.2 presents the setup used to obtain the results presented in this discussion. The QP values and Group of Pictures (GOP) structure are the ones recommended in the common test conditions document published by JCT-VC [23].

The Bjøntegaard delta bitrate (BD-rate) metric will be used to compare both encoders in terms of compression efficiency [24]. The BD-rate value compares a reference and a test encoder, measuring the average BR difference between them for the same image quality. This metric is computed using the following equation:

$$BD - rate(E_{ref}, E_{test}) = \frac{\int_a^b BR_{ref} - BR_{test}}{b - a} \tag{2.1}$$

In (2.1), BR_{ref} and BR_{test} are the BR values obtained at each point of the integration interval (a, b). The rate-distortion curves are obtained using a piece-wise cubic Hermite interpolating polynomial with at least four (BR, PSNR) pairs, one for each QP value (22, 27, 32, 37).

2.3.1 Analysis of VVC and HEVC encoders

2.3.1.1 Compression efficiency

The first set of results presented in this section show how the VVC standard compares with its previous version in terms of compression. The results presented

Table 2.3 BD-rate and encoding time ratio between the HM and the VTM encoders

Resolution	Sequence	BD-rate Y (%)	BD-rate YUV (%)	Time ratio[a]
2,560 × 1,600	PeopleOnStreet	−12.76	−14.13	2.54
	Traffic	−18.5	−19.95	1.17
	SteamLocomotive	−14.82	−18.41	1.70
	NebutaFestival	−4.72	−10.59	2.27
1,920 × 1,080	BasketballDrive	−16.91	−19.02	2.39
	ParkScene	−14.23	−15.33	1.37
	Kimono	−11.87	−12.71	2.28
	Cactus	−13.65	−15.2	1.85
	BQTerrace	−11.35	−14.39	1.19
832 × 480	BQMall	−14.28	−15.45	1.91
	RaceHorsesC	−10.23	−11.4	2.51
	BasketballDrill	−16.89	−18.17	2.06
	PartyScene	−12.98	−13.91	1.99
416 × 240	BasketballPass	−13.56	−14.66	1.94
	BQSquare	−11.44	−12.44	1.19
	BlowingBubbles	−12.83	−13.93	1.66
	RaceHorses	−11.43	−12.29	2.35
	Average	−13.09	−14.82	1.91

[a]VTM using SIMD optimizations.

in Table 2.3 represent the BD-rate for the Y and YUV PSNR values, as well as the ratio between the VTM and the HM encoding time. In this analysis, the SIMD optimizations were enabled as the main goal was to assess the compression efficiency of both encoders.

It is possible to observe that the VVC encoder is still far from its 50% improvement goal, achieving an average BD-rate gain of 13.1% (Y) and 14.8% (YUV). One of the reasons for the higher gain in the YUV BD-rate, which is a weighted combination of luminance and chrominance PSNRs, is likely the fact that chrominance CTBs may have a different tree structure than luminance ones in intra-predicted frames [22].

Another observation from Table 2.3 is that the results do not seem to be greatly affected by spatial resolution, but scene characteristics are still playing a major factor in the efficiency of the new standard. This goes against the versatile characteristic that is being targeted by the JVET group, showing that novel solutions capable of achieving a similar coding performance under diverse input sequences are necessary.

Using SIMD optimizations, the VTM encoder takes 1.91× more time to encode sequences on average. Of course this is not a fair comparison, but it already shows that the compression-complexity trade-off remains a problem in this new encoder. The next section will elaborate on this analysis using an unoptimized VTM encoder.

2.3.1.2 Computational effort

This section will assess how encoding computational effort[1] scales from one encoder to another across the different encoding steps. The Valgrind tool set [25] was used to dynamically instrument the HM and VTM references, accounting the cycles spent in each function. The VTM parallelism tools were not accounted for in this analysis, as this would compromise the confidence of our measurements. The SIMD optimizations were also disabled.

Figure 2.4 shows a bar chart containing the main kernels and the estimated cycles spent using the HM and VTM encoders. The white-filled boxes indicate the complexity-increasing factor of each tool. To obtain these results, 30 frames of each sequence were encoded using the Random Access structure and a QP of 22. The reduced number of frames was necessary to obtain results in feasible time, because Valgrind increases processing time by a factor of 4 on average due to its dynamic instrumentation.

Observing Figure 2.4 we can conclude that the VTM encoder is more complex than HM in practically every encoding step. Overall, VTM took 3.4 and 5.4 more time to encode the BQSquare and the RaceHorses sequences, respectively.

The increase factors with values above the average were written in bold in Figure 2.4. It is possible to observe that the fractional ME (FME) interpolation, the integer ME (IME) search and the intra-prediction search had a significant increase in complexity in this new standard.

The encoding time difference between the BQSquare (left part of Figure 2.4) and the RaceHorses (right part) sequences is a product of several coding tools, but the chart evidences that most of it lies in the IME search, which includes the fast test zone search (TZS) and the bi-prediction refinement. For the BQSquare sequence, the encoding complexity of this step increased by 4.28× compared with HM, whereas the increase was 18.77× when encoding RaceHorses. One of the reasons for such disparity comes from the fast search algorithm implemented in both HM and VTM, which takes longer to converge in scenes with complex motion.

The intra-prediction search should also be pointed out, even though this is not visually noticeable in the chart. By comparing HM and VTM, intra-prediction search takes 8.09× and 11.05× more cycles to be computed in the latter for the BQSquare and RaceHorses sequences respectively. Though this does not seem to pose any harm in this particular example, as the intra-prediction is responsible for less than 3% of the overall encoding time, some applications require all frames to be intra-coded (e.g., surveillance footage), so the VTM intra-prediction is clearly an important optimization target as well.

The reference implementations [14,21] used in this analysis serve as basis to assess the coding efficiency of the supported methods but are not necessarily followed in practical applications, since standards only define the decoding process.

[1]In this chapter, computational effort and complexity will be used interchangeably, both meaning processing time of the encoder applications.

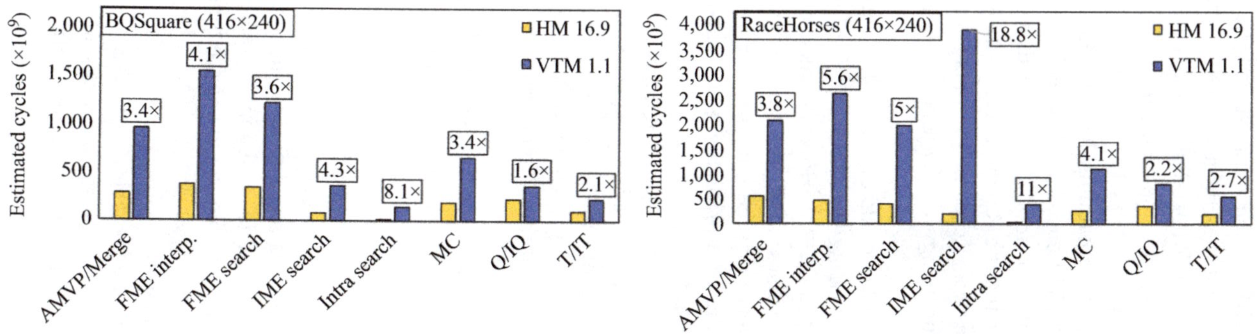

Figure 2.4 *Cycle estimation of the main encoding steps for the HM and VTM encoders (obtained with the Valgrind profiler) when encoding two sequences with distinct motion characteristics (QP = 22, 30 frames, Random Access)*

The following sections will discuss how video coding researchers exploit this fact to design efficient hardware and software solutions to mitigate encoding processing and energy requirements while keeping coding efficiency close to its reference.

2.4 Rate–distortion optimization

RDO is applied at the encoder to decide the coding modes that minimize the rate and the distortion of coded video [26]. RDO is formulated as an optimization problem to minimize the cost J, considering rate R and distortion D of each coding mode, as shown in the following equation:

$$J = D + \lambda \cdot R \tag{2.2}$$

Ideally, R and D are obtained at the output of entropy coding and in-loop filter, respectively. This process is known as full RDO. However, it has a huge computational complexity, due to the high number of combinations of block partitioning and prediction modes of current video encoders. In practice, some coding modes are skipped from *RD*-cost calculation, and D is obtained from the prediction step, simplifying the calculation of J.

The following sections will present two important coding tools related to this metric: the block-partitioning decision and the distortion computations (used to obtain D), highlighting the important contributions that optimize each one.

2.4.1 Block-partitioning decisions

As explained in Section 2.2.2, HEVC supports flexible coding structures to find the best encoding mode of different regions in the frame. This was necessary to achieve the superior coding performance of this standard, but it also conveys an optimization problem: finding the best CU, PU and TU partitioning in terms of *RD*-cost given all possible modes. The solution for this problem is not trivial, as there is no known algorithm capable of predicting the *RD*-cost efficiently.

The best solution can only be obtained by computing the actual cost of every possibility by applying the inter- and intra-frame prediction, transforms and quantization and their inverse counterparts, and entropy coding. This brute-force process, known as full RDO, requires a significant amount of computations. To mitigate this, several speed-up heuristics are implemented in the current HM version, skipping some CU, PU and TU partitions depending on specific *RD* tests. However, real-time HEVC encoding remains an open problem despite the current software optimizations. Therefore, solutions that further reduce the complexity of the HEVC partitioning decisions are still being studied.

The main approach adopted in related works consists in early termination techniques for the CTU quadtree splitting process. When a certain condition is met, the current CU is not further split, eliminating the computations that would be spent to process its four sub-CUs. Early terminating the quadtree nodes might remove important CU partitions that would increase the coding efficiency. Therefore, the challenge is designing efficient solutions to maximize the encoding time savings while minimizing the *RD* loss obtained from wrong early terminations.

In [27], data mining and machine-learning techniques are used to train classification models that decide whether the current CU being processed should be further split or not. By analyzing encoding process in HM, it was observed that *RD*-cost and best mode of the current CU are highly correlated with the final split decision. A decision tree was trained using these variables for each CU size. The reported results show a 36.7% encoding time reduction with a coding performance loss of 0.28% in BD-BR.

The work in [28] employs a classification-based CU-split decision using support vector machines (SVM). The classifiers are trained using encoding-domain variables obtained offline. The SVM model loading and prediction routines were implemented in the HM software to obtain encoding time savings and compression efficiency results. The results presented point to an average reduction of 48% in the encoding time, with a BD-BR loss of 0.48%.

In the following sections, we present VLSI architecture for the most computing-intensive encoding steps. Note, however, that the algorithmic solutions presented in this section can be jointly applied with the hardware architecture discussed hereafter, further reducing the overall encoding time.

2.4.2 Distortion metrics

Three metrics are usually employed in reference implementations to estimate the distortion D between two blocks: sum of absolute differences (SAD), sum of absolute transformed differences (SATD) and sum of squared differences (SSD) (a.k.a. sum of squared error). Although the most efficient metric to estimate distortion is SSD, as it approximates the PSNR better than other metrics, it is rarely found in practical hardware implementations because it would require higher complexity hardware modules. Hence, we focus on the most common metrics found in hardware implementations: SAD and SATD.

2.4.2.1 Sum of absolute differences

SAD is calculated as given in (2.3), in which O is the original block, R is the reference/candidate block, m and n are the dimensions of both blocks in samples.

$$SAD = \sum_{i=0}^{m-1} \sum_{j=0}^{n-1} |O_{i,j} - R_{i,j}| \tag{2.3}$$

An example of SAD hardware architecture with four original (O) and four reference (R) input samples is shown in Figure 2.5. It is composed by subtractors, absolute operators and an adder tree. An accumulator is placed at the output to calculate the final SAD for a block of $N \times 4$ samples in N cycles. Different SAD architecture with 8-input, 16-input or other number of input samples and pipeline levels can be designed depending on the needed throughput and the available memory bandwidth of DPB and current frame buffers.

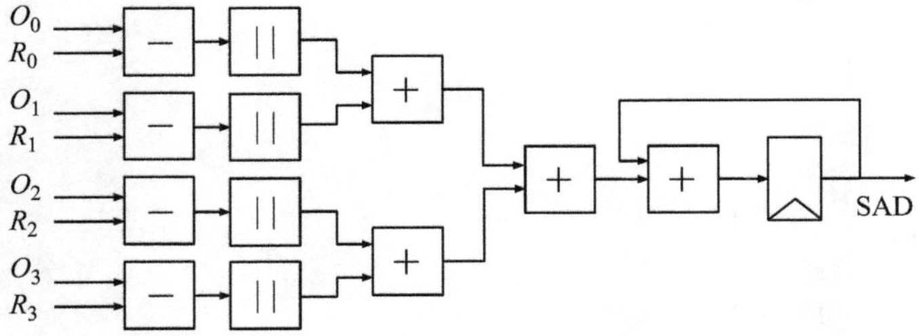

Figure 2.5 SAD hardware architecture with 4-input samples

Table 2.4 Results of state-of-the-art SAD architecture

	Technology node (nm)	**Maximum frequency (MHz)**	**Throughput (resolution @ frame rate)**	**Gate count**	**Area (μm^2)**	**Total power (μW)**
[29]	45	300	1,920 × 1,080@30fps	1,838	1,470	996.15
[30]	65	133	1,920 × 1,080@30fps	1,760	3,665	488
[31]	65	300	3,840 × 2,160@30fps	NA	NA	707.27

Recent works propose SAD architecture focusing on high-throughput and low-power dissipation. The adder tree can be implemented with different adder compressor topologies. Silveira *et al.* [29] propose an 8-input sample SAD architecture with 8-2 adder compressors implemented in different ways. Different absolute operators are proposed in [30]. The absolute operator designed with XNOR gates achieved the lower power dissipation compared to eight proposed solutions in [30] and the *abs* macro-function implemented by the synthesis tool. Abreu *et al.* [31] propose a partial distortion elimination optimization on SAD units which avoids the computation of candidates that will surely not be selected in the ME search. Results of the hardware implementations mentioned above are shown in Table 2.4. Power results in [29,31] are averaged of difference video sequences, while those reported in [30] are obtained with one 1,920 × 1,080 video sequence. Results are given in terms of the number of 2-input NAND gates that would fit in the same area in CMOS technology circuit.

2.4.2.2 Sum of absolute transformed differences

SATD calculation is given in (2.4), in which *HT(i,j)* is the Hadamard transform (HT) of the residual block *W*, as shown in (2.5). HM encoder implements SATD with the option to use HT with three different Hadamard matrix sizes: 2 × 2, 4 × 4 and 8 × 8, as shown in (2.6), (2.7) and (2.8).

$$SATD_{n\times n} = \sum_{i,j} |HT_{n\times n}(i,j)| \tag{2.4}$$

$$HT_{n\times n} = H_{n\times n} \cdot W \cdot H^T_{n\times n} / \left(\frac{n}{2}\right) \tag{2.5}$$

$$H_{2\times2} = \begin{bmatrix} 1 & 1 \\ 1 & -1 \end{bmatrix} \tag{2.6}$$

$$H_{4\times4} = \begin{bmatrix} 1 & 1 & 1 & 1 \\ 1 & -1 & 1 & -1 \\ 1 & 1 & -1 & -1 \\ 1 & -1 & -1 & 1 \end{bmatrix} \tag{2.7}$$

$$H_{8\times8} = \begin{bmatrix} 1 & 1 & 1 & 1 & 1 & 1 & 1 & 1 \\ 1 & -1 & 1 & -1 & 1 & -1 & 1 & -1 \\ 1 & 1 & -1 & -1 & 1 & 1 & -1 & -1 \\ 1 & -1 & -1 & 1 & 1 & -1 & -1 & 1 \\ 1 & 1 & 1 & 1 & -1 & -1 & -1 & -1 \\ 1 & -1 & 1 & -1 & -1 & 1 & -1 & 1 \\ 1 & 1 & -1 & -1 & -1 & -1 & 1 & 1 \\ 1 & -1 & -1 & 1 & -1 & 1 & 1 & -1 \end{bmatrix} \tag{2.8}$$

Two common architectural templates to implement SATD are found in the literature. Figure 2.6(a) shows a parallel solution for SATD architecture employing a 2×2 HT. The subtractors calculate the difference between original (O) and reference/predicted (R) samples that are further processed by a 2×2 HT and

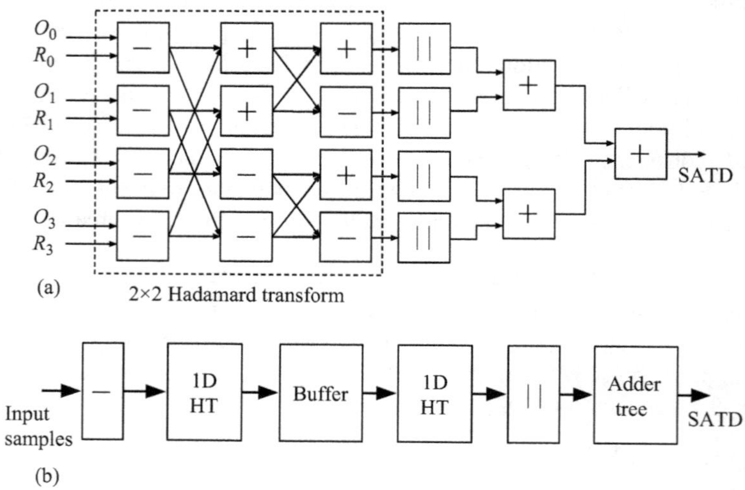

Figure 2.6 SATD hardware architecture: (a) parallel solution using 2 × 2 HT and (b) generic template of a semi-parallel HT solution

Table 2.5 Results of state-of-the-art SATD architecture

	Matrix sizes	Technology node (nm)	Maximum frequency (MHz)	Throughput (resolution @ frame rate)	Gate count	Area (μm^2)	Total power (μW)
[36]	4 × 4	45	650	NA	7,553	NA	4,535.6
[32]	4 × 4	45	352	NA	NA	NA	283.27
[32]	32 × 32	45	264.5	NA	NA	NA	2,424.4
[33]	2 × 2, 4 × 4, 8 × 8	45	125	NA	NA	70,321	7,412.6
[34]	8 × 8	45	333	1,920 × 1,080 @30fps	20,900	16,729	5,001.8
[35]	4 × 4, 8 × 8	45	32	NA	NA	14,978	467.8

absolute operators. An adder tree calculates the final SATD value. For larger HT sizes, it is common to employ the semi-parallel solution (Figure 2.6(b)) which use the separability property of HTs to design two 1D HTs and a buffer in between the HTs. The buffer guarantees that the second 1D HT is applied after all coefficients are processed by the first 1D HT.

In the same context of SAD, recent works propose SATD architecture focusing on high-throughput and low-power dissipation. Soares *et al.* [36] propose an approximate SATD architecture based on coefficient-pruning algorithm of 4 × 4 HT. The approximate computing strategy reduces power compared to precise SATD with negligible BD-PSNR reduction. Seidel *et al.* [32] propose different SATD architecture with HT sizes ranging from 4 × 4 to 32 × 32. It also evaluates two types of buffers (linear and transposition), but transposition buffer results in low operating frequency that is less useful for high-resolution video coding. Silveira *et al.* [33] propose an SATD architecture supporting multiple HT size (2 × 2, 4 × 4 and 8 × 8) and adder compressors to reduce power. Since the calculation of HTs of different sizes shares common operations, it is possible to reuse hardware of the architecture while supporting different HT sizes. The work in [34] proposes an SATD architecture with 8 × 8 HT using adder and subtractor compressors to reduce power in 14.4% compared to baseline SATD architecture. Monteiro *et al.* [35] demonstrate how previously computed HT can be reused in the computation of larger blocks. A generic SATD hardware design based on this demonstration is also shown in [35]. Results of the solutions mentioned above are shown in Table 2.5.

2.4.3 Challenges on rate–distortion optimization for VVC encoder

Most of the solutions for fast HEVC partitioning decisions focus on the binary CU split problem, and that there is a growing trend of using learning-based techniques to build a solution. In VVC, this problem is harder to solve, because the CU and PU are now part of the same structure, increasing the number of possible outcomes.

CUs in VVC can assume five different types of splitting, one of which is the quad-split also supported in HEVC (see Figure 2.3).

One possible approach is to train multi-class classifiers that are capable of predicting all of the possible partitioning modes supported in VVC. It is not possible to accurately predict if this will be an easier or harder problem compared with a binary framework, because it depends on how the attributes used for training will be distributed in each class (splitting mode).

A second approach would be relaxing the problem by putting all splitting modes in the same bag, so the decision to be made is still binary, i.e., either the current CU will not be split or it will be (in some of the five possible ways). This might be easier than the first approach, but it also has less potential of achieving significant time savings, because, when a split decision is made, every mode will have to be tested.

Some blocks will certainly be evaluated by SAD and SATD. State-of-the art SAD and SATD architecture target the performance requirements of HEVC encoder. The performance requirements of SAD and SATD increased in VVC encoder compared to HEVC encoder, as shown in the time increases of IME and FME search in Section 2.3. Note that VVC and HM use fast ME algorithms, as discussed in Section 2.5. VTM implementation uses different HT sizes to support the new block sizes.

Future solutions for RDO will need joint algorithm-architecture optimizations by combining fast-block partitioning algorithms with high-throughput hardware architecture employing intelligent reusing schemes, fast and low power arithmetic operators and newer CMOS technology node. New algorithms and architecture for inter-frame prediction is the key to reduce the number of blocks evaluated by SAD and SATD, as we discuss in Section 2.5.

2.5 Inter-frame prediction

As discussed in Section 2.2, inter-frame prediction is composed by ME and MC steps. Since current video coding standards support fractional-precision motion vectors, ME stage is split into two stages: the IME and the FME. The following subsections detail both of these stages. MC design is simplified in video encoders since the blocks can be reconstructed and saved inside the ME module. Another important consideration is the high memory bandwidth required to fetch reference and current blocks from buffers to SAD accelerator inside the ME architecture. Therefore, we have also discussed about dedicated memory design for ME.

2.5.1 Integer motion estimation

IME is responsible for searching on reference previously encoded frames for similar blocks to the current block to be encoded. The search occurs inside a search area smaller than the frame itself, because image patterns tend to slightly displace from the area where they were in temporally close frames. This process generates an integer-precision motion vector (x,y) pointing to the upper left corner of the best block found in the search. In the IME stage, the similarity between two blocks, i.e., the reference block and the current block, is calculated with the SAD metric in HEVC-compliant encoder software (such as HM and x265 [37]), as well as in VTM, in their default presets. SATD is used for FME and intra-frame prediction, as we will discuss next.

There are several fast block-matching algorithms (BMA) to search for similar blocks in the IME stage, since it is not normalized by standards. They reduce the number of candidate blocks to be searched, compared to exhaustive search (called full search, which is not implemented in practice due to huge complexity). Fast BMAs do not lead to the optimal block inside the search area, but have acceptable *RD*-cost results. Two well-known algorithms are analyzed in this chapter, since they are found in HEVC and VVC encoders: TZS, which is used in HM and VTM, and Hexagon search (HS), employed in x265. Both algorithms start their executions from a motion vector which was decided before the IME stage, based on the resulting motion vectors of the temporally and spatially neighboring blocks of the one being executed. The analysis presented in this chapter considers the TZS implementation from the HM encoder software, and the HS from x265.

Both TZS and HS are divided into four interdependent stages. Such dependencies between stages complicate pipeline-based architectural implementations that may waste a high amount of additional cycles due to pipeline bubbles. These stages are explained in the following items:

- Search vector initialization: This stage is responsible for checking whether other motion vectors may be better than the initial chosen one to start the search in the next steps. This stage checks for the (0,0) vector and the resulting vector from the $2N \times 2N$ prediction of the same CU, to analyze if any of these are better than the initial motion vector.
- First search: In TZS, a diamond-shaped search is performed, starting from the position defined by the previous stage. The diamond patterns test four, eight or sixteen candidates around the center, depending on the distance between samples, which is incremented after each diamond step. In HS, a six-point search in a hexagon shape is done around the center of the search. This stage ends when no better candidates have been found for three diamond steps (in TZS) or for one Hexagon iteration (in HS).
- Raster/Square search: This stage searches through the entire search area (in TZS), only executing if the current best vector is in a specific distance from the center, or in an eight-point area, after the first search (in HS).
- Refinement search: Considering TZS, the last stage works similarly to the first search. The main difference here is that the algorithm may be iterated several times with the center being set to the current best center in a new iteration. The algorithm stops when no better candidates have been found in an entire iteration. In HS, this last stage checks whether a set of candidates decided before the IME has any better candidates.

To achieve real-time throughput with an IME architecture, we have analyzed TZS and HS algorithms on default presets of HM and x265. Simulations were performed to obtain the number of 8 × 1 SAD calculations required to encode 1 s of five full-high definition (1,080p) sequences using the TZS and HS algorithms, as well as the quality difference (BD-BR) of HS when compared to TZS. The 8 × 1 SAD calculation is used to normalize the comparison for both algorithms, because of the different block sizes evaluated by SAD. The results are presented in Table 2.6.

The analysis presented in Table 2.6 shows that HS requires about 4× less 8 × 1 SAD computations when compared to TZS. The results are easily understandable, considering that the x265 is a faster encoding software and, therefore, needs to apply a faster BMA in its IME. However, when considering the quality results in terms of BD-BR, the HS obtains an average reduction of 0.63%, which indicates a considerable reduction in quality.

Therefore, similar analysis must be conducted in order to decide which BMA to implement in a hardware architecture, considering the demands of throughput and quality of a given application or service.

The design of an IME architecture highly depends on the SAD architecture used within it. More specifically, the input width and the number of pipeline levels of the SAD architecture are two crucial factors to be considered in the design of the IME architecture. When using a high input width, SAD for high blocks can be calculated with less chunks accumulations. However, this requires a higher number of bits to be sent to the architecture in a given cycle, which may be an issue when dealing with memory communication. Additionally, the importance of the number of pipeline levels is due to the fact that the recent BMAs have several dependencies between the stages, which requires the pipeline to be emptied at every transition between stages, which may compromise the effective throughput of the architecture. For instance, when considering the HS BMA, the square refinement stage cannot begin until the last SAD chunk of the Hexagon stage is sent to the SAD accumulator, given that the last calculated candidate may be the best one and the algorithm may not switch to the square refinement, continuing for another iteration instead.

Table 2.6 Number of 8 × 1 SAD requests and BD-BR values for 1 s of different video sequences, comparing the TZS and HS BMAs

Video sequence	**8 × 1 SAD calculations (1 s)**			**BD-BR (%)**
	TZS (×10^6)	**HS (×10^6)**	**Reduction (%)**	
Basketball Drive	14.11	2.39	83.1	−0.11
BQTerrace	6.36	2.24	64.8	−0.56
Cactus	6.37	1.81	71.6	−1.27
Kimono	5.24	1.14	78.2	−0.66
ParkScene	1.81	0.82	54.7	−0.55
Average	6.78	1.68	75.2	−0.63

Along with the SAD architecture, some additional registers are an important part of generic IME architecture. Registers that store the best SAD are crucial so that this SAD value can be compared with the one currently being calculated, to check whether the current candidate has a smaller SAD value than the current best one.

A kernel (control unit) that generates the vector requisitions for the memory is also required. The values of the vectors are represented by *x* and *y* coordinates. Considering the two aforementioned BMAs, vectors are generated based on their current center. Therefore, it is recommended that registers for the current center (which may change with every single iteration) are used, so that the control unit may generate the current vector to be sent to the memory based on the current center and on the current candidate position, relative to the center. This control unit is also responsible for checking whether the new SAD chunk coming to the SAD register should be added to the current one or loaded as a new SAD. The number of accumulations depends on the size of the PU being encoded, as well as on the input width of the SAD architecture.

An IME architecture was implemented considering the HS algorithm. The full architecture is shown in Figure 2.7. The architecture considered an 8×1 SAD module (64 bit for each block chunk), due to the fact that, in the default preset of the x265 software, the smallest PU dimension is 8, and using larger SAD modules could lead to unused hardware when calculating SAD for smaller PUs. The architecture did not employ pipeline in the SAD module, given that the HS contains several data dependencies between each of the Hexagon iterations, which would generate pipeline bubbles. The control block required to manage the new block, the accumulator register and the best SAD register are also shown in Figure 2.7.

We estimated the throughput required for the IME architecture to work in real-time based on the 8-byte requests from the x265 software, for the same Full HD

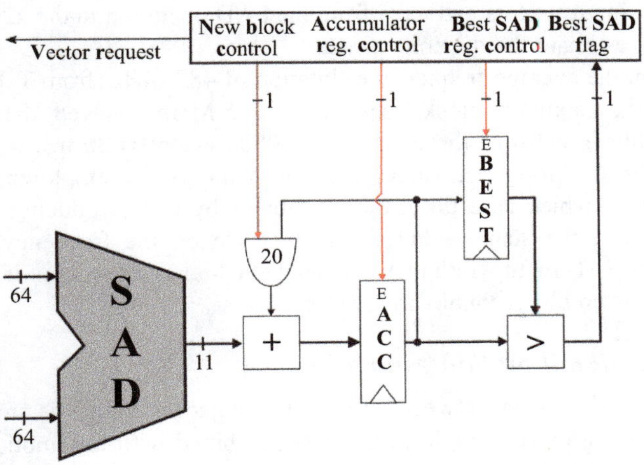

Figure 2.7 IME architecture and its control unit

Table 2.7 Frequency estimation (1,080p@30fps) for different video sequences

Video sequence	**Frequency (MHz)**
BasketballDrive	561.61
BQTerrace	402.93
Cactus	441.19
Kimono	641.61
ParkScene	388.07
Average	487

Table 2.8 Area and power results of the IME architecture (in 65nm CMOS)

Circuit area		**Power (μW)**		
μm^2	**kGates**	**Leakage**	**Dynamic**	**Total**
20,367.9	11.25	15.936	6,950.626	6,966.562

video sequences from the previous analysis, for 30 frames of each video (equivalent to 1 s of a 30 fps video). In our analysis, the number of 8-byte requests of the SAD operation in the HS is equivalent to the required frequency. Table 2.7 shows the estimated frequency for the five videos, as well as the average case, which corresponds to our frequency goal.

The architecture was synthesized for ASIC, with ST 65 nm CMOS standard cells library, for the maximum frequency of 555 MHz. The area and power results of the IME architecture are shown in Table 2.8. The power results were obtained by applying real input vectors extracted from the SAD operation in the x265 software, to accurately estimate the dynamic power.

Based on the average frequency estimation of 487 MHz (from Table 2.7), and considering the maximum clock frequency of 555 MHz achieved, the architecture is able to achieve real-time throughput for 1,080p videos at 30 fps.

Even higher throughput values may be achieved by employing hardware-friendly BMAs, which ease the use of pipeline—by not introducing high dependencies between the stages—due to the increase on the frequency. Moreover, increasing the SAD input width may also lead to a higher throughput if the memory is able to dispatch block chunks in fewer cycles.

2.5.2 Fractional motion estimation

In some cases, the offset between an object in adjacent frames is smaller than a pixel. This is a product of high frame rates combined with low motion intensity. Since the motion vectors produced in the IME step are not capable of representing fractional offsets, a second search is required to include this possibility. This search is called FME and is composed of two main steps: (i) an interpolation of fractional pixels, since the only ones available are in integer precision and (ii) a block-matching search to find the direction in which the sub-pixel offset occurred. The second step generates a fractional motion vector that is added to the integer one, producing the final vector encoded in the bitstream.

In HEVC, the FME is two-fold: In the first search, half-pixel samples are interpolated, then a search is applied in the interpolated area to find the half-pixel motion vector; in the second search, a quarter-pixel area is interpolated using both the integer and half-precision pixels as input, and a new search is performed.

For a block with $W \times H$ pixels, the area of both searches is limited to a $(2W + 1) \times (2H + 1)$ region surrounding the IME block, corresponding to eight fractional candidates. To generate this region in the first interpolation, a block of $(4 + W + 4) \times (4 + H + 4)$ integer samples must be fetched from memory (the 4-pixel padding on each side is required to interpolate border samples), so memory access is also an important concern for designers.

After the half search is completed, another $(2W + 1) \times (2H + 1)$ search occurs in the region surrounding the best block found in the half search, now using quarter-interpolated samples. It is common to interpolate half and quarter samples before searching, so the quarter regions surrounding all possible half-pixel candidates are generated speculatively, yielding a $(4W + 3) \times (4H + 3)$ area. Figure 2.8 shows a 2×2 integer block (shaded cells) and the surrounding half and quarter pixels that must be interpolated to perform both searches. Note that there are actually 4-pixel padding zones surrounding this block that were omitted for visual purposes.

The interpolation of fractional pixels in HEVC is computed using finite impulse response (FIR) filters with different input sizes (usually referred to as number of taps) depending on the precision (half or quarter) and on the color channel being processed (luminance or chrominance). An N-tap FIR filter consists

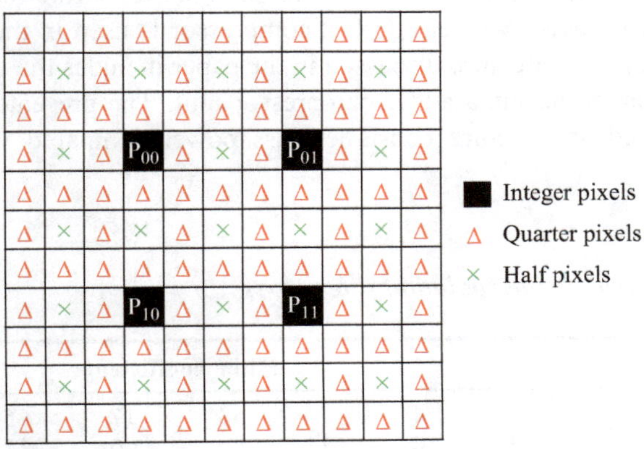

Figure 2.8 Interpolated area surrounding a 2 × 2 block (padding integer samples omitted)

of multiplying the input samples by a set of N coefficients and adding up the partial results.

Luminance pixels are interpolated using 8-tap filters for half pixels and 7-tap filters for the quarter ones. Table 2.9 shows the coefficients used to interpolate luminance pixels for the quarter (1/4 and 3/4) and half (2/4) precisions.

As pointed out in Section 2.3.1, interpolation is the most complex FME step due to its multiplications and large block sizes supported in the HEVC standard. Although hardware multipliers can be replaced by add/shift operations in this case (one of the terms is a constant coefficient), the throughput required to process fractional samples in real-time is enormous. Therefore, solutions that accelerate this process are demanded.

Diniz *et al.* [38] propose an HEVC interpolation architecture that implements luma and chroma filters using configurable design techniques. The architecture processes 12 pixel in parallel and configures its datapaths to support different filter types. The configuration logic also adapts to the different execution scenarios that require interpolation: ME or MC.

A second approach by Diniz *et al.* [39] employs dynamic reconfiguration in its design. Two engines are proposed to process luma and chroma samples with different throughput capabilities (up to 12 pixel in parallel) depending on the current data path configuration. This work also features a component that predicts the number of filter calls based on observations from previous frames and knowledge obtained from analyses. The output of this prediction is used in conjunction with the current block size in an adaptive scheduling algorithm that selects the smallest amount of data paths capable of achieving the required throughput. It is capable of achieving a 2,560 × 1,600@30fps throughput with a frequency of 283 MHz in the luma engine and 184 MHz in the chroma one.

The work proposed in [40] employs adder compressors to replace regular ones in the add/shift-based multiplication part of the interpolation. Adder compressors are power-efficient parallel adders that can be used when the partial sums are not required, which is the case with interpolation. For luma filtering, a 7-2 adder compressor is employed, whereas an 8-2 compressor is used in the chroma interpolation. A second architecture discussed in the paper includes the final adder used for accumulation in the luma adder compressor unit. The presented results show that the proposed architecture consumes less power compared with competing solutions.

Table 2.9 Luminance interpolation filter types

Filter types				**Filter coefficients**				
1/4	−1	4	−10	58	17	5	1	
2/4	−1	4	−11	40	40	−11	4	−1
3/4	1	−5	17	58	−10	4	−1	
Integer sample index	−3	−2	−1	0	1	2	3	4

Table 2.10 Results of state-of-the-art interpolation filter architecture

	Platform	Max. frequency (MHz)	Throughput resolution@fps[a]	Resource utilization	Area (μm^2)	Total power (μW)
[38]	TSMC 150 nm	312	3,840 × 2,160@30fps	30,209 gates	N/A	N/A
[39]	Virtex-5 65 nm	283/184	2,560 × 1,600@30fps	5,017 LUT + 2,550 register	N/A	89,000
[40]	Nangate 45 nm	707	N/A	11,201 gates	21,098	9,967
[41]	Nangate 45 nm	471	7,680 × 4,320@60fps	82,090 gates	65,508	33,790

[a]Considering motion compensation only.

A multi-standard interpolation architecture capable of processing the HEVC, AVS and MPEG-2 interpolations is proposed in [41]. The authors found that the different standards shared similarities despite the distinct filters employed in each of them. This similarity was used to design luma and chroma architecture that deliver the interpolated pixels of a given standard depending on the input coefficients. Resource utilization was optimized by reusing operators common among standards. Using a 45 nm standard cell technology, the proposed architecture is capable of achieving a 7,640 × 4,320@60fps throughput in HEVC mode using a target frequency of 471 MHz.

Table 2.10 shows a summary of the references aforementioned. Note that the throughput results of every work only consider the MC requirements.

2.5.3 Dedicated memories for motion estimation

Video codecs introduce very high pressure on memory hierarchy, leading to undesirable energy and performance overhead [42]. Besides employing novel complex coding tools, an HEVC encoder requires a significant amount of data from the off-/on-chip memories due to more memory intensive reference frames transmission for the prediction steps.

On average, the memory demand of HEVC is 2×–3× higher compared to the H.264/AVC [43]. Thus, high off-/on-chip memory bandwidth along with larger on-chip video memories (to support bigger resolutions) leads to increased energy consumption in the HEVC encoders. The memory bottleneck in HEVC encoders is related to the access to already processed (and reconstructed) video frames, called reference frames. In this aspect, up to 50% of off-/on-chip memory accesses are required for reference frames reading and writing leading to high memory-related energy consumption during HEVC encoding [42].

These aforementioned memory problems of HEVC tend to be aggravated when considering the next-generation of video encoders. In the case of current VVC draft, several adopted coding tools may intricate the memory bottleneck. Initially, the novel strategy for CTUs partitioning, which is based on a quad-tree with more possibilities to be analyzed by encoder control, will lead higher memory demand to inter-frame prediction. Precisely, the ME must be extensively executed to analyze more possibilities of motion vectors from several reference frames. Furthermore, refined FME precision (up to 1/16 pixel) will lead to extra on-chip memory to keep stored the generated fractional pixels. Still, another integrated coding tool may contribute to complicate the memory management for next-generation VVC encoders, like features of larger CTUs sizes and accurate motion vectors resolutions [20].

Therefore, there is a strong need of dedicated memory infrastructure (in both on-chip and off-chip perspective) and application-driven memory management to reduce the bandwidth requirements of current (HEVC) and next-generation (VVC) video coding applications.

To provide on-chip data-storage support in ME architecture in HEVC encoders, several works proposed on-chip video memory architecture along with memory management strategies [42,44,45]. The main goal is to reduce memory energy consumption by adopting novel memory organizations (combined private/shared memories with mixed cache/scratchpad design) and emerging memory technologies (SRAM along with MRAM/STT-RAM[2] cells).

2.6 Intra-frame prediction

Intra-frame prediction is included in the latest video encoders such as H.264/AVC, HEVC and the upcoming VVC. Intra-prediction is employed to exploit pixel correlation within a video frame. The intra-prediction is performed in the spatial domain by using neighbor samples already coded and reconstructed to predict the current block. The neighbor samples are localized in regions left, up, left-down and up-right from the current block and they can be interpolated in different ways to generate the predicted block.

The number of possible ways to perform the interpolation of the neighbor samples, called intra-prediction modes, is increasing with the evolution of the video coding standards, since they need to improve compression rates targeting high-resolution videos and immersive video. Although in the H.264/AVC there were nine intra-prediction modes, HEVC increases this number to 35, and for the VVC, this number is currently in 67 modes. Figure 2.9 shows the available intra-prediction modes in the HEVC.

The high number of intra-prediction modes together with all possible block sizes available in current standards makes the intra-prediction mode selection one of the main challenges of the video encoding. The following subsections present the intra-prediction mode selection and a hardware implementation of the intra-prediction for the HEVC standard.

[2]Magnetic/Spin Transfer-Torque RAM memory technology

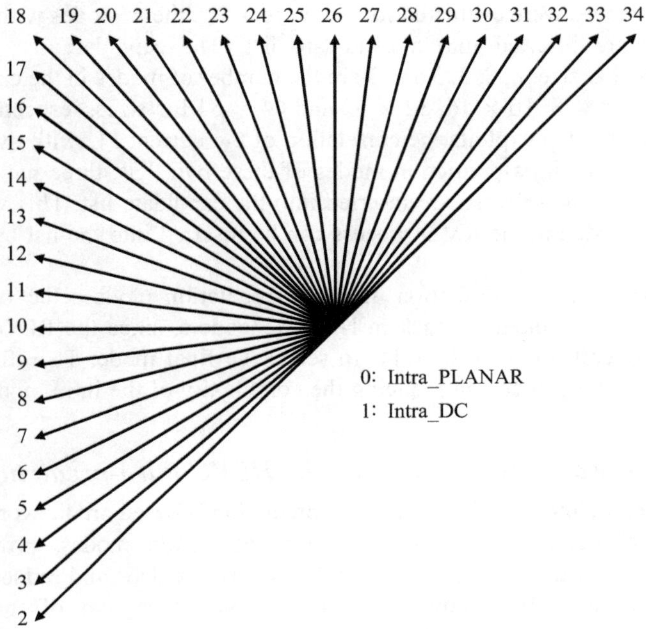

Figure 2.9 Intra-prediction modes in H.265/HEVC [46]

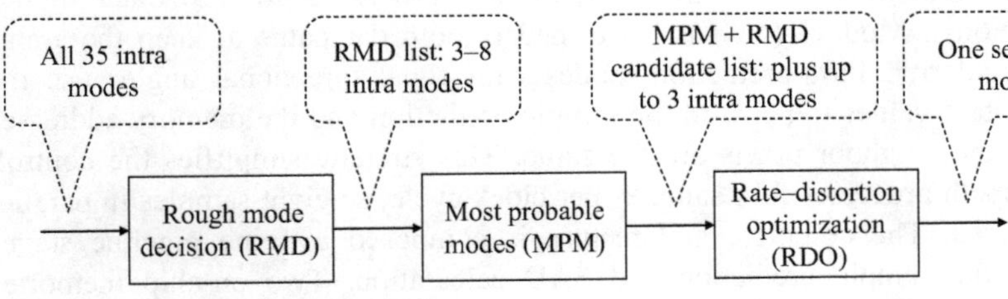

Figure 2.10 Intra-prediction mode decision in HEVC

2.6.1 *Intra-prediction mode decision in H.265/HEVC*

The main challenge of the intra-prediction in HEVC is to choose which mode will be used for each PU. RDO (Section 2.4) is used to deal with this decision; however, it cannot evaluate all possible intra-prediction modes, since it would be very time-consuming. The solution to deal with this problem evaluates only a subset of the total number of modes. In HM, the intra-prediction mode selection is performed using a three-step mechanism composed by rough mode decision (RMD), most probable mode (MPM) and RDO. Figure 2.10 shows this three-step mechanism.

RMD goal is to reduce computational complexity of RDO. In the RMD, the predicted PU is generated for all 35 modes and SATD is calculated to each mode.

Once all 35 modes were evaluated for the current PU, the *N* modes with the lowest RMD costs are inserted into a candidate list. The candidate list size varies according the PU size: 8, 8, 3, 3 and 3 are the number of modes in the candidate list for the 4 × 4, 8 × 8, 16 × 16, 32 × 32 and 64 × 64 PU sizes, respectively. After RMD, the MPM step exploits the correlation of the current PU with two neighbor PUs. Based on the intra-prediction modes of these two PUs, three modes that are most probable to be selected are inserted into the candidate list. This way, modes that were discarded in the RMD process can be inserted into the list by the MPM evaluation.

Even with this simplification, the intra-prediction mode selection is still a high computational intensive task in HEVC encoders, since the RDO process is performed several times in the CTU to select the final mode. Therefore, several works, such as [46], focus on reducing the complexity of the intra mode selection process.

2.6.2 Hardware architecture for the HEVC intra-prediction

The hardware architecture for intra-frame prediction is based on the work proposed in [47]. The architecture supports all intra-prediction modes, provides high throughput to real time processing of high-resolution videos and reduces memory accesses. Figure 2.11(a) shows a high-level block diagram of the designed architecture.

To achieve high throughput, the architecture is divided into two datapaths (called datapath above and datapath left). The datapath above processes all vertical modes (17 directions) while the datapath left processes all horizontal modes (17 directions). Mode 3 (45°) is performed in both datapaths to keep the architecture regularity. Intra-prediction modes with equal directional angles use the same constant values to perform the sample prediction and the memory addresses to access the neighbor pixels are the same. This strategy simplifies the control. Each datapath processes four samples per clock cycle, so eight samples in parallel are delivered. The complete architecture is composed by nine pipeline stages including the sample prediction and SAD calculation. Two on-chip memories (neighbor and original memory) were designed to locally store the input samples for intra-prediction.

First, all neighbor samples are fetched from the neighbor memory to neighbor buffer, once for each PU size. With all input samples available after the loading step, the pipeline strategy improves the throughput and the architecture usage, since after the latency of loading all samples, the architecture will always have valid input data to process. Figure 2.11(a) shows the timing schedule of the designed architecture. The time needed to load these neighbor samples varies from 9 to 129 clock cycles depending on the target PU size. The loading of the four original samples for each data path takes only one-clock cycle. Finally, it is possible to process all modes for the PU being predicted to calculate and store the SAD partial values. The architecture takes 73,682 clock cycles to process an entire CTU, i.e., all possible modes for all possible PU sizes. Table 2.11 presents the results of intra-frame prediction architecture.

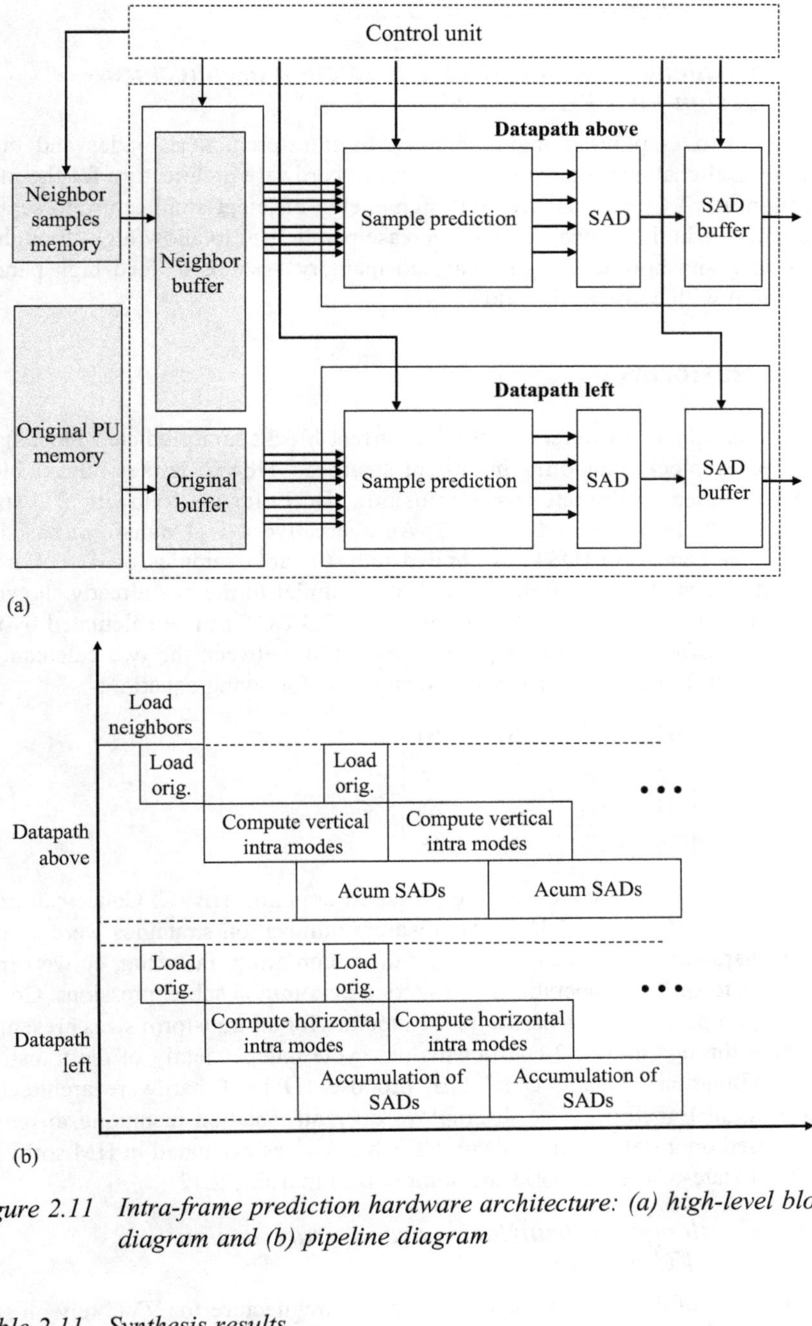

Figure 2.11 Intra-frame prediction hardware architecture: (a) high-level block diagram and (b) pipeline diagram

Table 2.11 Synthesis results

Technology node (nm)	**Gate count**	**Max. Freq. (MHz)**	**Throughput (resolution @ frame rate)**
65	36,734	500	1,920 × 1,080@13fps

2.6.3 Challenges on intra-frame prediction architecture design for VVC encoder

VVC standard employs a higher number of intra-prediction modes and block sizes. Several challenges arise when designing hardware architecture for the intra-prediction in VVC, such as how to design energy-efficient architecture to support more intra-prediction modes, how to increase parallelism to allow high-throughput architecture and how to design dedicated memory systems to feed high parallel architecture with constant data flow.

2.7 Transforms

The residual blocks (difference between current blocks and predicted blocks) are processed by block transforms in current standards. HEVC defines integer block transforms based on discrete cosine transform (DCT) for the following TU sizes: 4×4, 8×8, 16×16, and 32×32. An alternative 4×4 transform based on discrete sine transform (DST) is applied only to luma samples of 4×4 intra-predicted blocks. The transform calculation is similar to the one already shown in (2.5) for HTs in Section 2.4.2.1. Similar to HT, 2D DCT can be calculated by two 1D DCT transforms with a transposition operation between the two calculations. The 4×4 DCT transform matrix is shown in the following equation:

$$C_{4\times4} = \begin{bmatrix} 64 & 64 & 64 & 64 \\ 83 & 36 & -36 & -83 \\ 64 & -64 & -64 & 64 \\ 36 & -83 & 83 & -36 \end{bmatrix} \tag{2.9}$$

State-of-the-art DCT architecture are proposed targeting HEVC. Conceição *et al.* [48] propose a 16×16 2D DCT. Hardware minimization strategies were used to simplify hardware architecture, e.g., operation reordering, factoring, convert multiplications to shift-add operations and sharing of common sub-expressions. Goebel *et al.* [49] propose 1D DCT hardware architecture for all transform sizes presenting a constant throughput of 32 coefficients per cycle independently of the transform sizes combination. Bonatto *et al.* [50] propose 1D DCT hardware architecture supporting all transform sizes. It employs operand isolation technique to reduce power based on a statistic analysis on DCT block sizes evaluated in HM software. Results of state-of-the-art works are summarized in Table 2.12.

2.7.1 Challenges of transforms architecture design for VVC encoder

The challenge of designing transform hardware architecture for VVC rely on supporting more DCT sizes than in HEVC, ranging from 2×2 to 64×64. In addition to that, VVC unifies transform and prediction in the same units with same block size, which leads to the transforms to handle non-square blocks such as 8×4, 4×8 and so on. It results in an increase on the performance requirements, especially to large transform sizes (e.g., 32×32 and 64×64). To handle to such requirement increase, higher throughput hardware architecture design of DCT is desired.

Table 2.12 Results of state-of-the-art DCT architecture

	Dimension	Transform sizes	Technology node (nm)	Maximum frequency (MHz)	Throughput (resolution @ frame rate)	Gate count	Total power (mW)
[48]	2D	16 × 16	65	742.02	7,680 × 4,320 @238fps	14,954	135.4
[49]	1D	All	45	50	7,680 × 4,320 @32fps	97,300	24.2
[50]	1D	All	65	54.56	3,840 × 2,160 @32fps	124.8	30.56

2.8 In-loop filter

The in-loop filter is applied to reconstructed samples before delivering them to DPB to improve the *RD* results of inter-frame prediction. In HEVC, a deblocking filter (DF) is applied only to boundary samples of blocks larger than 4 × 4 samples in order to reduce blocking artifacts caused by strong quantization. Sample adaptive offset (SAO) filter is applied after DF to all samples of the image, by adding an offset of some pixels classified based on pixel intensity and edge properties. A few hardware solutions are found for DF and SAO, e.g., [51,52]. Adaptive loop filter (ALF) was considered to be included in HEVC but was removed prior to finalization of the standard. In-loop filter in VVC includes DF, SAO and ALF again. Since in our analysis (Section 2.3) these filters spend together less than 1% of encoding time of VTM, we foresee that their performance requirements will easily be targeted by current software/hardware solutions.

2.9 Entropy coding

The entropy coding is the final encoding step hybrid video coding standards, such as H.264/AVC, HEVC and VVC [11,12,18,19]. All data previously generated by the video encoder process is categorized into Syntax elements (SEs) that undergo the lossless entropy encoding process. In HEVC, the CABAC algorithm is used to perform this task.

CABAC is a binary arithmetic coding (BAC) algorithm [53], which means that the coding decisions are binary (i.e., the choice of the algorithm is always between two possibilities). The advantage of using a BAC comes from the fact that the alphabet of symbols is reduced to 2, simplifying the design of solutions compared with a multiple-symbol AC [54].

The control variables of this process are range, which corresponds to the interval in which the overall probabilities distributions for all encoded symbols are represented, and low, which is the smallest possible value.

CABAC is composed of three steps: binarization, context modeling and BAE [54]. The details of these blocks are presented next.

During binarization, the generated SEs are converted to specific binary representations, and each bit of this new representation is called a bin. A bin can be of three different types: regular, bypass and terminate. Most encoded bins fall into the regular and bypass categories.

Regular bins are divided into two groups: the most probable symbol (MPS) or the least probable symbol (LPS). This is the only type of bin that undergoes the context modeling step. Bypass bins, on the other hand, are categorized as their value (i.e., "0" or "1"), because each value is randomly distributed throughout the encoding flow. Thus, such bins do not require the context modeling step.

The context modeling updates the occurrence probabilities for regular MPS and regular LPS bins. Every time a regular bin occurs, the likelihood of occurrence of the next MPS or LPS for the same SE is updated. The new probability is indicated by a variable called State, which is used in a look-up table fashion to decide the occurrence probability of the LPS bin (and by consequence the MPS bin, as will be further explained).

The BAE is the central part of the CABAC block. To avoid the multiplications involved in the AC step, CABAC utilizes a precomputed Range for the LPS bin (i.e., the rLPS variable), which is addressed by the State variable coming from the context modeling [54]. Hence, the update of the remaining control variables follows (2.10) for regular bins. For bypass bins, the Range is not updated (since each bypass value is equiprobable), and the control variables follow (2.11). The Range and Low variables may be re-normalized if they fall below or above specific values [12].

$$\begin{cases} rMPS = Range - rLPS \\ Range = rMPS \quad \text{and} \quad Low = Low, & \text{when} \quad bin = MPS \\ Range = rLPS \quad \text{and} \quad Low = Low + rMPS, & \text{when} \quad bin = LPS \end{cases} \tag{2.10}$$

$$\begin{cases} Range = Range \\ Low = Low, & \text{when} \quad bin = 0 \\ Low = Low + Range, & \text{when} \quad bin = 1 \end{cases} \tag{2.11}$$

2.9.1 Upcoming challenges related to entropy encoding

As already mentioned, VVC will possibly make use of the CABAC as entropy encoding algorithm. The usage of CABAC implies that the BAE has to be the same, whereas the binarization and context modeling may change in case new SEs are added to the new standard, or different types of binarization are chosen for an already existing SE. The BAE is the critical part in terms of throughput bottleneck [55]. All recent CABAC alternatives propose multi-bin processing, where a multicore structure is a baseline for the BAE block, so the other blocks have to deliver the minimum required quantity of bins that the BAE can process per clock cycle. As reported in [55,56], a four-core BAE offers the best trade-off concerning the achievement in terms of throughput.

The first alternative consists in decreasing the critical path. The basic approach is to use a pipeline structure within the BAE. A possible four-stage pipeline architecture is depicted in Figure 2.12. The first stage corresponds to the rLPS

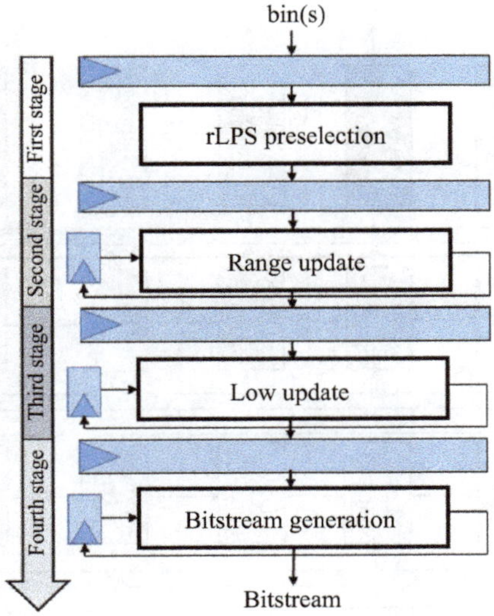

Figure 2.12 Baseline four-stage pipeline approach for the BAE

preselection, the second to the Range update, followed by the low (and auxiliary variable OB) update and finally the bitstream generation in the final stage [56].

The second option is to reduce the logic of the Range update. It is possible to move the rLPS variable renormalization to the first stage and process it speculatively, as proposed in [57], increasing maximum frequency by 9.6%. This solution was named pre-normalization rLPS (PNrLPS).

Another alternative is to use heterogeneous cores that can process only a particular type of bin, instead of all of them. The approach presented in [55], named Lookahead Range for the Least Probable Symbol (LH rLPS), uses a structure where seven distinct cores are employed in the four-stage pipeline BAE design. The designed cores are capable of processing either LPS or MPS bins. In doing so, some cores require less hardware resources and a reduced critical path. This technique improves frequency by 29% in frequency [55].

In [57], a solution that completely separates the flow of regular and bypass bins, named bypass-bin splitting (BPBS), is presented. Hence, the merge of the streams occurs only at the Low update stage, which now has five cores, each one able to process a single regular bin of any type or a single bypass bin. This solution increases the bins/cycle throughput in around 9.5%.

This idea was improved by processing more than one bypass bin per low core in [58,59]. The proposed scheme processes two bypass bins per core in parallel, with an average increase of 12.76% in throughput.

A final remark is that the techniques presented herein can be jointly applied to achieve even better results, which is depicted in Figure 2.13. The usage of all techniques combined leads to an eight-stage pipeline structure, where the first stage is the rLPS preselection [56] integrated with PNrLPS [57]. The first and the second

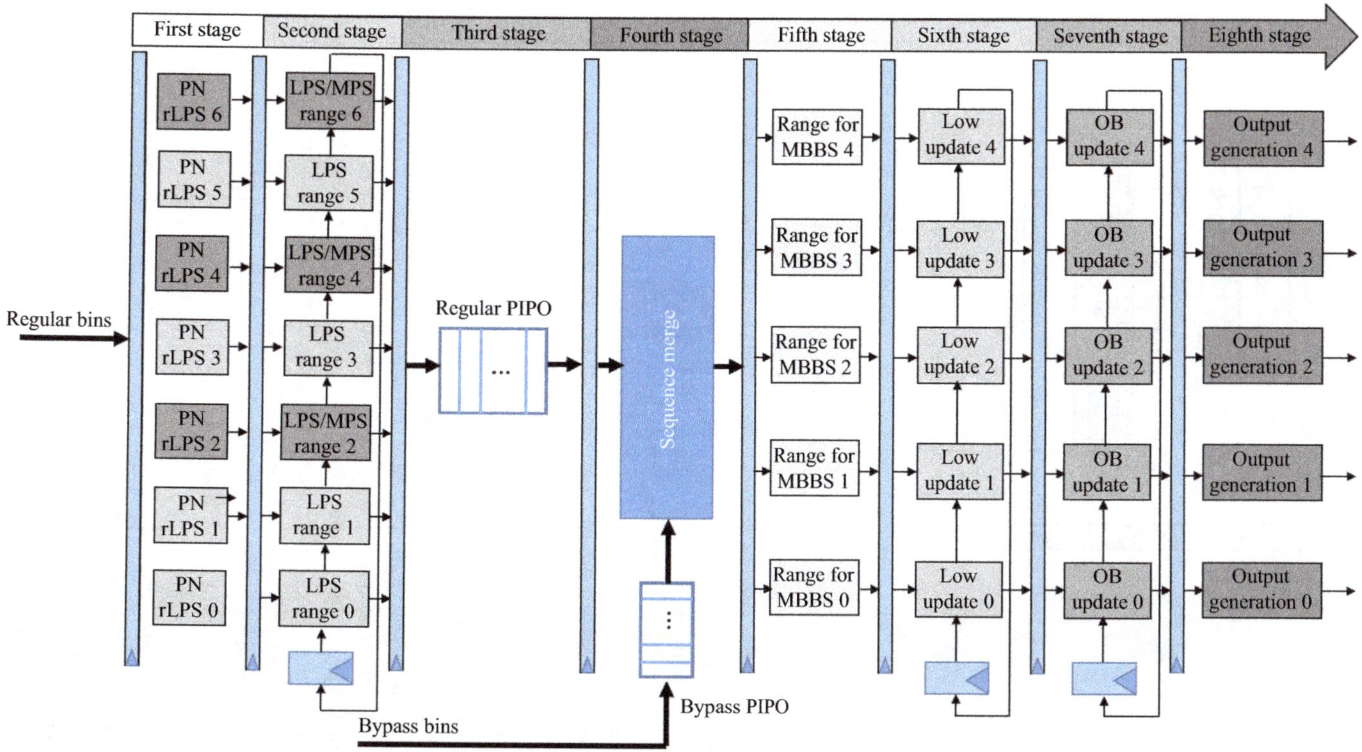

Figure 2.13 BAE architecture with all related techniques integrated

stages use the LH rLPS technique [55], necessary only for regular bins, since the usage of BPBS [57] separate the flows, which is shown in the same figure. Finally, the Multiple-Bypass Bins Scheme (MBBS) [59] technique is integrated into the fifth, sixth, seventh and eight stages.

2.10 Conclusions

This chapter overviews the joint effort of ITU-T and ISO/IEC to develop the upcoming VVC standard that will outperform HEVC standard by 50% of compression efficiency for similar video quality. Our analysis on current video encoder of VVC reference software (VTM 1) shows that one order-of-magnitude increase in encoding time is expected for VVC encoder compared to HEVC encoder. The complexity growth numbers are expected to increase since the standardization process is still in the beginning and the reference software includes only a small set of coding tools, while others are under test. The BD-rate reduction of VTM 1 compared to HEVC are far from the goal of 50%, but other sophisticated (and complex) coding tools are under test to reach this goal. Monthly meetings are happening to define and test coding tools to be included in each VTM version. The final standard is planned to 2020.

This chapter also presented some state-of-the-art solutions on joint algorithm-architecture design of video coding modules, mainly targeted to HEVC, and discussed the challenges on adapting or designing new solutions for VVC standard. The new block partitioning is a key tool that affects many modules, such as RDO, inter- and intra-frame prediction, and transforms. It has also increased the performance requirements because of the rise of possible combinations of block partitioning and coding modes. We foresee that the future video coding systems will need joint algorithm-architecture optimizations by combining fast block partitioning algorithms with high-throughput hardware architecture employing intelligent reusing schemes, fast and low power arithmetic operators implemented in newer technology nodes. These types of challenges on implementing high-throughput hardware modules need to be tackled by industry and academic communities to make the future VVC video-coding standard usable for a wide range of applications.

Acknowledgements

The authors would like to thank CNPq, CAPES and FAPERGS Brazilian funding agencies for financial support to their research.

References

[1] Vanne J, Viitanen M, Hamalainen TD, *et al.* Comparative Rate-Distortion-Complexity Analysis of HEVC and AVC Video Codecs. IEEE Transactions on Circuits and Systems for Video Technology. 2012;22(12):1885–1898.

[2] CISCO. Cisco visual networking index: Forecast and methodology 2017–2022; 2018.
[3] Khan MUK, Shafique M, Grellert M, *et al.* Hardware-software collaborative complexity reduction scheme for the emerging HEVC intra encoder. In: Proceedings of the Conference on Design, Automation and Test in Europe. EDA Consortium; 2013. p. 125–128.
[4] Shafique M, and Henkel J. Hardware/software architectures for low-power embedded multimedia systems. Springer Science & Business Media; 2011.
[5] Zatt B, Shafique M, Henkel J, *et al.* 3D video coding for embedded devices. Springer; 2013.
[6] ITU-T. Recommendation H.261, version 1. Video Codec for Audiovisual Services at 64 kbit/s; 1988.
[7] ISO/IEC. 11172-2 – MPEG 1. Coding of Moving Pictures and Associated Audio for Digital Storage Media at up to About 1.5 Mbit/s – Part 2: Video; 1993.
[8] ITU-T/ISO/IEC. Recommendation H.262 and ISO/IEC 13818-2 (MPEG 2 Video). Generic Coding of Moving Pictures and Associated Audio Information – Part 2: Video; 1994.
[9] ITU-T. ITU-T Recommendation H.263. Video Coding for Low Bit Rate Communication; 1995.
[10] ISO/IEC. Coding of Audio-Visual Objects – Part 2: Visual, ISO/IEC 14496-2 (MPEG-4 Visual version 1); 1999.
[11] ITU-T/ISO/IEC. ITU-T Recommendation H.264 and ISO/IEC 14496-10 (MPEG-4 AVC). Advanced Video Coding for Generic Audiovisual Services; 2003.
[12] ITU-T/ISO/IEC. ITU-T Recommendation H.265 and ISO/IEC 23008-2. High Efficiency Video Coding; 2013.
[13] Sullivan GJ, Ohm JR, Han WJ, *et al.* Overview of the High Efficiency Video Coding (HEVC) Standard. IEEE Transactions on Circuits and Systems for Video Technology. 2012;22(12):1649–1668.
[14] JCT-VC, HEVC Test Model (HM) v. 16.9 [Online]. Available at: http://hevc.hhi.fraunhofer.de.
[15] Alliance for Open Media (AOM). AV1 Bitstream & Decoding Process Specification, version 1.0.0 [Online]. Available at: https://aomedia.org; 2018.
[16] Nguyen T, Marpe D. Future video coding technologies: A performance evaluation of AV1, JEM, VP9, and HM. In: 2018 Picture Coding Symposium (PCS); 2018. p. 31–35.
[17] Laude T, Adhisantoso YG, Voges J, *et al.* A comparison of JEM and AV1 with HEVC: Coding tools, coding efficiency and complexity. In: 2018 Picture Coding Symposium (PCS); 2018. p. 36–40.
[18] ITU-T/ISO/IEC. Joint Call for Proposals on Video Compression with Capability beyond HEVC. Joint Video Exploration Team (JVET); 2017.
[19] ITU-T. Requirements for future video coding (FVC). In: ITU-T SG16/Q6 VCEG 56th Meeting, Torino, Italy, Jul. 2017, Doc. VCEG-BD03 (available at http://wftp3.itu.int/av-arch/video-site/1707_Tor); 2017.

[20] Bross B. Versatile video coding (Draft 1). In: 10th Meeting of Joint Video Experts Team (JVET): San Diego, US, 10–20 Apr. 2018, Doc. JVET-J1001-v2 (available at http://phenix.it-sudparis.eu/jvet/); 2018.

[21] JVET, VVC Test Model (VTM) v. 1.1 [Online]. Available at: http://jvet.hhi.fraunhofer.de.

[22] Chen J, and Alshina E. Algorithm description for Versatile Video Coding and Test Model 1 (VTM 1). In: 10th Meeting: San Diego, US, 10–20 Apr. 2018, Doc. JVET-J1001-v2 (available at http://phenix.it-sudparis.eu/jvet/); 2018.

[23] Bossen F. Common test conditions and software configurations. In: JCT-VC Document No. L1100; 2013.

[24] Bjøntegaard G. Calculation of Average PSNR Differences between RD-Curves. Video Coding Experts Group from ITU-T; 2001. VCEG-M33.

[25] Nethercote N, and Seward J. Valgrind: A framework for heavyweight dynamic binary instrumentation. In: ACM SIGPLAN Notices. vol. 42. ACM; 2007. p. 89–100.

[26] Sullivan GJ, and Wiegand T. Rate-Distortion Optimization for Video Compression. IEEE Signal Processing Magazine. 1998;15(6):74–90.

[27] Correa G, Assuncao PA, Agostini LV, *et al.* Fast HEVC Encoding Decisions Using Data Mining. IEEE Transactions on Circuits and Systems for Video Technology. 2015;25(4):660–673.

[28] Grellert M, Zatt B, Bampi S, *et al.* Fast Coding Unit Partition Decision for HEVC Using Support Vector Machines. IEEE Transactions on Circuits and Systems for Video Technology. 2019;29(6):1741–1753.

[29] Silveira B, Paim G, Abreu B, *et al.* Power-Efficient Sum of Absolute Differences Hardware Architecture Using Adder Compressors for Integer Motion Estimation Design. IEEE Transactions on Circuits and Systems I: Regular Papers. 2017;64(12):3126–3137.

[30] Abreu B, Paim G, Grellert M, *et al.* Exploiting absolute arithmetic for power-efficient sum of absolute differences. In: 2017 24th IEEE International Conference on Electronics, Circuits and Systems (ICECS); 2017. p. 522–525.

[31] Abreu B, Santana G, Grellert M, *et al.* Exploiting partial distortion elimination in the sum of absolute differences for energy-efficient HEVC integer motion estimation. In: 2018 31st Symposium on Integrated Circuits and Systems Design (SBCCI); 2018. p. 1–6.

[32] Seidel I, Bräscher AB, Güntzel JL, *et al.* Energy-efficient SATD for beyond HEVC. In: 2016 IEEE International Symposium on Circuits and Systems (ISCAS); 2016. p. 802–805.

[33] Silveira B, Ferreira R, Paim G, *et al.* Low power SATD architecture employing multiple sizes Hadamard Transforms and adder compressors. In: 2017 15th IEEE International New Circuits and Systems Conference (NEWCAS); 2017. p. 277–280.

[34] Silveira B, Abreu B, Paim G, *et al.* Using adder and subtractor compressors to sum of absolute transformed differences architecture for low-power video

encoding. In: 2017 24th IEEE International Conference on Electronics, Circuits and Systems (ICECS); 2017. p. 409–493.

[35] Monteiro M, Seidel I, and Güntzel JL. On the calculation reuse in Hadamard-based SATD. In: 2018 IEEE 9th Latin American Symposium on Circuits Systems (LASCAS); 2018. p. 1–4.

[36] Soares LB, Diniz CM, da Costa EAC, *et al.* A novel pruned-based algorithm for energy-efficient SATD operation in the HEVC coding. In: 2016 29th Symposium on Integrated Circuits and Systems Design (SBCCI); 2016. p. 1–6.

[37] Multicoreware, HEVC x265 Encoder [Online]. Available at: https://bitbucket.org/multicoreware/x265/.

[38] Diniz CM, Shafique M, Bampi S, *et al.* High-throughput interpolation hardware architecture with coarse-grained reconfigurable datapaths for HEVC. In: 2013 IEEE International Conference on Image Processing; 2013. p. 2091–2095.

[39] Diniz CM, Shafique M, Bampi S, *et al.* A Reconfigurable Hardware Architecture for Fractional Pixel Interpolation in High Efficiency Video Coding. IEEE Transactions on Computer-Aided Design of Integrated Circuits and Systems. 2015;34(2):238–251.

[40] Diniz CM, Fonseca MB, da Costa EAC, *et al.* Evaluating the Use of Adder Compressors for Power-Efficient HEVC Interpolation Filter Architecture. Analog Integrated Circuits and Signal Processing. 2016;89(1):111–120.

[41] Penny W, Goebel J, Paim G, *et al.* High-Throughput and Power-Efficient Hardware Design for a Multiple Video Coding Standard Sample Interpolator. Journal of Real-Time Image Processing. 2019;16(1):175–192.

[42] Sampaio FM, Zatt B, Shafique M, *et al.* Hybrid Scratchpad Video Memory Architecture for Energy-Efficient Parallel HEVC. IEEE Transactions on Circuits and Systems for Video Technology. 2018:1. doi: 10.1109/TCSVT.2018.2870825

[43] Sampaio F, Shafique M, Zatt B, *et al.* dSVM: Energy-efficient distributed Scratchpad Video Memory Architecture for the next-generation high efficiency video coding. In: 2014 Design, Automation Test in Europe Conference Exhibition (DATE); 2014. p. 1–6.

[44] Sampaio F, Shafique M, Zatt B, *et al.* Approximation-aware multi-level cells STT-RAM cache architecture. In: 2015 International Conference on Compilers, Architecture and Synthesis for Embedded Systems (CASES); 2015. p. 79–88.

[45] Sampaio F, Shafique M, Zatt B, *et al.* Content-driven memory pressure balancing and video memory power management for parallel high efficiency video coding. In: 2014 IEEE/ACM International Symposium on Low Power Electronics and Design (ISLPED); 2014. p. 153–158.

[46] Palomino D, Cavichioli E, Susin A, *et al.* Fast HEVC intra mode decision algorithm based on new evaluation order in the coding tree block. In: 2013 Picture Coding Symposium (PCS); 2013. p. 209–212.

[47] Palomino D, Sampaio F, Agostini L, *et al.* A memory aware and multiplierless VLSI architecture for the complete intra prediction of the HEVC emerging standard. In: 2012 19th IEEE International Conference on Image Processing; 2012. p. 201–204.

[48] Conceição R, Souza J, Jeske R, *et al.* Low-Cost and High Throughput Hardware Design for the HEVC 16 × 16 2-D DCT Transform. Journal of Integrated Circuits and Systems. 2014;9(1):25–35.

[49] Goebel J, Paim G, Agostini L, *et al.* An HEVC multi-size DCT hardware with constant throughput and supporting heterogeneous CUs. In: 2016 IEEE International Symposium on Circuits and Systems (ISCAS); 2016. p. 2202–2205.

[50] Bonatto LVM, Ramos FLL, Zatt B, *et al.* Low-power multi-size HEVC DCT architecture proposal for QFHD video processing. In: Proceedings of the 30th Symposium on Integrated Circuits and Systems Design: Chip on the Sands. ACM; 2017. p. 41–46.

[51] Diniz CM, Shafique M, Dalcin FV, *et al.* A deblocking filter hardware architecture for the high efficiency video coding standard. In: 2015 Design, Automation Test in Europe Conference Exhibition (DATE); 2015. p. 1509–1514.

[52] Rediess F, Conceição R, Zatt B, *et al.* Sample adaptive offset filter hardware design for HEVC encoder. In: 2014 IEEE Visual Communications and Image Processing Conference; 2014. p. 299–302.

[53] Moffat A, Neal RM, and Witten IH. Arithmetic coding revisited. In: IEEE Data Compression Conference (DCC); 1995. p. 202–211.

[54] Marpe D, Schwarz H, and Wiegand T. Context-Based Adaptive Binary Arithmetic Coding in the H.264/AVC Video Compression Standard. IEEE Transactions on Circuits and Systems for Video Technology. 2003;13(7): 620–636.

[55] Zhou D, Zhou J, Fei W, *et al.* Ultra-High-Throughput VLSI Architecture of H.265/HEVC CABAC Encoder for UHDTV Applications. IEEE Transactions on Circuits and Systems for Video Technology. 2015;25(3):497–507.

[56] Fei W, Zhou D, and Goto S. 1 Gbin/s CABAC encoder for H.264/AVC. In: European Signal Processing Conference (EUSIPCO); 2011. p. 1524–1528.

[57] Zhou J, Zhou D, Fei W, *et al.* A high-performance CABAC encoder architecture for HEVC and H.264/AVC. In: IEEE International Conference on Image Processing (ICIP); 2011. p. 1568–1572.

[58] Ramos FLL, Zatt B, Porto M, *et al.* Novel multiple bypass bins scheme for low-power UHD video processing HEVC binary arithmetic encoder architecture. In: Symposium on Integrated Circuits and Systems Design (SBCCI); 2017. p. 47–52.

[59] Ramos FLL, Zatt B, Porto M, *et al.* High-throughput binary arithmetic encoder using multiple-bypass bins processing for HEVC CABAC. In: IEEE International Symposium on Circuits and Systems (ISCAS); 2018. p. 1–5.

Chapter 3

High-throughput architectures for high-resolution video coding: system architecture analysis

Grzegorz Pastuszak[1]

The need for high video resolutions in nowadays applications lead to the development of new compression standards such as H.264/advanced video coding (AVC) [1], H.265/high efficiency video coding (HEVC) [2], audio video standard (AVS), and VP9. The best compression efficiency is achieved by encoders implementing the second standard. However, the compression gain involves a significant increase in the algorithm complexity. Compared to its predecessors, the H.265/HEVC format offers more coding options and modes. Moreover, their computational complexity usually exceeds those inherited from previous standards (e.g., transforms on larger blocks). In order to benefit from new possibilities, it is indispensable to find what set of modes and options should be selected for particular pixel units to obtain a better video quality. Since this task is accomplished by the encoder, its complexity is much higher than that of the decoder.

Apart from the algorithm complexity, the amount of operations depends on supported video resolutions. Provided that the aspect ratio is preserved, the doubling of the pixel number in line imposes the same change in the vertical dimension. As a consequence, the number of pixels in a frame is increased four times. To obtain better quality, it is also desired to increase the frame rate, which affects the number of pixels processed in assumed time periods. Thus, the growth in spatio-temporal resolutions also puts higher requirements on coding performance/throughput. To meet the requirements, the encoder should be implemented on the platform featuring a significant amount of computational resources. In practice, the hardware acceleration is indispensable. The analysis of hardware framework constraints is conducted in Section 3.1. Section 3.2 reviews general hardware optimization methods. In Section 3.3, the timing constraint analysis is conducted. The following sections discuss particular parts of the H.265/HEVC encoder in terms of their mapping onto hardware resources.

[1]Institute of Radioelectronics and Multimedia Technology, Warsaw University of Technology, Warsaw, Poland

3.1 Hardware vs. software encoders

The compression efficiency of developed standards is demonstrated with software reference models. During the standardization process, successive versions of Joint Model [3] and HEVC Test Model (HM) [4] have been elaborated for H.264/AVC and H.265/HEVC, respectively. Although their compression efficiency is significant, their coding speed is very low. Generally, the speedup using multicore processors or dedicated hardware accelerators is difficult due to the irregular dataflow and control-intensive operations. Therefore, the software models can be used only in offline applications.

There are other software implementations able to achieve higher performance, e.g., x264, x265, Kvazaar, and Nvidia Encoder (NVENC). Depending on the selected computation profile, they can provide different throughputs. When executing on typical general-purpose processors, the coding can be performed in real time for lower video resolutions. The implementation on the graphic processors (NVENC) allows the support even for the 4K resolution. In fact, such a platform employs hardware acceleration. Generally, the increase in the performance involves quality/rate losses in the compressed domain. Therefore, the fastest computation profiles have significantly lower compression efficiencies.

In order to increase the throughput, the software encoding can be decomposed into a number of threads executed on the general-purpose processors. Another enhancement consists of the processing with single-instruction-multiple-data operations. Such optimizations have been conducted in the Kvazaar encoder [5]. Its hardware-oriented C source code enables more straightforward hardware/software partitioning than other eligible open-source H.265/HEVC encoders. The intraultrafast encoding for 2,160@30 fps has been achieved on the 22-core Intel Xeon E5-2699 v4 processor. If the intercoding is used, the throughput is reduced at least by half.

The software implementation can be partly mapped to hardware resources to increase the encoder performance or to save power consumption. This approach is demonstrated for the Kvazaar encoder in [6]. The encoder is decomposed into two parts as depicted in Figure 3.1. The first is executed on 14 processor dual cores and includes control-intensive operations (e.g., context adaptive arithmetic coding), communication interfaces, and chroma processing. The data-intensive coding tools (intra-prediction, transforms, quantization, and dequantization) are implemented in field programmable gate array (FPGA) Arria 10 device placed on the peripheral component interconnect (PCI) card. The hardware accelerator is accessed by the software driver which controls the communication on the PCI interface. The synchronization of both parts is performed for successive coding tree units (CTUs), which are 64×64 pixel in size. The FPGA capacity allows the implementation of two intracoding paths. Additionally, the device embeds some buffers to queue CTU data forwarded to and returned from the paths.

When the Kvazaar encoder is executed only on the processor cores, its average throughput achieves 20 fps for the $2,160 \times 3,860$ resolution. The exploiting of one and two intracoding paths allows the increase to 30 and 40 fps, respectively.

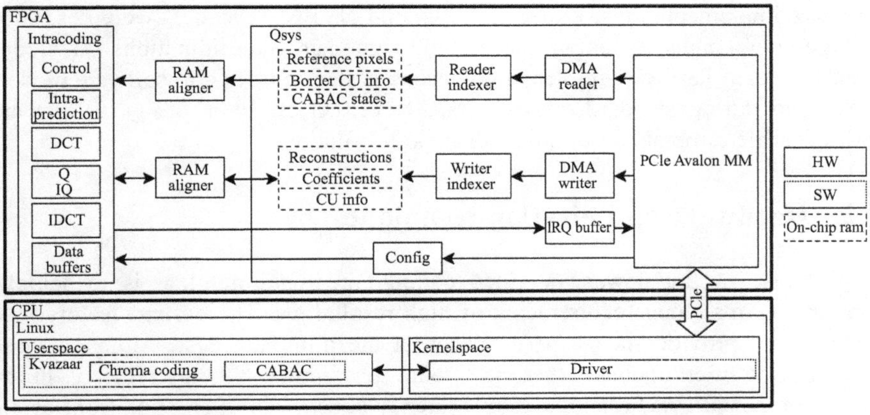

Figure 3.1 System architecture of the hardware–software Kvazaar encoder [6]

These results are achieved for the fast profile, which involves significant losses in the compression efficiency. Lower resolutions enable the medium profile with better efficiencies. In the presented hardware–software design, the achieved throughput is not constant and varies to some extent with video content. Thus, the stable processing requires a somewhat higher throughput than that stemming from the resolution. The same rule also applies to software implementations.

Since the use of the C code optimized for parallel processing, the intracoding tools of the Kvazaar encoder are easily mapped to hardware resources with the use of a high level synthesis tool. The synthesized design consumes 308k ALUTs and 862 DSP units. It is clocked at the common frequency of 125 MHz. Compared to dedicated hardware implementations, the resource cost is much higher, even if it is normalized to the throughput. An additional cost is related with the processor implementing the software part. Although the FPGA design can be optimized at the module level, the hardware–software codesign is limited by the communication bandwidth between both parts. It becomes apparent when comparing throughputs achieved for single and dual-core hardware acceleration (30 and 40 fps, respectively). Doubling of the amount of hardware resources increases the throughput by about 33%.

The advantage of hardware architectures stems from the ability to provide highly optimized solutions, which can be adapted to a specific framework. The optimality depends on a number of parameters such as throughput, delay, silicon area, power consumption, flexibility, accuracy, available interfaces, amount of external resources, and additional (parameterized) functionality. In design of integrated circuits, the parameters are implicitly correlated due to design rules, algorithm complexity, and dataflow dependencies. Therefore, hardware designs involve some constraints on the implemented algorithm due to data dependencies and the dataflow regularity. These constraints are not present or are limited in typical software implementations. Kvazaar, x.265, and NVENC encoders were described in high-level languages taking into account some hardware limitations. As a consequence, their hardware implementations are easier compared to software models developed during standardization works. Nevertheless, some design efforts are indispensable to obtain an optimized hardware solution.

3.2 Hardware optimization techniques

In the case of the nowadays video coding, the main problem is to achieve throughputs matching requirements of high resolutions. The desired feature of a device is to provide such a performance at the minimum cost. Although these demands are often opposite, there is a room to minimize this discrepancy. In the design of digital synchronous circuits, latency constraints often arise from cyclic dependencies, feedback paths, and/or performance requirements. Ad hoc architectural modifications are typically used to improve the performance of circuits.

Usually, algorithms can be decomposed into a number of tasks that can be performed independently. This property allows parallel processing by mapping each task onto a dedicated hardware unit. Since design for high-speed coding involves using parallel processing, a number of techniques have been developed to efficiently transform algorithms from the sequential form to the parallel one. These techniques include pipelining, retiming, loop unrolling, folding/unfolding, look-ahead, and resource sharing. Generally, the techniques trade higher speed for increased area. Power consumption can be optimized concurrently with speed or the two quantities can be traded for one another.

A specific functionality can often be realized through several alternative algorithms. Different algorithms usually exhibit various characteristics regarding performance, accuracy, and power consumption. Algorithms can have both a direct and an indirect influence on architecture design. In particular, complexity and operation count have a straightforward and direct impact. Other issues such as available computational concurrency, regularity, and modularity (locality) have a more indirect influence. Concurrency exhibits opportunities for parallel processing, whereas regularity facilitates using the same hardware resources to perform different tasks. Maximizing the modularity in an algorithm ensures that efficient use can be made of the distributed processing structure.

Once a particular algorithm has been chosen, its dataflow and the representation of internal variables can be modified to emphasize some features allowing efficient hardware implementations. These modifications should not affect the final outcome of the original algorithm. The set of possible modifications includes the following decisions: operation reordering, algebraic transformations, and changes of data representation. They are described below:

1. Algebraic transformations enable equivalent variants of complex algebraic expressions. In particular, they can change the number, types, and the order of operations. The transformations can lead to a dataflow that is either more simplified or more complex with respect to the original one. Although the

simplification is usually beneficial, some complex variants can be more suitable for hardware due to their regularity.

2. Data representation relates to the format and the required word length for the data-path. Regarding the format, the designer typically has several different alternatives from which to choose: e.g., fixed-point vs. floating-point, sign-magnitude vs. two's-complement, and unencoded vs. encoded data. For example, certain data representations such as sign-magnitude have an inherently lower activity than two's complement when signal transitions between positive and negative values occur frequently.
3. Data structure relates to an organization of information, usually in memory, for better accessibility, savings in storage resources, and/or conceptual unity. The set of possible data structures includes queue, heap, linked list, tree, dictionary, and stack.
4. Reordering of operations allows explicit changes in the dataflow, providing the path for concurrency and/or regularity.

3.3 Timing constraints on pixel units

The real-time video encoder should have the throughput sufficiently high to support required resolutions. In practice, the required throughput depends on the video resolution specified in time and pixel domains. They are measured in fps and pixel area, respectively. Subsampling of chroma components can additionally affect the performance. Since the video compression processes pixels in $N \times N$ pixel units, it is convenient to use their number per second to specify the throughput. The performance of a given architecture depends on the clock frequency. In particular, the throughput increases proportionally with the frequency. Table 3.1 summarizes average throughputs for different unit sizes required to support particular resolutions and the average number of clock cycles allocated to each unit at selected frequencies. The results included in the table are obtained with the assumption that incomplete pixel units located at the bottom order take the same number of cycles as the remaining ones. In practice, they require less computation since their bottom part is outside the frame and includes empty pixels. Nevertheless, the savings are slight.

In practice, the hardware encoder performance should have a computation margin to compensate for wait states introduced by control and communication operations, e.g., initializations of probability models, rate control, the fullness of the output stream buffer, and shared access to external memories. Therefore, numbers of clock cycles given in Table 3.1 should be slightly decreased. For example, it is useful to assume that the number of available cycles for each 16×16 unit and the resolution of 1,080p@30 fps is equal to 400 rather than 408. Since cycle budgets for higher resolutions are scaled almost proportionally, the division by the appropriate power of two will provide the target values of the number of clock cycles. Similarly, the frequency scaling translates directly onto the budget. Thus, numbers of available cycles are four times greater in the case of the 400 MHz clock.

Table 3.1 Summary of timing requirements for different video formats

Resolution	**64 × 64 unit**			**16 × 16 unit heads**		
	Throughput (unit/s)	**Cycles at**		**Throughput (unit/s)**	**Cycles at**	
		100 MHz	**400 MHz**		**100 MHz**	**400 MHz**
1,080p@25	12,750	7,843	31,372	204,000	490	1,960
1,080p@30	15,300	6,535	26,143	244,800	408	1,633
1,080p@50	25,500	3,921	15,686	408,000	245	980
1,080p@60	30,600	3,267	13,071	489,600	204	816
2,160p@25	51,000	1,960	7,843	810,000	123	493
2,160p@30	61,200	1,633	6,535	972,000	102	411
2,160p@50	102,000	980	3,291	1,620,000	61	246
2,160p@60	122,400	816	3,267	1,944,000	51	205
4,320p@25	204,000	490	1,960	3,240,000	30	123
4,320p@30	244,800	408	1,633	3,888,000	25	102
4,320p@50	408,000	245	980	6,480,000	15	61
4,320p@60	489,600	204	816	7,776,000	12	51

High resolutions involve decreased number of clock cycles assigned to each pixel unit. For the 4,320p resolutions, the number becomes so small that the finishing of necessary computation is a challenging task. Although most operations can be parallelized and pipelined, the video coding algorithm makes the processing of a unit dependent on the readiness of results for preceding one. Since the processing path introduces a delay, the conditions may be mutually exclusive. The limitation is more serious when considering smaller pixel units, i.e., 8 × 8 and 4 × 4. Corresponding cycle budgets are 4 and 16 times smaller than in the case of 16 × 16 units. As a consequence, the smaller sizes cannot be supported if the data dependencies between successive units are utilized.

In order to address the problem of short-time periods assigned to small pixel units, video encoder can apply the simultaneous processing of data-independent units. This approach extends the periods proportionally to the number of processing paths. In practice, each path should embed almost all encoder modules. Therefore, the amount of resources is also scaled in similar way as the performance. To support high resolutions, it becomes possible to employ several instances of the encoder designed for a lower resolution.

In all video compression standards, it is possible to code units taken from different frames. This approach can be easily exploited when frames are coded without mutual references, i.e., all frames use only intra-mode. In the case of inter-modes, there is an additional constraint on the location of pixel units in frames. The inter-prediction can refer to some reconstructed pixels of the previous-in-decoding-order frame. The reference area of a given unit is usually centered around the collocated unit in the previous frame. To preserve the data consistency, the processing of units including pixels from this area should be completed prior to the coding of the unit in the following frame. Provided that motion vectors (MVs) are limited, it is sufficient to introduce the almost-constant location shift between units taken from successive prediction-dependent frames, as shown in Figure 3.2. The shift can temporally change when passing from the last unit located at the right or bottom border to the first in the following line or frame, respectively. The number of unit locations between coded units can be increased or decreased until the condition on the availability of reference areas is satisfied for all of them.

The parallel coding of units taken from separate frames introduces a significant latency. Moreover, the size of input and output buffers is large since they should save data corresponding to several frames. In order to minimize these drawbacks, the simultaneous processing can be performed on pixel units taken from the same frame under the condition that their independence is kept. This is possible when applying the division of frames into slices and/or tiles. Each slice/tile groups some successive pixel units. Tiles form rectangular groups, whereas slices include an arbitrary number of successive units taken from one or more rows within one tile. An example of the frame division into tiles and slices is shown in Figure 3.3. Owing to the data independence, the predictions on tile/slice borders must not be used. Moreover, the probability models adapted based on coded data are

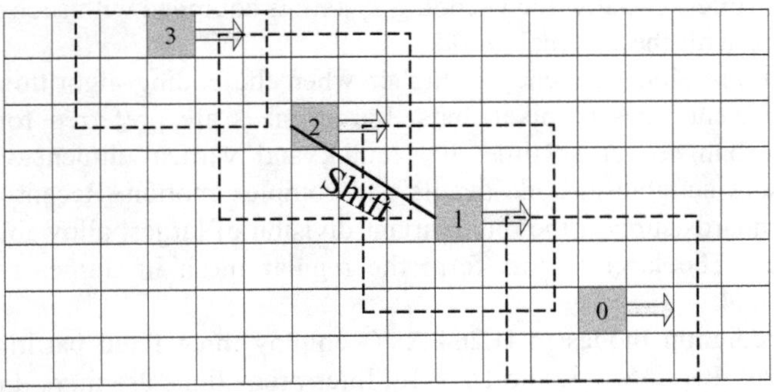

Figure 3.2 Parallel coding of units taken from successive frames. Numbers point the frame order. Dashed lines indicate reference areas

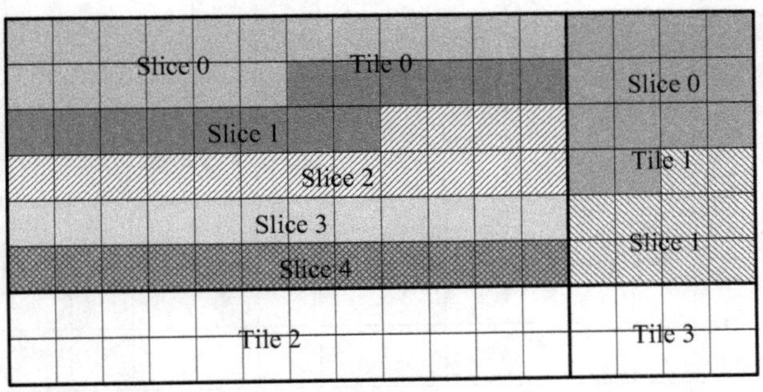

Figure 3.3 Example division into tiles and slices

reinitialized at the beginning of each slice. These limitations lead to lower compression efficiencies. Although this impact is slight in terms of objective quality measures, the borders become subjectively visible at low bit rates.

The H.265/HEVC standard provides the wavefront parallel processing as a new coding method for simultaneous computations on pixel units. Each unit row is started with a delay of at least two units with respect to the row above, as depicted in Figure 3.4. As a consequence, the method can take advantage of the prediction between neighboring units. Also, the adaptation of probability models is used between processing paths. Wavefront parallel processing requires an additional inter-core communication to transfer the reference data for the prediction and states of the probability models.

The common feature of the parallel unit coding is the need for separate buffers of output streams. This requirements stem from the fact that streams produced in each processing path should form the final one by their concatenation rather than the switching on the unit basis. Due to a significant amount of data, the buffers should be sufficiently large. In practice, the use of external memories is necessary. The buffering introduces a delay, and the generation of the final stream should be performed with an additional computation module, e.g., general-purpose processor or microcontroller. Usually, the encoding system is equipped with such a module to communicate with the external world.

The compression efficiency is higher when the coding algorithm can select between different sizes of pixel units. Larger sizes are preferred for fragments with small changes across time, horizontal, and vertical dimensions. Smaller sizes are more suitable to code details and complex motion. Recent video compression standards support the multivariant division of largest allowable units into smaller ones. The largest units form the regular mesh in frames as shown in Figures 3.2–3.4.

Intra-prediction modes in H.264/AVC employ three fixed partition schemes with unit sizes 4×4, 8×8, and 16×16. Inter-predictions use more unit sizes and

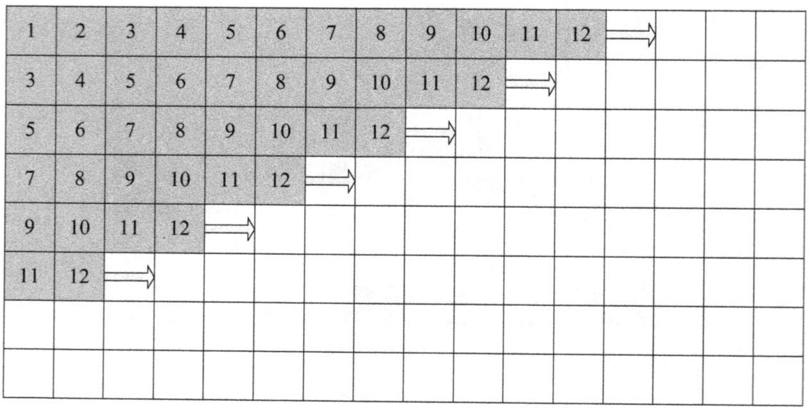

Figure 3.4 Wavefront processing order

the quad tree structure. Each macroblock can be coded as one 16 × 16 unit or partitioned into 16 × 8, 8 × 16, or 8 × 8 subunits. If the last mode is selected, each 8 × 8 subunit can be coded as one unit or partitioned into 8 × 4, 4 × 8, or 4 × 4 subunits. Such partitions form the second level of the quad tree. The concept of the quad tree subdivision is extended in H.265/HEVC, as shown in Figure 3.5. The standard takes advantage of CTUs, typically 64 × 64 pixel in size. At each tree level, the $2N \times 2N$ can be left without the subdivision or partitioned into units $2N \times N$, $N \times 2N$, and $N \times N$. Inter-predictions can use asymmetrical partitions formed by the division along one dimension with the ratio of 3:1. As a consequence, four additional partition modes are specified. The smallest unit sizes are 4 × 4 and 8 × 4/4 × 8 ones for intra-mode and inter-modes, respectively.

Each CTU can be divided into square coding units (CUs) according to the quad tree scheme. CUs are processed with either intra/inter-predictions. Inter-mode CU can be divided into two symmetrical or asymmetrical prediction units (PUs). Intra-mode 8 × 8 CUs can be divided into four 4 × 4 PUs. The PUs partitioning is indicated by dashed lines in Figure 3.5. Apart from the PU subdivision, CUs can be divided into square transform units (TUs).

The plurality of the subdivision combinations leads to highly complex encoder algorithms able to select the best one. In practice, the problem is decomposed into decisions at each tree node. The complexity depends on the number of nodes, which is determined by the tree depth. It is useful to support all depths to achieve high compression efficiencies. Although the number of nodes at small depths (large units) is small, they require much more computations compared to nodes at large depths. With the assumption that the amount of computations is proportional to the unit size, the encoder complexity increases proportionally with the number of analyzed partition modes. This stems from the fact that each pixel is analyzed

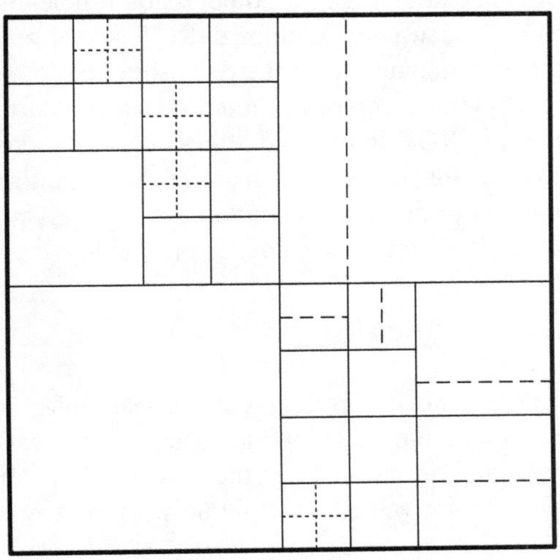

Figure 3.5 Example of quad tree subdivision

several times for allowable partition modes at different depths. In reality, more computations are necessary for larger TUs due to the property of the discrete cosine transform (DCT). Therefore, if the same transform size was used at each node of the quad tree, the complexity would be roughly proportional to the number of analyzed partition modes.

The highest compression efficiency is achieved when partition modes are evaluated for all possible CU nodes. However, this approach involves the highest complexity. It turns out that the complexity can be reduced with slight impact on the compression efficiency when skipping some smaller CUs. The HM reference software applies the top-down partition search with the early termination. There are some conditions which disable the evaluation of smaller CUs. For example, such savings are possible when all quantized coefficients are zeros. Although the reduction of computations can be significant, the irregular dataflow makes the conditioned top-down search unsuitable for the hardware encoder. Owing to the regularity, the hardware design must evaluate all assumed partition modes. Therefore, the bottom-up search order is suitable. The architecture can employ separate resources for particular PU/TU sizes. Alternatively, the same resources can be shared by PUs/TUs with different sizes provided that the scheduling meets timing requirements. In both cases, time slots utilized by units should not exceed limits inferred from the video resolution and the unit size, as described in previous subsection. Therefore, one $2N \times 2N$ and four $N \times N$ units should be evaluated within the same period. This rule must be simultaneously preserved for all tree levels supported by the encoder, e.g., N can be 4, 8, 16, or 32 for H.265/HEVC. In the case of the shared processing path, time slots should be shorter since available clock cycles must be distributed to more units of different sizes. The architecture composed of paths separate for each N allows longer slots.

The transform sizes are selected with the residual quad tree (RQT) for each CU. The varying transform size allows higher compression efficiency. The compression gain is about 1% [7]. The use of additional RQT levels almost proportionally increases the amount of computations in the reconstruction loop. In the case of the intracoding, the size of prediction blocks should match that of transform. From the hardware perspective, the RQT implies additional computations with tight timing constraints similar to those for smaller PUs. It seems that the intracompression gain is relatively small compared to the increase in the complexity. On the other hand, the use of at least two-level RQT for inter-modes is profitable.

3.4 Mode decision tradeoffs

The encoder must determine the coding cost of particular subdivision modes, including the case without the partitioning. The cost (J) estimation takes into account distortions (D) and bit rates (R). In the classical R–D optimization (RDO), they are combined into a joint value by weighting with the Lagrangian multiplier λ as follows:

$$J(R, D) = R \cdot \lambda + D$$

It is desired to design real-time encoders able to find modes which minimize the cost *J*. In general, the estimation of rates and distortions is a complex task. Thus, the design for high-resolution videos must take into account many tradeoffs between the compression efficiency and the hardware complexity. The following subsections discuss particular aspects of the modes decision process in terms of the tradeoffs.

3.4.1 Reconstruction loop

The estimation of rates and distortions requires some operations to determine quantized transform coefficients and reconstructed pixels. These operations include the intra/inter-prediction generation, the residual computation (difference between original pixels and their prediction), the forward transform, the quantization, the dequantization, the inverse transform, and the reconstruction. This chain is usually called the reconstruction loop and constraints computations between pixel units. In the hardware framework, the chain determines the encoder architecture, as shown in Figure 3.6, each operation in the loop introduces a delay. The number of clock cycles corresponding to the delay determines if a given unit size is supported for the assumed video resolution and the clock frequency.

Apart from the loop delay, the throughput of the hardware encoder depends on the number of different candidate modes analyzed in RDO for a given unit. In particular, different intra-prediction and inter-prediction modes should be evaluated to select the best one. H.264/AVC employs nine intra-prediction modes, whereas H.265/HEVC specifies 35 intra-prediction modes (DC, planar, and 33 angular modes). Examples of intra-modes are depicted in Figure 3.7. For interframes, the number of candidates can be huge since each possible MV can be analyzed in RDO for each CU partition. In practice, a limited number of intra-prediction and inter-predictions is evaluated with the *R–D* cost. Nevertheless, the number cannot be too small to avoid significant losses in the compression efficiency. Each intra/intercandidate block

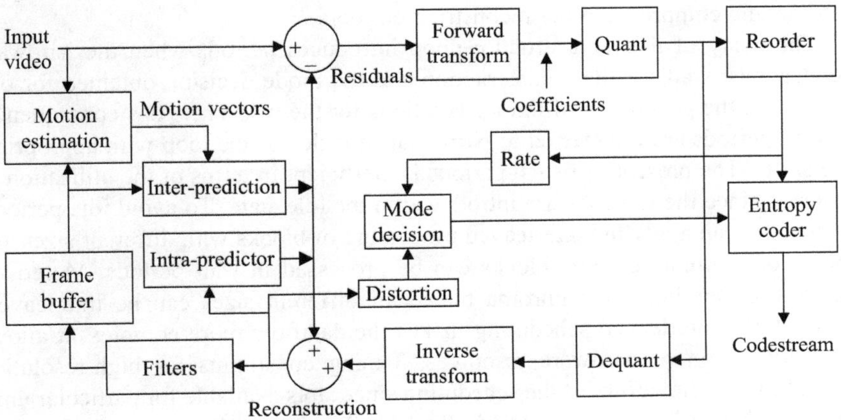

Figure 3.6 General video encoder architecture

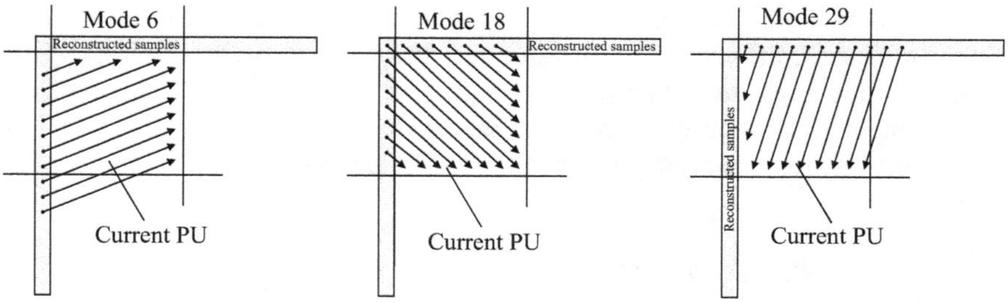

Figure 3.7 Examples of intra-prediction modes

forwarded to the reconstruction loop takes some clock cycles in particular modules. Provided that each candidate takes the same number of cycles M, the minimal period required for a given CU size can be estimated with the following equation:

$$\text{Period} = N \cdot M + \text{Delay},$$

where N is the number of candidate modes.

The CU period can be decreased by the parallel processing of samples and transform coefficients in the reconstruction loop. The parallelism of all operations performed for one candidate block (scaling on M) makes it possible to decrease the number of clock cycles required by each block to one. A further reduction can be achieved with several instances of the loop (scaling on N). In general, the scaling of the throughput of the reconstruction loop strongly affects the consumption of hardware resources. Usually, the increase of the throughput and the amount of resources are proportional. Unfortunately, the throughput of the mode decision process is not scaled proportionally with that of the reconstruction loop. The delay of the reconstruction loop imposes the limit on the minimal number of clock cycles needed to process all prediction modes for a block. For higher parallelisms, the increase in performance saturates. Thus, it is beneficial to minimize the delay of each module comprised in the reconstruction loop.

The delay of the reconstruction loop introduces periods when the hardware encoder must wait for the reconstruction and the mode decision obtained for one unit to start the generation of intra-predictions for the next one. As a consequence, the wait periods are transferred to particular modules of the loop with appropriate time shifts. The presence of such periods is inefficient in terms of the utilization of resources since the modules are introduced in the idle state. To avoid idle periods, the encoder can apply the interleaved processing of blocks with different sizes and types. For example, chroma blocks can be processed in wait periods. Moreover, timing slots for luma and chroma blocks of different sizes can be interleaved. Although the interleaved scheduling makes the dataflow more complex, it allows better utilization of hardware resources. Timing constraints for high-resolution videos limit the flexibility of the scheduling since slots available for particular units are small. Nevertheless, wait periods for luma blocks can be assigned to chroma ones with separate processing paths designed for each CU/PU/TU size.

3.4.2 Transforms

In H.264/AVC, transform blocks are 4×4 and 8×8 in size. Additionally, transforms can be implemented without multiplications with shift and add operations. As a consequence, their hardware complexity is moderate and lower than the (de) quantization based on multiplications. The complexity of the transforms in H.265/HEVC is much higher since also larger TU sizes are supported (16×16 and 32×32). This feature increases the size of the transposition buffer and the depth of the cascaded arithmetic operations. Integer basis functions specified in transform matrices better approximate original DCT. However, the extended range of values included in the transform matrix (from -90 to 90) involves the use of multipliers. Even with the hierarchical decomposition of the transform operations and the use of multiple constant multipliers, the amount of consumed hardware resources is significant. For high-throughput implementations (the parallelism greater than 32), the cost of the transposition buffer becomes smaller than that of each of two 1D transforms.

All results of the first 1D transform must be written to the buffer before the second transform stage starts. The amount of written data increases with the transform size. Provided that the throughput of the 1D transform submodule is fixed, the number of clock cycles taken to fill the buffer also increases. Therefore, the need for the buffering during the transposition makes the delay of the reconstruction loop dependent on the transform size. Although the increased delay is of less importance in the case of larger transform sizes, it negatively affects the delay for smaller ones if the interleaved processing is applied. In particular, the transformation of smaller blocks can be started after the finishing of delayed processing for larger ones.

To save hardware resources, the architecture can be composed of the submodules able to compute the 1D transform for one row/column in each clock cycle. The throughput of the submodule can vary for different transform sizes. Higher throughputs can be obtained with multiple instances of the submodule. One submodule can be employed to design the folded architecture, where two 1D transformations are executed sequentially. In the first step, the submodule takes samples/coefficients from the input and writes the result to the transposition buffer. In the second step, the submodule reads coefficients from the buffer and produces 2D-transformed outputs. The throughput of the folded architecture is equal to the half of that offered by the 1D transform submodule. The use of this architecture is inefficient. First, they have insufficient throughput to support the encoding for high resolutions. Second, the resource cost of the transposition buffer is significant compared to that of the submodule. Third, the discontinuity of output data deteriorates the hardware utilization of since the quantization must wait for valid data during the first step. Therefore, the use of separate submodules assigned to both transform steps/stages is beneficial.

If the transform sizes supported in the same module are various, the use of separate submodules for each one is inefficient in terms of resource consumption. Since DCT of larger sizes includes cores for smaller ones, it is beneficial to share the same resources for transforms of different sizes. However, throughputs match corresponding transform sizes in terms of samples/coefficients processed in one clock cycle. To keep the throughput of the 1D submodule equal to that for the largest transform size, additional resources are necessary. For example, if the maximal transform size is 32, the support of 16-point transform with the same throughput (32 samples/coefficients) can be implemented with an additional 16-point transform core. More details on transform architectures can be found in Chapter 1.

The processing of whole rows/columns in the same clock cycle has an impact on the encoder dataflow. In particular, the organization of samples generated by the intra-predictor should be adapted to the transform input interface. The adaptation can be performed in the predictor by the selection of samples from the same row. Alternatively, the buffering and the multiplexing are indispensable combine contributions generated in various clock cycles. For example, four 8×8 prediction blocks must be buffered to obtain eight 32-sample rows. With the preselection described in the previous subsection, the delay and the resource cost are not increased considerably.

If the RQT is enabled for inter-modes, the same residual block is submitted to the reconstruction loop for different transform sizes. Provided the same throughput is allowed for all cases, the delay introduced by the transposition buffer is constant. Thus, the time slot assigned to each CU is increased only by the number of cycles need to resubmit the same blocks. In the case of the intra-modes, smaller TUs impose predictions on smaller blocks. As a consequence, the RQT cannot take advantage of the parallel processing on TUs. To maximize the throughput, several rows of a residual block should be combined to form one 32-sample row submitted to the reconstruction loop in each clock cycle. Such a rearrangement makes the architecture more complex. First, the assignment of reference samples to processing elements in the intra-predictor has more variants leading to larger multiplexers. Second, additional multiplexers are required to write (or read) 1D-transformed coefficients to (from) the transposition buffer in the processing order modified compared to the 32-point DCT dataflow.

For high-resolution video encoding, separate reconstruction loops for particular transform block sizes are attractive as some disadvantages of the interleaved processing order are not present. First, the separate intra-predictors and transform modules assigned to each path can be optimized in terms of the dataflow, which leads to the reduction of hardware resources and/or higher clock frequencies. Second, the delay of each reconstruction loop is kept constant more easily since transposition takes the same number of clock cycles and there is no interleaving. Although separate reconstruction loops increase the complexity of the encoder dataflow, the resource consumption can be traded for throughput by the adaptation of the architecture to desired parallelism within each loop.

3.4.3 Mode preselection

As described above, the number of intra-prediction and inter-prediction forwarded to RDO is limited to improve the encoder throughput. In the HM reference model, the default number of selected intra-predictions is 3, 3, 3, 8, and 8 for PU sizes 64,

32, 16, 8, and 4, respectively. More candidates are used in RDO process for smaller PU sizes. Additionally, the mode candidate list is supplemented with most probable modes which are at most three non-contained prediction modes. They are determined based on final prediction modes selected for neighboring pixel units from the left and the top. As a consequence, this dependence forms an additional constraint on timing between successive units apart from the readiness of reconstructed samples used in the prediction.

Only one candidate is analyzed in RDO for each allowable inter-PU. Although this is a strong restriction, eight partition-mode candidates must be evaluated for CU sizes 16, 32, and 64. This set includes partition modes shown in Figure 3.8. The smallest 8 × 8 CU has only three partition modes since asymmetrical divisions are not allowed. The number of intracandidate and intercandidate can be changed to be adapted to specific timing constraints. The change affects the compression efficiency. In general, the efficiency falls and rises if less and more candidates are processed, respectively.

The preselection of candidates submitted to the reconstruction loop and the RDO has a significant impact on the encoder architecture. First, the throughput of the intra-predictor should be much higher compared to the reconstruction loop and the RDO. Second, memory buffers are needed to keep residual blocks during the preselection. It is desired to generate blocks for all modes in the intra-predictor to select the subset with lower coding cost. There are some techniques limiting the number of generated modes to some extent. If the reduced number matches the throughput of the reconstruction loop and the RDO, the buffering is not needed. In particular, residual blocks can be directly forwarded to the reconstruction loop. Otherwise, the preselection for a given CU must be performed. To achieve a high clock rate, each block should be evaluated with the sum of absolute differences (SAD) or sum of absolute transformed differences (SATD) in successive pipeline stages. To avoid the repetition of the intra-prediction process for preselected candidates, residual blocks should be written to memories. The submission to the reconstruction loop and the RDO can start when at least one candidate from the preselected set is known. This dependence introduces a delay which adds to that of the reconstruction loop.

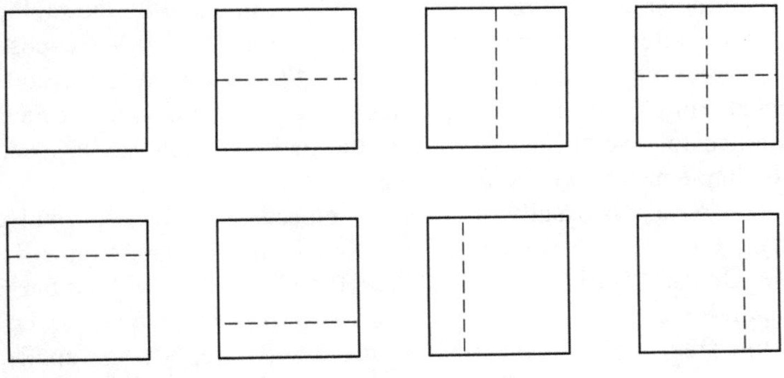

Figure 3.8 Inter-CU partition modes

In video coding, the mode decision must be taken for luma and chroma blocks. The H.265/HEVC standard makes the decision on chroma blocks dependent on that on the corresponding luma block in terms of their sizes and prediction modes. For the 4:2:0 format, the chroma block width and height are equal to the half of corresponding luma block dimensions. The prediction mode selected for luma can be inferred by chroma components. However, it is possible to select between four intra-prediction modes regardless of the luma one. The chroma mode dependence affects the design dataflow. The best mode decision is taken when all components are evaluated to preselect candidates for RDO and to decide the final mode. However, this strategy considerably extends time slots utilized for particular CUs since the chroma processing must follow the luma one. Alternatively, all luma and chroma blocks can be generated by the intra-predictor and some of them submitted to the reconstruction loop with the RDO. In this case, the increased number of blocks leads to higher resource cost or longer CU processing. To save computations and time, the CU mode can be selected only based on its luma block(s). Thus, the chroma processing can be delayed without the impact on the throughput. This simplification involves some losses in the compression efficiency (about 1% for intracoding). In the case of the inter-prediction, chroma blocks can be determined after the MVs for luma are known.

3.4.4 Cost estimation

Due to the high complexity of the RDO, only a preselected set of candidates is submitted to the reconstruction loop. The preselection is usually performed with a simplified *R–D* cost estimated from the difference between original and prediction samples. The SAD has the smallest complexity and limited reliability. If the differences are additionally transformed with 2D Hadamard transform (SATD), the reliability is better. On the other hand, the hardware complexity is higher. Moreover, the transform involves an additional delay. The residual SAD/SATD cost is increased by a value estimated from the number of mode bits. Thus, the final simplified cost is expressed as follows:

$$J = R_{\mathrm{mode}} \cdot \sqrt{\lambda} + \mathrm{SA(T)D}$$

Although the mode rate contribution is originally computed by the multiplication with the value after the square root, the corresponding hardware cost can be significantly reduced. Since λ is depended on QP, the square root results can be precomputed and placed in the look-up table (LUT). The number of mode bits is usually limited to several bits. Therefore, the multiplication has low complexity and can be implemented with adders.

In H.264, SATD is usually used for mode decision because it involves lower computation cost than *R–D* cost and the loss in the compression efficiency is acceptable. On the other hand, in the H.265/HEVC encoding without the *R–D* cost calculation, it leads to the loss of 10%–15% for intra-frame encoding and more than 40% for interframes [7]. In the HM reference software, SAD is applied in the integer-pel motion estimation (ME), whereas SATD is used in the fractional-pel ME and intra-mode preselection (raw mode decision). Both measures can be simplified by the consideration of the part of samples from the original block and its prediction. Another method consists in the reduction of bit accuracy of samples taken to the cost estimation. Although these simplifications save hardware resources, they have a negative impact on the compression efficiency.

The complexity of the accurate *R–D* cost calculation is high due to two reasons. First, all or at least the part of operations included in the reconstruction loop is required. Second, context adaptive arithmetic coding (CABAC) is used to estimate the rate based on a LUT. In a real-time hardware encoder, the rate and distortion estimation should not introduce a significant delay since it could negatively affect the minimal period required for a given CU size. Moreover, the throughput of the reconstruction loop and the rate/distortion estimation should be balanced to avoid wait states. It is difficult to meet these requirements taking into account timing constraints imposed by high resolutions. The main limitation stems from the fact that CABAC processes residual and mode data on the bit/bin level after the binarization stage. Even with the estimation of rate contributions from the LUT, the adaption of particular CABAC probability models for each bin is complex and data dependent. In particular, the model selected from the set can be accessed and modified for successive bins. As a consequence, the parallelization of this process is difficult without simplifications.

In order to allow the parallel rate estimation for whole transform blocks or their parts, the adaptation of probability models should be disabled. Fixed probabilities involve estimation inaccuracies, which negatively affect the compression efficiency. Provided that the models are fixed after initialization, losses in terms of BD-Rate are 0.79%, on average [8]. Such losses seem to be acceptable when taking into account the reduction of the algorithm complexity. Nevertheless, a great number of LUTs is indispensable to support the simultaneous processing of transform blocks and subblocks. The LUT cost can be removed with the assumption that all bins have equal probability. This simplification decreases the compression efficiency by 2.11% [8]. In fact, this assumption eliminates the CABAC probability modeling in the rate estimation. As a consequence, the binarization stage can skip the generation of codewords. It means that only corresponding bin-rate contributions and their sum should be estimated.

Since the binarization is based on 4×4 (sub)blocks, it is convenient to design the rate estimator able to process one such block in each clock cycle. Higher throughputs can be achieved by the reuse of such a module. As the binarization in H.265/HEVC uses the adaptive rice code, high-frequency quantized transform coefficients have an impact on the binarization of the following (lower frequency) ones within each 4×4 subblock. This dependence creates a critical path if all coefficients are analyzed at the same pipeline stage. To maximize the clock frequency, the process should be distributed into several stages. On the other hand, it is important to limit the number of stages to avoid the long delay which may extend the minimal period assigned to particular CUs.

The distortion estimation is much less complex compared to the rate estimation. In the straightforward approach, it can follow the reconstruction stage in the loop. However, this order negatively affects the delay of the mode decision since it becomes longer than that of the reconstruction loop. When the distortion is computed from reconstructed residuals, the delay becomes slightly smaller, however, the problem remains. The solution is to perform estimation in the transform domain. Although the transform in H.265/HEVC is not strictly orthogonal, the energy of quantized coefficients after the normalization is almost proportional to that of corresponding residuals in the pixel domain. Therefore, the difference between values of coefficients before the quantization and after the dequantization is a good approximation of introduced errors. To obtain the distortion, these errors should be squared and summed. If the high parallelism is assumed, a number of multiplications are necessary to perform square operations. However, it is possible to limit the hardware cost by simplifications based on decreased accuracy or LUTs.

The estimation of rates and distortions in the transform domain allows the encoder to complete both tasks in similar moments provided that their pipeline depths are similar. Moreover, the *R–D* cost for successive modes can be computed in parallel with the inverse transform and the reconstruction. As a consequence, the simplified RDO may add no clock cycles to the delay of the reconstruction loop. The buffering of the quantized coefficients makes it possible to save computations in the inverse transform since only transform blocks for selected modes must be processed to provide the reconstruction.

3.5 Motion estimation and compensation

The ME involves the largest part of computations in video encoders. It aims to provide MVs used as parameters in the inter-prediction, as shown in Figure 3.6. The reference area pointed by MV is taken as the inter-prediction block. The compensation amounts to the computation of the difference between original and prediction samples. Due to the multiplicity of possible partition modes in recent video compression standards, the number of candidate MVs required for particular pixel units is significant. Moreover, the use of multiple reference frames also increases the complexity proportionally to the number of frames. The considerable improvement in the compression efficiency is achieved when the fractional-pel ME is enabled. In this case, reference samples should be computed in fractional positions with interpolation filters. As a consequence, the ME complexity increases due to the filtering and the evaluation of fractional MVs.

The ME is based on the block-matching algorithm, which searches various positions of blocks in the reference frame. The output MV should point the block most similar to the original one in terms of a selected measure. Usually, the SAD measure is employed to quantify the similarity owing to its simplicity. Slightly better results can be obtained with the SATD measure. In the HM software, SAD and SATD are applied in the integer-pel and the fractional-pel ME, respectively. For each evaluated MV, one SAD/SATD must be computed. The corresponding complexity depends on the number of samples in the block.

In terms of high video resolutions, the ME involves a huge amount of computations. First, there are more pixel units in frames. Second, the reference area should be extended in both horizontal and vertical dimensions to reflect the same motion in the lower resolution. Third, the requirement of the high efficient compression involves the evaluation of more partition modes. Therefore, to support the real-time video encoding, the hardware acceleration is indispensable. If the ME algorithm assumes the regular dataflow, the throughput is proportional to the amount of hardware resources, and both factors can be easily scaled. However, the huge number of MVs determined in shorter time periods leads to complex architectures. Usually, the high complexity is not acceptable in nowadays and future applications due to limitations on silicon area and power dissipation. Therefore, ME architectures should be able to support algorithms with a reduced number of evaluated MVs and a high reusability of partial results.

3.5.1 Search strategy

The search performed for all MVs in the allowable range provides the best results in terms of the compression efficiency. Moreover, it can be easily parallelized and computationally scaled due to its regularity. On the other hand, the complexity of such a full search is often unacceptable, especially in the case of high resolutions. A significant complexity reduction is achieved with the hierarchical search. In this approach, the full search is performed on the subsampled reference area. Usually, every second or fourth sample in each dimension is taken to form the subsampled representation. As a consequence, both the number of MVs and the number of arithmetic operations in SAD are decreased inversely proportional to the square of the subsampling factor. If the factor is 2, the decrease of each is 4 folded. Since the complexity can be expressed as the multiplication of both numbers, the reduction is 16 folded. In fact, the ME on the subsampled area is equivalent to that on the corresponding lower resolution. The MV found in the subsampled representation should be projected to the original one and refined with the local search. The decreased accuracy of the subsampled search may lead to worse matching results compared to the basic full search. To limit losses, more than one candidate can be selected in the subsampled search and refined at the original resolution. For the subsampling factor equal to 4, better accuracy can be achieved when the hierarchy is composed of three levels. In this case, the local search is performed at two finer levels.

Apart from the hierarchical search on different resolution/subsampling levels, there are intermediate methods for the complexity reduction with a small impact on the quality. In the partial computation of the cost function, the subsuming of blocks taken to SAD/SATD can be combined with the evaluation of all integer-pel MVs. If the subsampling rate is 2, the complexity reduction is 4 folded. The same reduction is achieved when the full cost is computed for one rather than four adjacent MVs, i.e., the subsampling in the MV domain. If samples taken to SAD/SATD have a reduced accuracy (less significant bits are discarded), the hardware complexity decreases since it is almost proportional to the number of remaining bits.

There are some fast search algorithms able to find good matching using a reduced number of MVs. The set of such algorithms includes square three/four-step search, kite-cross/cross/diamond/octagon/hexagon search, and variants with modified test patterns. Three of them are shown in Figure 3.9 as examples. Some implementations

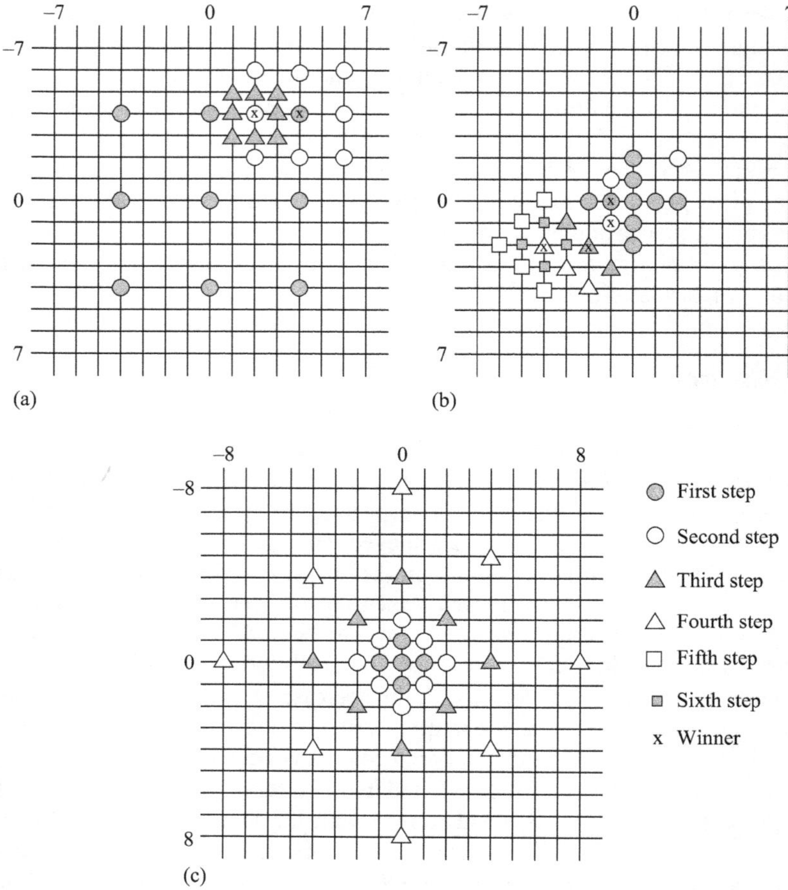

Figure 3.9 Example patterns of fast search algorithms: three steps (a), cross-diamond (b), and test zonal (c)

take advantage of more than one pattern. This is the case of the test zonal search used in the HM reference software. The effectiveness of fast algorithms stems from the assumption on the unimodality of the matching cost in the MV space. In particular, SAD values should become smaller when evaluated MV is closer to the best one. In reality, this assumption is not always true, and the search process tends to trap in local minima. Nevertheless, such situations are statistically less probable, and the worse matching also can be used for the inter-prediction with slight losses in the compression efficiency. Most of the fast algorithms were developed for low resolutions. Therefore, their transfer for high resolutions should not be straightforward to limit losses. In particular, it is important to avoid the sparse distribution of MVs, which exposes the search to trap in local minima.

The main disadvantage of fast search algorithms is the requirement of random access to reference samples. This requirement is difficult to satisfy in the hardware design based on processing elements organized into the systolic array due to the irregular dataflow. However, the 2D array of memory modules with the cyclic position shifters allows block access to reference samples for arbitrary MVs [9]. In each clock cycle, one MV can be evaluated for the block of size matching that of the memory array. Larger blocks take more cycles. When SAD is computed for a larger block, it is possible to simultaneously obtain SAD for its partitions with the same MV. For example, the reuse of SADs for 4×4 blocks allows the composition of cost for all partitions by the selective additions. This property saves computations when the search pattern is common for all partitions. In reality, they are different in the case of the complex motion, which negatively affects the implementation efficiency.

ME result for smaller partitions can be reused to save computations for larger ones within the same unit (e.g., CTU) [9]. Although smaller partitions are more suitable to reflect more complex motion, their MVs can be good candidates when a similar motion occurs for larger objects in the video scene. Indeed, the candidates for different partitions included in a given unit are substantially redundant, which allows the reusability. Thus, MVs found for particular smaller partitions can be evaluated for that including them. For example, MVs obtained for four 8×8 blocks can be checked for 16×16 block mode they form. The method can be used in a hierarchical way for 64×64 and 32×32 blocks evaluated with MVs found for 32×32 and 16×16 ones, respectively. The reuse of MVs of two adjacent blocks makes it possible to get good candidates for rectangular partitions (e.g., 16×8 from two 8×8). According to tests, the BD-Rate loss in the compression efficiency is equal to about 2.9%, on average. However, the loss rises when high-resolution videos are taken into account. It stems from the fact that MV bit costs for larger blocks have a lower impact on the total cost. As a consequence, MVs for smaller blocks tend to be sometimes shorter than optimal MVs for collocated larger blocks.

3.5.2 Fractional-pel motion estimation

The fractional-pel ME allows a significant improvement of the compression efficiency. Evaluation results of the H.265/HEVC encoding are summarized in Table 3.2 for low-delay profile and sequences specified in available repositories JCT-VC [10], UVG [11], and SVC [12]. As can be seen, the skipping of the fractional-pel ME would lead to unacceptable losses. However, the losses tend to decrease when the resolution becomes higher. Although the compression gain of fractional-pel ME is relatively small for 4K sequences, it is still considerable.

In the HM software model, the fractional-pel ME employs SATD to estimate costs associated with particular MVs. The need for the Hadamard transform makes hardware architectures more complex. As shown in Table 3.2, the replacement of SATD by SAD has a limited negative impact on the compression efficiency. Moreover, losses are very low or even there is a gain for 4K sequences. Therefore, it is beneficial to remove the Hadamard transform from the cost computation. On the other hand, the Hadamard transform can be replaced with the discrete cosine

Table 3.2 Impact of the fractional-pel ME on BD-Rate for low-delay H.265/HEVC encoding

Class	No. of fractional-pel ME	Only half-pel ME	SAD instead of SATD
A	33.07	10.14	0.20
B	57.05	12.30	0.38
C	81.87	15.56	0.48
D	127.49	23.18	0.60
E	84.85	14.79	0.70
4K-UVG	9.56	1.77	−0.42
4K-SVT	14.80	3.06	0.22
Average	58.38	11.54	0.30

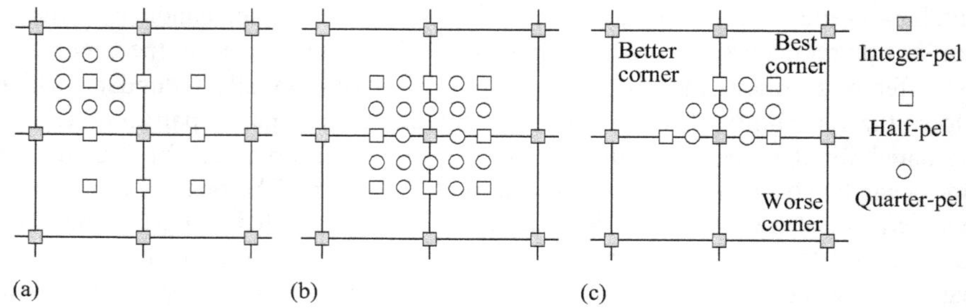

Figure 3.10 Fractional-pel patterns: two-step conventional software oriented (a), fixed hardware oriented (b), and reduced hardware oriented (c)

one to improve the compression efficiency. As reported in [13], the improvement is about 1% (BD-PSNR equal to 0.03 dB) for the H.264/AVC encoder and low-resolution sequences. The complexity of the transform is decreased by the selection of the half of coefficients after each stage with the negligible impact on the compression efficiency. Taking into account the results in Table 3.2, it is expected that the efficiency gain for 4K sequences should be smaller.

The conventional fractional-pel ME assumes the two-step processing, where the quarter-pel step follows the half-pel one. In each step, eight MVs around the currently best one are evaluated, as depicted in Figure 3.10(a). Thus, the total number of fractional-pel MVs is 16. If any new MV has a lower matching cost, it becomes the best MV. The two-step search is inconvenient in terms of hardware design. Therefore, the search pattern should be fixed. To avoid compression efficiency losses, more MV should be evaluated. For example, the fixed search pattern shown in Figure 3.10(b) is composed of 25 positions. The increase in the number of evaluated MVs negatively affects the throughput and/or the complexity. The reduction of the number of evaluated MVs can limit the problem. The elimination of MVs from the search pattern can be performed based on results of the integer-pel ME. In particular, the algorithm can evaluate only those fractional-pel MVs which are located between several best integer-pel MVs. This method is applied in the design described in [14], and its search pattern is shown in Figure 3.10(c). Since the number of MVs is decreased to 12, the average loss in the compression efficiency is about 1%.

The complexity of fractional-pel ME can be reduced by skipping it for some partition modes. Evaluations reported in [15] show that losses in the compression efficiency are in the range from 0.5% to 2.5% if only one CU size has no fractional-pel ME. The smallest and the greatest impact occur when the CU size is 64 $\times$ 64 and 16 $\times$ 16, respectively. Another method used to reduce the complexity consists in the estimation of fractional-pel matching costs based on the integer-pel ones. Instead on the sample, the interpolation is performed on the matching function. Achieved accuracies allow better results than in the case of the skipping. In particular, losses are decreased about two times [15].

The H.264/AVC standard specifies the six-tap luma filters for half-pixel positions, followed by the linear interpolation for quarter-pixel positions. Chroma samples are computed by the weighed interpolation of four closest integer-pixel samples. In H.265/HEVC, seven-tap and eight-tap filters are used for the luma interpolation of half-pixel and quarter-pixel positions, respectively. Chroma samples are computed using four-tap filters. Since the processing for high resolutions involves the use of a number of interpolation filters, their complexity plays an important role.

The straightforward implementation of interpolation filters with multipliers is inefficient. Since the filters are specified with constant multipliers, it is possible to design dedicated circuits with the use of adders and subtractors. Such decomposition should significantly reduce the hardware complexity. The hardware efficiency is better when samples are simultaneously computed by filters dedicated to particular fractional positions [16]. However, this approach restricts the design flexibility since interpolated samples must be either written to on-chip memories or directly submitted to compute SA(T)D for a number of fractional-pel MVs in parallel. Reconfigurable filters allow more flexible design at the cost of resources. In particular, such filters are reconfigured by additional multiplexers and the number of adders/subtractors is usually increased compared to dedicated filters.

The complexity of the fractional-pel ME can be reduced by the simplification of the quarter-pel filters. Their impact on the total complexity is dominant since they compute 12 out of 15 fractional-pel samples corresponding to each integer-pel position. In [14], the number of taps in quarter filters is reduced to two, and they take integer-pel and half-pel samples as inputs. As a result, the number of additions for 15 positions is decreased from 165 to 39 at the average quality loss of 0.02 dB (BD-Rate about 0.7%). Although the complexity reduction is significant, quarter-pel samples can be used only to estimate MVs due to their inaccuracy. Thus, inter-mode candidates for the best MVs should be computed with separate filters when the motion compensation is performed prior to the forwarding to the reconstruction loop.

3.5.3 Access to memories

Apart from the computation complexity, design for high resolutions is also challenging in terms of reading reference samples from the external memories. Although available bandwidths become even higher in successive generations of external memories, the growth in requirements is greater. In particular, the bandwidth scaling is similar to the complexity scaling when changing the resolution. If both the frame width and height are doubled, the reference area should be increased four times to match the same displacements after the projection to the higher resolution. Since the number of pixel units is also four times greater, the growth of the required bandwidth is 16 folded. It means that the transfer of ME algorithms from lower to higher resolutions cannot be straightforward.

The efficiency of memory access depends on the continuity of data transfers. The efficiency rises when pixels are placed in successive cells, i.e., when addresses are changed incrementally. For smaller blocks, address jumps are relatively frequent. Therefore, larger blocks and block-oriented pixel order facilitate access. Since the latest compression standard H.265/HEVC supports larger pixel units (64×64 CTUs) than previous standards (16×16 macroblocks), their transfer from external memories is easier. In particular, the reference area can be composed of CTUs reconstructed for the previous frame.

Reference areas of neighboring pixel units overlap to a great extent, as shown in Figure 3.11(a). Therefore, the reuse of common parts enables the bandwidth reduction. If each reference pixel/blocks is read only once, the bandwidth is minimal. However, the amount of data needed to buffer in on-chip memories is unacceptable for high resolutions. Therefore, blocks composing the reference area should be read several times, each time for different CTUs row. It increases the bandwidth proportionally to the number of repeats. The number depends on the vertical size of the reference area. If the vertical range of MVs is $[-128, 127]$, each 64×64 block of reference samples is read five times. If the horizontal search range is the same, the size of on-chip memories should be sufficient to keep at least 30 CTUs. This case is depicted in Figure 3.11(b). It is assumed that one CTU column is assigned to the write port, whereas the remaining ones can be used in the ME and compensation. To reduce the memory size, only luma samples need to be kept. Chroma samples should be fetched from the external memories for selected MVs.

The reduction of the bandwidth and the on-chip memory capacity is possible with the hierarchical search. The wide range is necessary only at the subsampled level corresponding to the lowest resolution. MVs found at this level can be applied to fetch narrow-range reference areas in the target resolution. Since MVs for particular partition modes are usually similar, their reference areas should mostly overlap. Therefore, the caching and the reuse of reference pixels should limit the average bandwidth. However, reference areas for particular partition modes can sometimes diverge leading to temporal peaks in the amount of data fetched from the external memory. If the bandwidth is limited, the peaks may increase access time assigned to pixel units, which in turn negatively affects the scheduling in the encoder. To compensate occasional delays, the scheduling should assume safety margins enabling the encoder to speed computations up.

In the local search, the number of fetched pixels should be significantly greater than the area of particular partitions. First, the local search at the original resolution extends the range. Second, interpolation filters used in the fractional-pel ME require an additional extension by 3 or 4 pixel in each of four directions. Third,

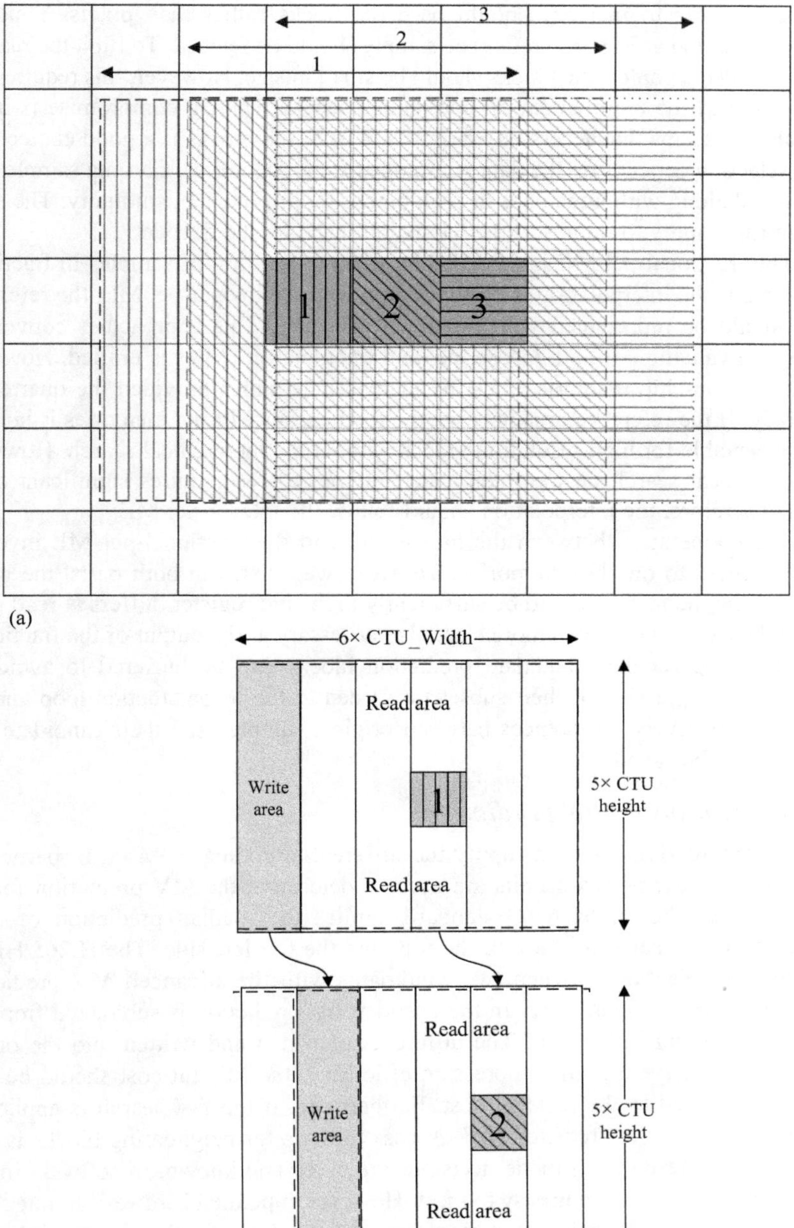

Figure 3.11 Example reference areas of successive pixel units in the frame domain (a), and their moving assignment to on-chip memory subspaces (b)

efficient access to memory should base on blocks rather than pixels. Thus, all blocks including at least one reference sample should be fetched. To limit the reading of not-needed samples, the blocks should be small in size. However, this requirement is opposite to the requirement on continuous access. Thus, a compromise is indispensable. It seems that access based on 4×4 or 8×8 blocks is a good choice. The final selection depends on the data width of external memories. Chroma samples can be fetched along with luma ones in such blocks to improve the continuity. The same design rules apply to minimize the bandwidth need by the decoder.

The fractional-pel ME is preceded by the interpolation of samples in fractional positions. If the interpolation is performed prior to the integer-pel ME, the reference area should be represented with fractional accuracy. This approach is convenient since the evaluation of both integer-pel and fractional-pel MVs is unified. However, the size of on-chip memories must be increased 16 times provided the quarter-pel accuracy. If the reference area is large, the total capacity of the memories is huge. It is unacceptable for high-resolution videos without the hierarchical search. However, even the local search on the interpolated reference area requires significant capacities. Therefore, the interpolation should follow the integer-pel ME.

The cooperation between the integer-pel and the fractional-pel ME involves shared access to on-chip memories. To avoid wait states in both parts, the bandwidth to the memories should be sufficiently high, and register buffers at read ports should be used. On-chip memories are also necessary at the output of the fractional-pel ME. In particular, candidate prediction blocks can be buffered to avoid the repetitive computation of their subset forwarded to the reconstruction loop and the RDO. Alternatively, differences between original samples and their candidate predictions can be saved.

3.5.4 Motion vector prediction

Video compression standards apply the differential coding of MVs. In particular, MVs of neighboring pixel units are used to determine the MV prediction for the current unit. The H.264/AVC standard applies the median prediction of three adjacent units located to the left, the top, and the top-left side. The H.265/HEVC enables the selection between two candidates with the advanced MV prediction scheme shown in Figure 3.12. In the encoder, the prediction is subtracted from the actual MV found in the ME. The difference is coded and written into the output stream. To achieve a high compression efficiency, the MV bit cost should be estimated and added to the residual cost. Furthermore, if the fast search is applied, it should start from the predicted MV. Access to MVs for neighboring blocks is possible when corresponding mode decisions are taken and known. In software implementations, the condition is easy to meet. However, pipelined hardware architectures introduce the delay, which makes the decision unavailable for the directly following unit during the search process. The delay is equal to the multiple of the period length each stage processes 1 pixel unit. For high-resolution videos, the unit size should be maximal for the sake of the compression efficiency. In the case of the H.265/HEVC encoder, it is beneficial to design the pipeline to deal with 64×64 CTUs or 32×32 CUs. The mode-decision delay for smaller units is equal to that for the maximal ones

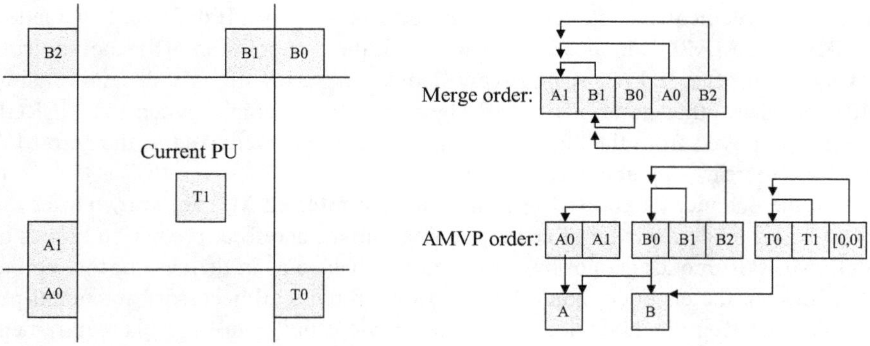

Figure 3.12 Selection of MV predictors from spatial (A0, A1, B0, B1, B2) and temporal neighbors (T0, T1). An unavailable candidate is replaced by the following one during the list construction

if the architecture employs ME and RDO stages assigned to successive units. Provided three stages are used (e.g., ME divided into fractional-pel and integer-pel parts), the first one can access the decision after the delay of two periods.

The unavailability of MVs of neighboring pixel units makes it impossible to implement ME algorithms in the form specified by software reference models. Thus, simplifications are indispensable. The negligence of the bit cost of MVs is an attractive solution due to the flexibility of the division into pipeline stages. However, losses in the compression efficiency can be considerable (about 1% for H.265/HEVC). Additional losses stem from fixing the search center corresponding to the zero MV (the increase to about 2% provided MERGE mode is supported). Although actual MVs can be accessed at the boundaries of unit rows, their use improves the efficiency in a small extent. A solution is to take final MVs found in the ME process. However, this approach should be completed with additional assumptions to specify the selection. First, if the ME selects more MV candidates for RDO, only the best can be considered for the MV prediction of the following units. Second, as partition modes of adjacent units are not decided, the design should fix one mode to take corresponding MVs. Third, if the search is performed simultaneously for different partition modes, the prediction of MVs cannot be determined on internal block boundaries. Thus, the MV prediction for the largest evaluated block can be applied in favor of smaller ones. Since the final partition and prediction modes are mostly not known at the ME stage, MVs used for prediction often differ between original and simplified algorithms. Nevertheless, losses are smaller compared to the case when the bit cost of MVs is neglected.

Video compression standards specify special inter-prediction modes when the MV of a given pixel unit is inferred from those of neighboring ones. The modes are named as SKIP and MERGE in H.264/AVC and H.265/HEVC, respectively. Their inferring scheme is similar to that applied in the MV prediction, as depicted in Figure 3.12. For such modes, prediction samples/pixels must be computed with the fractional-pel accuracy. Therefore, the hardware architecture must use interpolation filters either dedicated or shared with the fractional-pel ME. If dedicated evaluation of SKIP and MARGE modes is not supported, the compression efficiency decreases by about 50% and 6% respectively. However, losses strongly depend on simplifications in other parts of the ME algorithm. For example, when the MERGE mode is removed from the algorithm with the search center fixed to the zero MV, BD-Rate increases by about 10% (to 12.4%).

In the encoder dataflow, MVs taken for the inferred MV are known after the RDO-based mode decision of corresponding units. Therefore, prediction blocks of SKIP/MERGE modes cannot be computed in advance as in the case of the regular ME. Instead, the encoder should be able to interpolate all necessary fractional-pel samples just after the mode decision on the previous unit is taken. This requirement breaks the conventional pipeline organization where the ME and the RDO are placed in separate stages. In terms of the processing for high-resolution videos, the timing constraint on the interpolation process is strong. It means that the time slot available to finish this operation is shorter for smaller units. Moreover, since the H.265/HEVC standard allows the selection between up to five MERGE candidates, the interpolation speed should be significantly higher. Therefore, interpolation filters should be designed with a high degree of parallelism, which leads to a higher hardware complexity. Not all MERGE candidates should be computed in the time slot after the mode decision on the preceding unit. Candidates whose MVs are inferred from top neighbors can be computed before the decision on the left unit since their modes must be known earlier. This property allows more time-balanced processing in the interpellation filters.

Short time slots available to compute SKIP and MERGE predictions require very low delays in access to reference pixels submitted to the interpolation. Since the direct fetching from external memories introduces considerable delays, the pixels should be written in advance to on-chip memories. However, the uncertainty of predicted MVs during the fetching imposes a wide range of the reference area kept in on-chip memories. The wide range requires a larger capacity. To avoid additional resources, it is beneficial to share the memories with preceding ME stages.

3.6 Entropy coding

The performance of the entropy coding in recent video standards is limited by context adaptive binary arithmetic coding/context adaptive binary arithmetic decoding (CABAC/CABAD). The dataflow of the CABAC and CABAD is shown in Figures 3.13 and 3.14, respectively. The entropy coder receives syntax elements describing quantized transform coefficients, prediction modes, partition modes, quantization parameters, and parameters for sample adaptive offset (SAO). The syntax elements are first binarized. Bins are classified as either regular (context-coded) or bypass-mode symbols. For regular bins, context labels are determined based on the type of corresponding syntax elements, block sizes, the position in the binarized string, and values of previously coded syntax elements from the neighborhood. Symbols and corresponding context labels are submitted to the CABAC

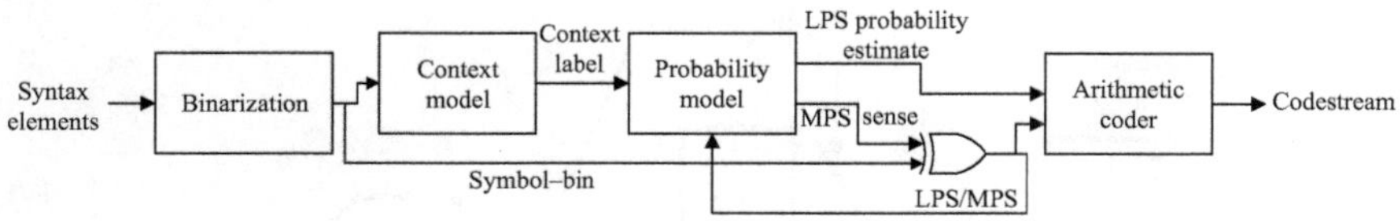

Figure 3.13 CABAC dataflow

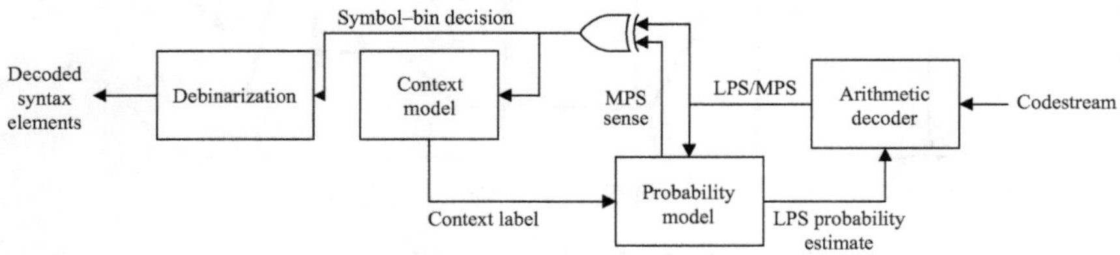

Figure 3.14 CABAD dataflow

engine, which produces the output stream. Since the general dataflow of the CABAC entropy coder has no feedbacks, the pipeline architecture can be employed at this level. As a consequence, critical paths can be minimized independently for each module.

At the decoder side, the stream is directed to the CABAD. Owing to the context and probability modeling, bin/symbols are restored in the loop composed of the arithmetic decoder core, the probability modeler, and the context model. Since the context label and/or the corresponding probability state for a symbol can be dependent on the preceding one, the loop imposes timing constraints in the hardware framework, which limits throughputs. Therefore, the CABAD architecture should integrate three modules to minimize the critical path.

If high-resolution videos are of interest, the number of bins submitted to the CABAC and/or received from the CABAD can be huge. However, there are some limits, which stem from standard specifications. In the H.265/HEVC standard, the maximum bit rate is specified as 800 Mbit/s at level 6.2 high tier [2]. With the assumption that the CABAC output/input compression ratio is about 0.8, the maximal bin rate (symbol rate) is approximately equal to 1 Gbin/s. In practice, the number of bins fluctuates strongly between frames and CTUs. Therefore, a real-time CABAC encoder should provide even higher bin rates. Even for the balanced distribution of bins across frames and CTUs, the CABAC/CABAD operating at the frequency of 200 MHz should process 5 bin in one clock cycle to match the limit imposed in the standard. In practice, bit rates of compressed videos are much lower. As a consequence, the required throughputs do not have to be so extreme. However, it is desired to maximize the throughput due to temporal peaks in the number of processed bins.

The ability to process several bins in one clock cycle affects critical paths of the entropy (de)coder. Main limitations on the throughput originate from the interval subdivision algorithm in the arithmetic (de)coder. The interval subdivision is depicted in Figure 3.15. The interval range is shortened and renormalized so that these

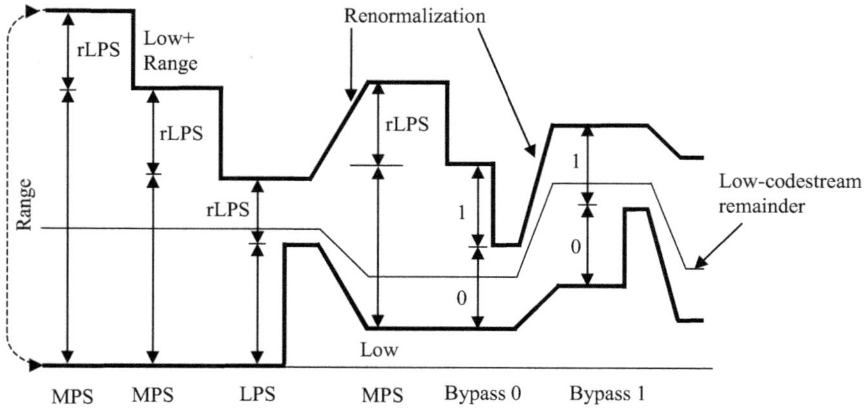

Figure 3.15 The interval subdivision

operations for a bin must be completed to start them for the next one. The sequence forms the critical path which limits the clock frequency. Although the higher parallelism decreases the frequency, the resultant throughput can rise if more bins are processed in the same clock cycle. The improvement is limited and tends to saturate.

Apart from the operations on the interval range, the conditional increase of the interval low border performed sequentially leads to longer paths in the hardware design. In the encoder, the low register is increased by the difference between the range register and the probability estimate of the less probable symbol. To avoid critical paths, it is useful to perform the operations on the low register in the following pipeline stage. Similarly, the probability estimates used in the interval subdivision should be placed in separate stages. Unfortunately, the pipelining cannot be applied in the CABAD due to the common dependence on bin/symbol decisions. Mutual dependencies between operations on the low (codestream remainder) register and the range one impose the use of a single stage. Moreover, the selection and the updating of probability estimates are conditioned by the decision on the preceding symbol if the same context models are used. These operations can be pipelined only when the look-ahead optimization is applied. In particular, two or four variants of the probability/context model can be computed in advance, and one of them is selected at the interval subdivision. Thus, instead of the multi-symbol processing, low-level optimizations are beneficial to maximize the CABAD clock frequency.

In the multi-symbol encoding, the frequency constraints can originate from the probability model. If the same context labels for two or more symbols processed in the same clock cycle are determined, the probability estimates (and sometimes MPS indicators) are different due to the model adaptation. In particular, the model selected by the context label is modified with the state transition table for successive bins/symbols. Such a table is easily mapped to LUT in the hardware design. In case when the same model is accessed and modified for several symbols, the architecture should be able to reconfigure corresponding LUTs into the cascade. Reconfiguration multiplexers and cascaded LUTs can form critical paths with timing similar to those in the interval subdivision. To shorten paths, the cascaded LUTs can be merged to support dual state transitions. The signal delay of merged LUTs is similar to basic ones.

For each CTU, the H.265/HEVC encoder can change parameters used by the SAO filter. The parameters are entropy coded at the beginning of each CTU prior to mode and residual data. However, the parameters are not available just after partition and prediction mode decisions for a CTU since reconstructed pixels undergo the filtering of edges in the deblocking filter. The output can be used to select SAO parameters. Usually, it requires evaluations of some parameter combinations. As a consequence, the parameters are determined with a delay with respect to CTU mode decisions. All quantized transform coefficients and modes must be buffered before the entropy coding even when the SAO filter is not present. However, the delay in the parameter determination significantly increases the buffer size. Provided that RDO modules, the deblocking filter, and the SAO one operate on successive CTUs, the buffer should be able to keep data for four rather than two CTUs.

3.7 Summary

Design for high-resolution video involves some challenges, mainly in encoder architectures. First, provided the same algorithm, the amount of computations increases proportionally to the number of processed pixels or to its square. Second, more compression efficient algorithms require more complex architectures. Third, mode and data dependencies between neighboring blocks introduce timing limitations in processing paths, especially in the reconstruction loop. Fourth, architectures operating at higher clock frequencies are indispensable. Fifth, simplifications introduced to algorithms implemented in reference models usually decrease the compression efficiency. Since there are many options and parameters in the encoder algorithm, the space of simplifications is huge. Therefore, designers should decide the algorithm specification subject to target design constraints.

References

[1] MPEG-4 Part 10, 'Advanced Video Coding (AVC)'. ISO/IEC Standard 14496-10, 2005.

[2] MPEG-H Part 2, 'High Efficiency Video Coding (HEVC)'. ISO/IEC Standard 23008-2, 2013.

[3] *AVC software repository—JM-16.0 reference model* [Online]. Available from http://iphome.hhi.de/suehring/tml/download/jm19.0.zip [Accessed 2018].

[4] *HEVC software repository—HM-16.6 reference model* [Online]. Available from https://hevc.hhi.fraunhofer.de/svn/svn_HEVCSoftware/tags/HM-17.0/ [Accessed 2018].

[5] Ylä-Outinen A., Lemmetti A., Viitanen M., Vanne J., and Hämäläinen T. D., 'Kvazaar: HEVC/H.265 4K30p Intra Encoder'. *Proceedings of EEE International Symposium on Multimedia*; 2017, pp. 362–363.

[6] Sjövall P., Viitamäki V., Oinonen A., Vanne J., Hämäläinen T. D., and Kulmala A., 'Kvazaar 4K HEVC Intra Encoder on FPGA Accelerated Airframe Server'. *Proceedings of IEEE International Workshop on Signal Processing Systems*; 2017, pp. 1–6.
[7] Sze V., Budagavi M., and Sullivan G. J. (eds.). *High Efficiency Video Coding (HEVC): Algorithms and Architectures*. Berlin, Germany, Springer; 2014.
[8] Pastuszak G., and Abramowski A., 'Algorithm and Architecture Design of the H.265/HEVC Intra Encoder'. *IEEE Transactions on Circuits and Systems for Video Technology*. 2016, vol. 26(1), pp. 210–222.
[9] Pastuszak G., and Trochimiuk M., 'Algorithm and Architecture Design of the Motion Estimation for the H.265/HEVC 4K-UHD Encoder'. *Journal of Real-Time Image Processing*. 2016, vol. 12(2), pp. 517–529.
[10] Bossen F., 'Document JCTVC-J1100: Common Test Conditions and Software Reference Configurations'. Stockholm: JCT-VC Meeting, Tech. Rep., 2012.
[11] T. Ultra Video Group. *4k test sequences* [online]. Available from http://ultravideo.cs.tut.fi/#testsequences [Accessed 2018].
[12] Haglund L. *The SVT high definition multi format test set* [online]. Available from https://tech.ebu.ch/hdtv/hdtv_test-sequences [Accessed 2018].
[13] Zhou D., Zhou J., He G., and Goto S., 'A 1.59 Gpixel/s Motion Estimation Processor With 211 to +211 Search Range for UHDTV Video Encoder'. *IEEE Journal of Solid-State Circuits*. 2014, vol. 49(4), pp. 827–837.
[14] He G., Zhou D., Li Y., Chen Z., Zhang T., and Goto S., 'High-Throughput Power-Efficient VLSI Architecture of Fractional Motion Estimation for Ultra-HD HEVC Video Encoding'. *IEEE Transactions on Very Large Scale Integration (VLSI) Systems*. 2015, vol. 23(12), pp. 3138–3142.
[15] Jou S. Y., Chang S. J., and Chang T. S., 'Fast Motion Estimation Algorithm and Design for Real Time QFHD High Efficiency Video Coding'. *IEEE Transactions on Circuits Systems for Video Technology*. 2015, vol. 25(9), pp. 1533–1544.
[16] Pastuszak G., and Trochimiuk M., 'Architecture Design and Efficiency Evaluation for the High-Throughput Interpolation in the HEVC Encoder'. *Proceedings of 16th Euromicro Conference on Digital System Design*; Sep 2013, pp. 423–428.

Chapter 4

High-throughput architectures for high-resolution video coding: hardwired oriented algorithms and VLSI architectures

Grzegorz Pastuszak[1]

High video resolutions along with more complex coding algorithms drive the need for new hardware accelerators featuring high throughputs. Simultaneously, it is desired to achieve this goal at the smallest possible cost of resources and power consumption. In the case of encoders, another important constraint is the allowable loss in the compression efficiency. The efficiency should be close to reference software implementations developed to demonstrate the compression possibilities of a coding standard. These opposite requirements must be satisfied with reference to hardware framework where design units usually support fixed throughputs and delays. Therefore, it is important to develop throughput-balanced processing paths with a high utilization of computation resources.

Provided the same performance, encoders are usually much more complex than decoders. Thus, this chapter focuses on hardware-oriented algorithms and architectures dedicated to encoders. However, some design techniques for particular modules can also be applied in decoders. There are some bottlenecks in the encoder dataflow as discussed in Chapter 3. They can be limited or removed by modifications introduced to coding algorithms. Usually, the modifications involve losses in the compression efficiency. The following sections review solutions to particular encoder tasks presented in the literature.

4.1 Reconstruction loop

The reconstruction loop processes residual blocks obtained for a preselected set of modes. The loop consists of the forward transform, the quantization, the dequantization, the inverse transform, and the reconstruction. The cascade of operations is easy to be pipelined. However, it introduces a significant delay, which can limit the encoder throughput as discussed in Chapter 3. The parallel processing of samples/coefficients included in a block decreases the delay at the cost of resources. On the

[1]Warsaw University of Technology, Institute of Radioelectronics and Multimedia Technology, Warsaw, Poland

other hand, the parallelism is limited by the block size. For example, 4 × 4 blocks can be processed with the maximal throughput of 16 samples/coefficients per clock cycle. Provided the same reconstruction loop is employed to blocks of different sizes, the delay increases for larger blocks.

There are several methods limiting the impact of the delay on the encoder throughput. The first focuses on decreasing the delay introduced by transform modules. The second employs more than one instance of the reconstruction loop. The third restricts the number of blocks processed in the reconstruction loop. The fourth applies the interleaved order of blocks of different sizes and types. The methods are described below in more detail with reference to architectures presented in literature.

4.1.1 Transform architectures

The buffering required to transpose blocks in transform operations increases the delay of the reconstruction loop. More cycles are needed to buffer larger blocks. Therefore, high-throughput architectures for the transformations are indispensable to support the high-resolution video coding. Larger sizes of transform blocks suggest the architecture able to receive and produce at least one row–column of samples/coefficients in each clock cycle. It means that the largest transform blocks allowable in the H.265/HEVC standard require the units with 32 inputs and 32 outputs. Such units can be optimized in terms of resources since there is a significant amount of computation redundancy existing inside and between the basis functions. Moreover, smaller transforms can be easily computed using the same unit due to the property of cosine basis functions. The straightforward architecture provides decreased throughput for smaller blocks. In order to obtain throughput fixed for all sizes, additional resources are needed. More detail can be found in Chapter 1.

The main disadvantage of transform architectures described above is the high cost of resources consumed by the transposition buffer. Furthermore, the delay increases significantly with the block size since the 1D-transformed block must be written to the buffer prior to the processing for the second dimension. In [1,2], the reconstruction loop employs transform modules with subblock interfaces. In particular, the forward transform reads and produces 4 × 8 subblocks. The order is illustrated in Figure 4.1 for the block size of 32 × 32. Arrows with numbers indicate the order of subblock contributions. In each clock cycle, two subblocks are read. They are taken from positions symmetrical to each other. For smaller block sizes, less input subblocks are accessed to compute particular output subblocks. As a consequence, the throughput is 32, 16, and 8 samples/coefficients per clock cycle for 8 × 8, 16 × 16, and 32 × 32 prediction blocks, respectively. The order of subblocks enables the reduced size of the transposition buffer and the decreased delay.

The architecture of the discrete cosine transform (DCT) module based on the subblock processing is shown in Figure 4.2. It consists of two similar parts responsible for horizontal and vertical 1D processing and the transposition buffer.

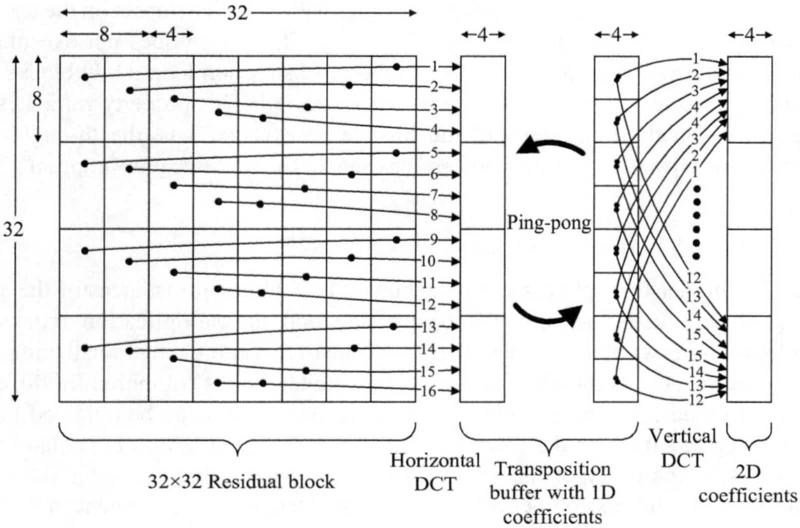

Figure 4.1 Column-oriented computation of the 2D 32 × 32 DCT [1]

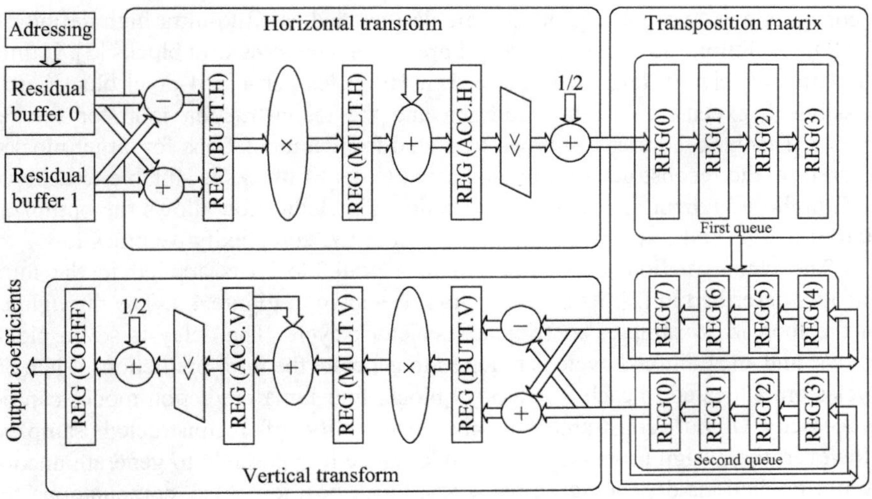

Figure 4.2 General DCT architecture [1]

The buffer comprises two register cyclic queues able to save eight columns of 8, 16, or 32 coefficients. Both 1D-transform engines multiply and accumulate contributions from successive input subblocks. The pipelining enables the design to operate at frequencies of 200 and 400 MHz when mapping into TSMC 90 nm and field programmable gate array (FPGA) Arria II technologies, respectively.

The reduced delay within transform modules has a positive impact on the encoder performance and the flexibility in the scheduling. The design does not execute the inverse transform on all blocks directed to the reconstruction loop. Only blocks corresponding to selected modes need to be reconstructed. This property mitigates the requirements on the throughput of the inverse transform. Thus, the throughput is decreased by half leading to proportional savings in the resource consumption.

4.1.2 Parallel loops

Generally, higher throughputs can be achieved by additional instances of the same processing path. In the case of the reconstruction loop, the multiplication of the same modules is also possible. However, the advantage of this approach is small compared to the resource cost. If parallel paths process residual blocks computed for different prediction modes, the mode decision and the reconstruction can be obtained faster. The speedup results from the shorter time needed to submit associated data to each loop. Usually, the delay of the path measured for single blocks is long and much greater than the submission period. As a consequence, the improvement in the total time assigned to a given block of original pixels is rather limited.

High video resolutions impose strong constraints on time slots assigned to blocks of input pixels. In the case of the smallest blocks of 4×4 samples, available number of clock cycles becomes too small to select the mode and to obtain the reconstruction. As a consequence, some designs dedicated to ultra high definition (UHD) resolutions do not support 4×4 prediction and transform blocks [3]. If time slots are sufficiently long (e.g., more than 10 cycles per 4×4-pixel block), it is possible to execute the intra-prediction and the reconstruction loop for several modes. This task requires resources separated from computations for larger blocks. Therefore, the reconstruction loop dedicated to 4×4 blocks is suitable to achieve both high-throughput and high video quality. The dedication allows the optimization of the general transform architectures since the complexity becomes low.

The reconstruction loop dedicated to 4×4 blocks is applied in the intra encoder described in [2]. Its block diagram is shown in Figure 4.3. The throughput of the loop is 16 samples/coefficients per clock cycle. The delay is seven clock cycles, and an additional cycle is required to generate the intra-prediction. When 50 cycles are allocated to each 8×8 pixel block, four luma prediction modes can be processed without constraints on the availability of reconstructed samples. However, the design processes more modes since it is possible to generate unconstraint predictions depending on the 4×4 block position. Also, computations for chroma modes are performed in the loop. The scheduling is depicted in Figure 4.4 for the 50-cycle slot assigned to each 8×8 unit.

The support for 4×4 luma and 4×4 chroma blocks involves the use of discrete cosine/sine transforms (DCT/DST), respectively. Both transforms can be designed separately and multiplexed depending on the color component type. However, the joint design allows some resource savings as the same arithmetic elements can be shared.

The concept of parallel loops can be extended to all transform sizes. In this approach, each transform size is assigned to a separate reconstruction loop. The loop

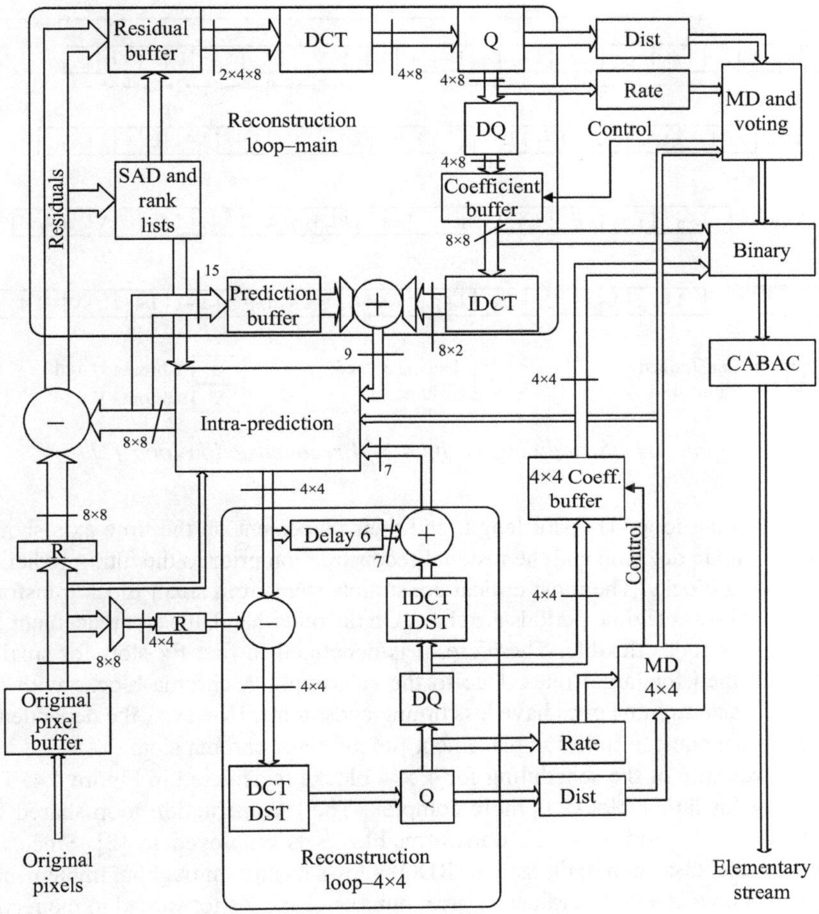

Figure 4.3 Architecture of the intra encoder proposed in [2]

delay is constant since there is no need to switch between transform sizes. Although the resource consumption increases significantly compared to the design with the shared loop, the growth of both the throughput and the utilization can be much higher. To balance the increased throughput, it is necessary to employ a high-throughput intra-predictor. The module can be decomposed into parts assigned to each loop. The parts should share the memory/registers to keep the consistency of reconstructed pixels.

4.1.3 Interleaved processing order

The interleaved processing order allows the use of the reconstruction loop to support different transform sizes and block types. The scheduling assumes that there are some time slots when residual blocks of a given type and size are forwarded and

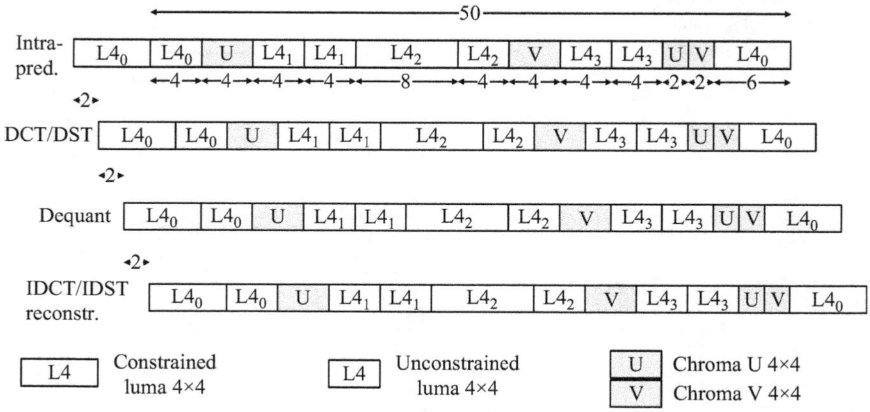

Figure 4.4 Scheduling in the 4 × 4 reconstruction loop [2]

processed in the loop. The slot length and their placement on the time axis should enable the mode decision and the residual reconstruction prior to the intra-prediction for following blocks. The most critical constraints stem from small block/transform sizes since allowable time periods are short. On the other hand, the slot placement for larger sizes is more flexible. Therefore, it is beneficial to first fix slots for smaller blocks and then for larger ones. Due to the subsampling, chroma blocks with the same block size as luma ones have less timing constraints. However, the dependence on the luma mode can limit the placement flexibility of chroma slots.

The example of the scheduling for 4 × 4 blocks is depicted in Figure 4.4. The scheduling for larger blocks is more complex. The reconstruction loop shared for 8 × 8, 16 × 16, and 32 × 32 transforms/blocks is employed in [2]. Since the simplified rate–distortion optimization (RDO) allows its high-throughput implementation, the encoder is able to evaluate a great number of modes forwarded to the reconstruction loop. Therefore, it possible to analyze some prediction modes for allowable block types (luma, chroma, and preselection based on original samples) and sizes. In particular, the processing path can be configured to process blocks of different sizes with scheduling based on the interleaving. The architecture has a significant flexibility to change the number of blocks of particular types and sizes under the condition that the number of clock cycles assigned to each 32 × 32 pixel unit is not exceeded.

4.2 Rate–distortion optimization

Due to the high computational complexity of the RDO, the designing of hardware encoders with the mode selection based on this method is a challenging task. Although there are many possibilities to achieve high throughputs of the reconstruction loop, the accurate rate estimation is a complex process which must keep track of context adaptive states of the probability models in context adaptive binary arithmetic coder (CABAC). Even when the adaptation is neglected and the probability models are fixed, the complexity is high since the estimation must take into account binarized syntax elements. Some of binary symbols have different fractional contributions to the rate.

There are two main groups of solutions to the complexity problem. The first group tries to find a statistical correlation between cost functions and signal features such as prediction error or texture gradients. The second group applies a simplified estimation of rates and distortions. The following subsections discuss both approaches with reference to designs described in literature.

4.2.1 RDO based on signal features

The coding mode selection based on signal features does not involve the complex processing in the reconstruction loop. However, the signal analysis can also be complex. Therefore, its advantage over the full RDO can be achieved when the amount of computations is kept at the moderate level, whereas losses in the compression efficiency are low.

The design proposed in [4] decomposes the mode selection process into two stages as shown in Figure 4.5. The stages process two successive coding tree units (CTUs). The first stage selects prediction block sizes in 8×8 and $16 \times 16/32 \times 32$ coding units (CUs). The selection is performed based on the analysis of texture features of original frames. The edge strength and the edge direction of each pixel are computed. At the CU level, the edge statistic features, such as the main edge direction, edge strength, and edge direction distributions, are obtained. According to the edge feature classification, linear relations between the pixel edge strength and its prediction error power are derived. Rates are estimated from error powers by the multiplication with weights dependent on the block size. The distortion includes the error power of coefficients greater than the threshold dependent on the quantization step.

The second stage incorporates two data paths for intra-prediction mode decision based on the RDO. The first path is assigned to $32 \times 32/16 \times 16$ CUs, whereas the second one supports 8×8 CUs. The paths do not perform the accurate estimation of rates and distortions. Instead, the design computes a new cost function which correlates rates with prediction errors. Fast rate models are computed by the binary classification and the linear regression method according to the features of $N \times N$ quantized coefficient block. The latency of rate estimation is reduced to three cycles. Although hardware-oriented modifications of the encoding algorithm reduce the amount of consumed resources, the consumption is still significant and equal to 2,269k gates. This result does not take into account entropy coder and memory buffers. For TSMC 90 nm, the encoder can operate at 357 MHz, which correspond to the throughput of 1,080p@44fps. According to evaluation results, the design has the compression efficiency lower by 4.53% (BD-Rate) compared to the HM reference software.

The use of convolution neural networks (CNNs) allows more efficient mode decisions based on the texture analysis. In [5], the architecture described above is modified to incorporate a CNN, as shown in Figure 4.6. Losses in the compression efficiency are lower, and BD-rate is equal to 2.67%, on average. The same network is employed to make the decision for 8×8 and $16 \times 16/32 \times 32$ CUs. Larger input

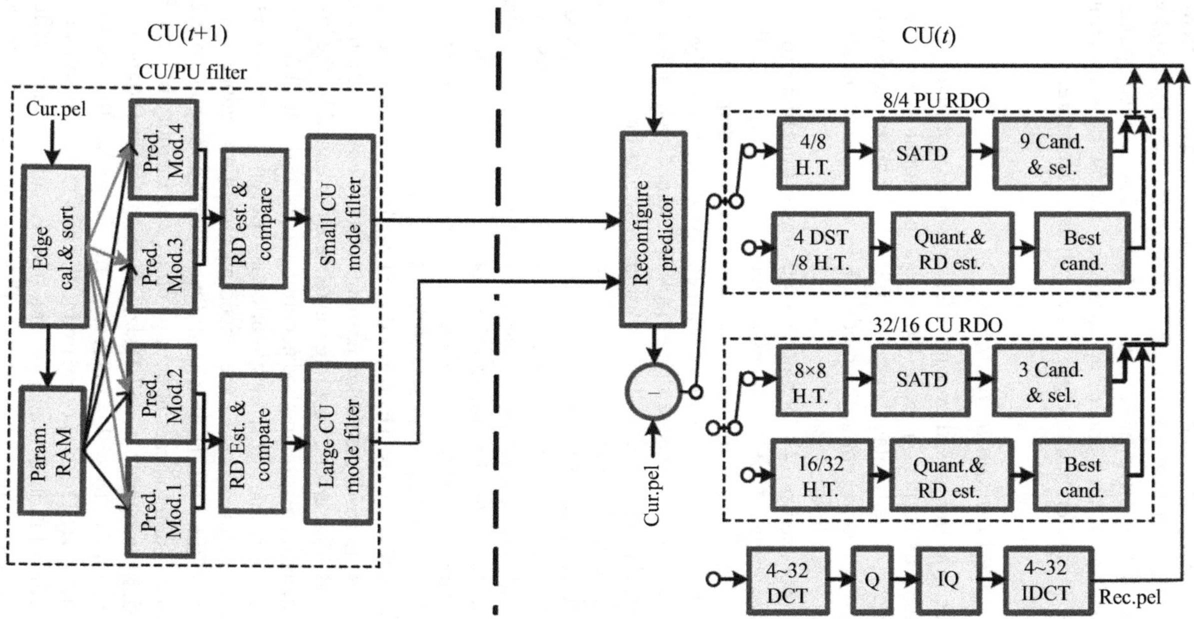

Figure 4.5 Architecture of the mode decision and the reconstruction loop proposed in [4]

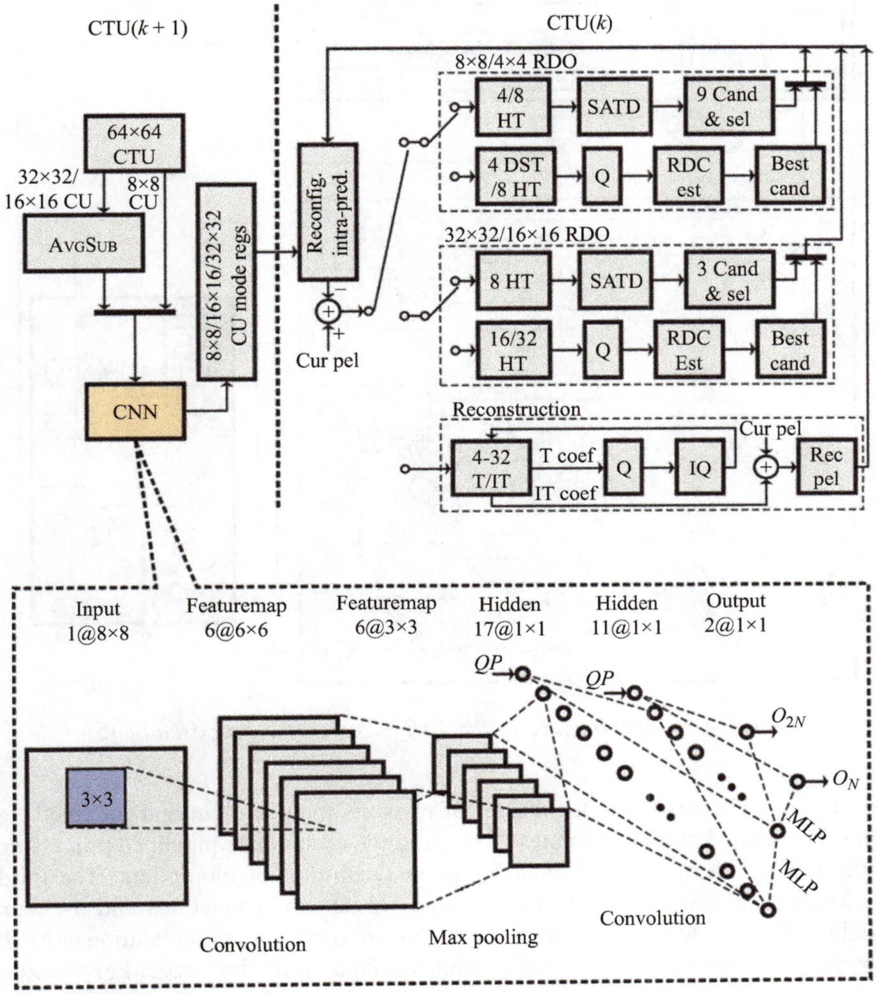

Figure 4.6 Architecture of the mode decision and the reconstruction loop proposed in [5]

blocks are scaled to the size of 8 × 8 samples by the averaging. The first hidden layer is a convolution layer producing six 6 × 6 feature maps. This layer extract signal features with 3 × 3 kernels. The second layer performs the subsampling and the max-pooling. The third layer performs the second convolution. Two last layers consist of fully connected multilayer perceptrons (MLP). The network is trained with gradient-based learning algorithms.

The architecture of the CNN accelerator is depicted in Figure 4.7. The design applies the propagate partial multiply–accumulate dataflow. It consists of three

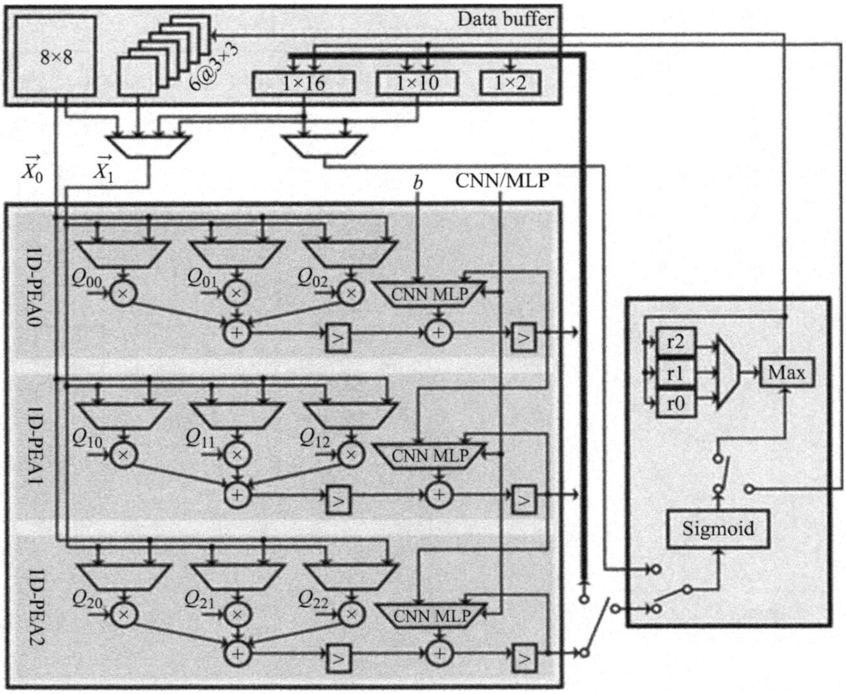

Figure 4.7 Architecture of the CNN accelerator proposed in [5]

basic parts. The first includes the set of registers to save input and intermediate data. The second part incorporates three multiply–accumulate pipelined processing units (PEA_i), each of which includes three floating-point multipliers. The third performs nonlinear operations such as sigmoid transition functions and the max pooling. Processing elements can be configured to perform the convolution or MLP operations by the selection of one multiplexer input at the last stage. For the convolution, the throughput is one operation per clock cycle. Due to the pipelining, the design is able to work at high frequencies, e.g., 714 MHz for the TSMC 65 nm technology. The network consumes 42.2k gates, which is quite small compared to encoder costs.

4.2.2 Simplified rate estimation

In RDO, rates and distortions are computed from quantized coefficients and reconstruction error, respectively. The rate estimation must follow the entropy coding dataflow, whereas the distortion requires the dequantization, the inverse transform, the reconstruction stage, and the computation of the sum of square errors. To reduce the complexity, computations can be simplified with additional assumptions on the adaptation of CABAC probability models and the orthogonality of transforms.

If probability models used to obtain predicted rate contributions for particular context-coded bins are fixed, estimated rates differ with actual ones to some extent. This difference introduces the inaccuracy of the cost function leading to the sub-optimal mode selection. On the other hand, the fixed models facilitate the implementation since the contribution of each context-coded bin depends only on its value. The fixing can be permanent or periodical. In the first case, initial model states are constant during whole frame coding. In the second case, the models are updated after the final mode decision for a given unit such as CTU or CU. The choice which method is more suitable depends on the optimization goal. The first method removes the need for probability model memory and the binarized representation of syntax elements at the BD-rate loss of 0.79%, on average. The periodical fixing has smaller loses; however, the probability model memory must be accessed and syntax elements must be binarized. After the final mode decision, probability states kept in the memory must be updated with all context-coded bins corresponding to the unit.

The periodical fixing of probability models is employed in the architecture proposed in [3]. The probability states are updated on the CTU basis. The architecture incorporates separate RDO paths for each of the six evaluated modes. The fixing allows the use of one probability model memory shared by all paths. The block diagram of the rate estimator is shown in Figure 4.8. The module employs six submodules, each of which consists of bit counters, multipliers, adders, and look-up tables.

As each bypass-mode bin contributes one bit to the compressed stream, simplifications are not necessary in this case. Thus, it is sufficient to apply the bin counting to estimate corresponding rate contributions. This method can be applied also to context-coded bins, which is equivalent to the fixing of all probability models to uniform distribution of zeros and ones. The inaccuracy in the rate estimation is higher compared to the fixing of models to initial states. According to evaluations on test sequences, BD-rate is equal to 2.11% [2].

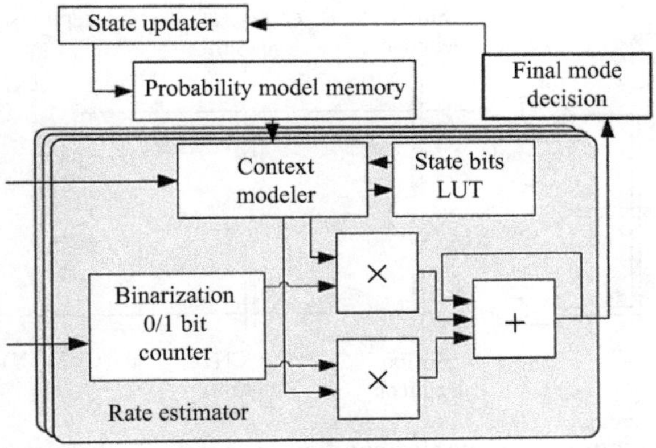

Figure 4.8 Rate estimator in [3]

Although the bin counting simplifies the rate estimation, the fast computation of numbers of bins for particular syntax elements is not trivial. It must be performed according to binarization procedures specified for the entropy coder. For quantized coefficients in each 4 × 4 block, the binarization applies adaptive rice codes. In this scheme, the suffix length is incremented for successive coefficients whenever predefined thresholds are exceeded. In turn, the current threshold depends on the last suffix length within the 4 × 4 block. This dependency leads to long chain of elementary operations when mapping to hardware framework.

The bin counting applied to all bins is used in the intra encoder described in [2]. The high throughput of the rate estimation is achieved by the deep pipelining and the representation of the suffix length with the unary code rather than the natural binary code. The later method is also applied for the counting of coefficients greater than 0 (GT0_CNT). The unary code applied in the chain allows the replacement of adders by two-input multiplexers, as shown in Figures 4.9 and 4.10. Moreover, threshold look-up tables (LUTs) are removed from the chain and replaced by simple LUT-based comparators. The architecture of the rate estimator is depicted in Figure 4.11. Apart from chain elements, there is an array of REM (remainder) subcircuits to compute remainders (REM decreases coefficients by 1, 2, or 3) and remainder bin counter LUTs. Additional LUTs are employed to compute bin contributions for signs, greater than zero indicators, and the position of the last nonzero coefficient. The module is able process one 4 × 4 block in each clock cycle. Higher throughputs are achieved by parallel instances of the module. Figure 4.12 shows the block diagram of the complete rate estimator with the throughput of two 4 × 4 blocks per clock cycle.

It is possible to approximate the bit rate without the quantization [6]. In this approach, transform coefficients are compared with thresholds derived from the quantization parameter (QP). In fact, the thresholds differ zero-valued and nonzero coefficients y_{ij} after the quantization. Only coefficients greater than the thresholds are considered to compute rates. Such coefficients are analyzed in terms of their

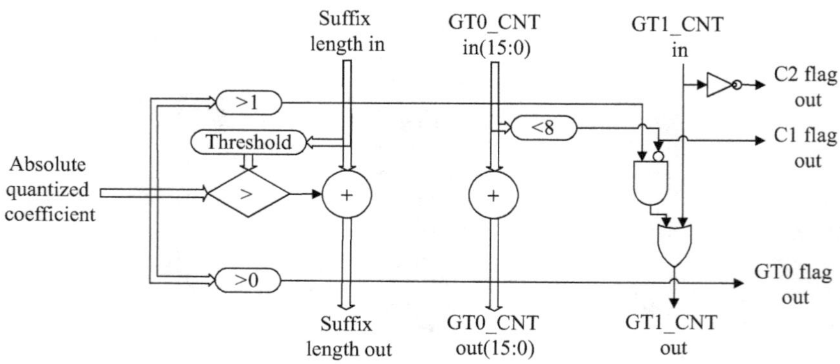

Figure 4.9 Straightforward implementation of the chain element for the rate estimation

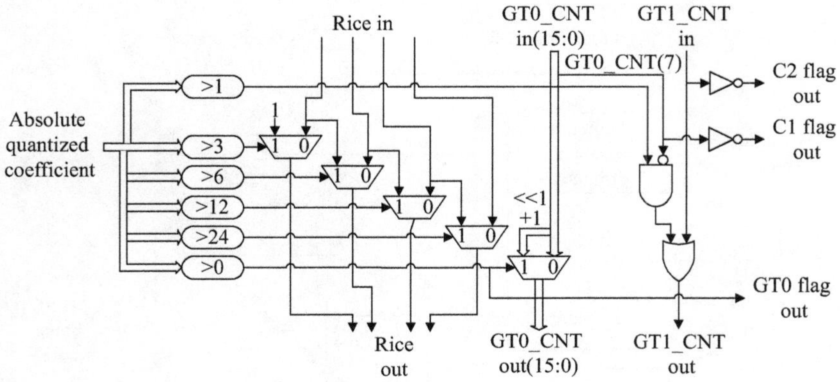

Figure 4.10 Optimized implementation of the chain element for the rate estimation

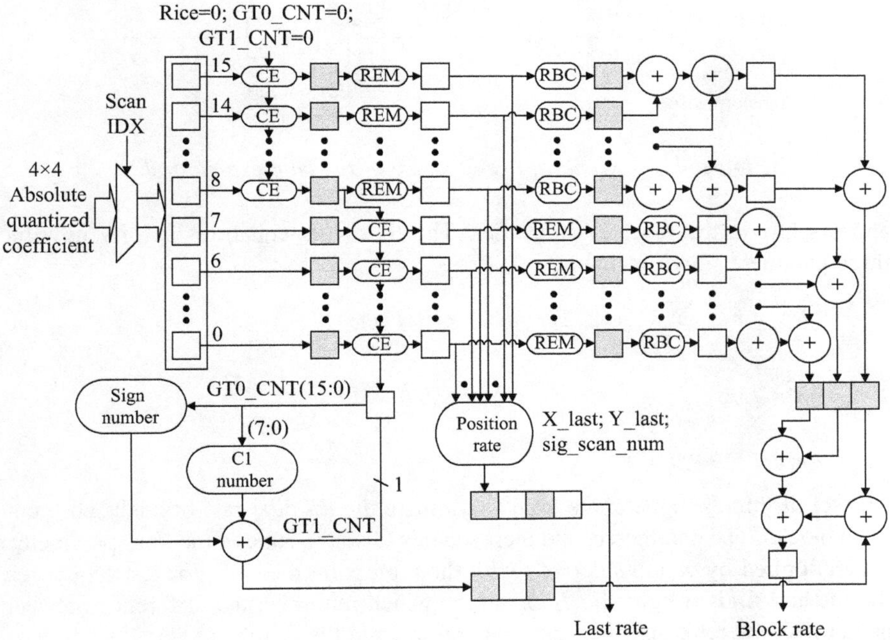

Figure 4.11 Rate estimator for 4 × 4 blocks [2]

absolute values and their position in the transform block. In particular, two sums are computed. The first (R_{value}) aggregates all absolute values, whereas the second (R_{posi}) takes their horizontal (i) and vertical (j) coordinates. An additional parameter is $R_{lastpos}$ is computed to take into account the position of the coefficient that

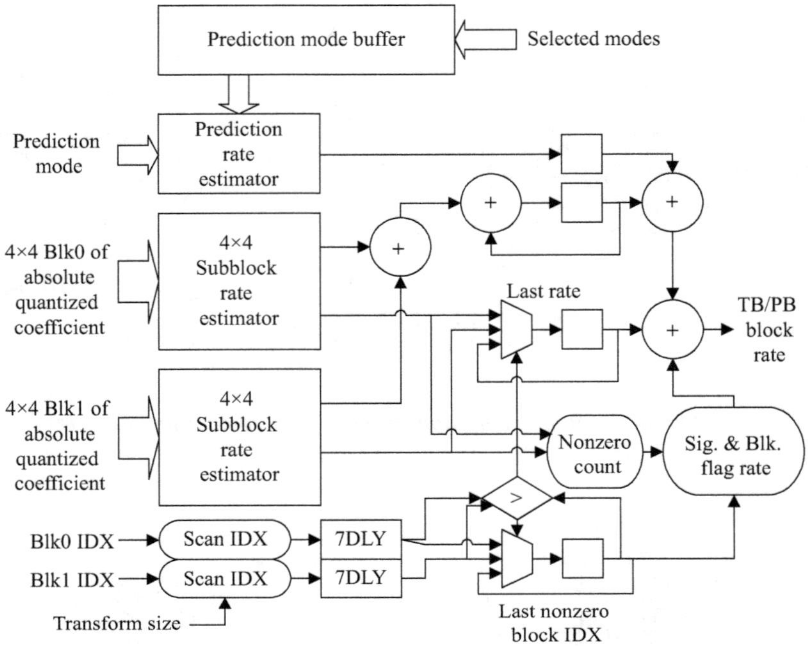

Figure 4.12 General architecture of the rate estimator [2]

appears last in the reverse scan order. The following equations specify how the three parameters are obtained:

$$R_{value} = \sum_{0 \le i,j < 2N} |y_{ij}|, \ Threshold(QP, TU)$$

$$R_{posi} = \sum_{0 \le i,j < 2N} (i + j), \ Threshold(QP, TU)$$

$$R_{lastpos} = i + j + 1$$

The set of three parameters is used to estimate the residual rate of analyzed transform blocks. The parameters and their squares (also the cube of the third parameter) are multiplied by weights derived with the regression method. For test sequences, the method finds weights α, β_i, δ_i, and γ_i which minimize the difference between actual residual rates and their approximation from the following equation:

$$R_{resi} = \alpha + \sum_{i=1}^{2} \beta_i \cdot (R_{value})^i + \sum_{i=1}^{2} \delta_i \cdot (R_{posi})^i + \sum_{i=1}^{3} \gamma_i \cdot (R_{lastpos})^i$$

Having fixed parameters for particular block sizes and QPs, the implementation is straightforward and can be easily pipelined. A total of 11 multipliers and two 16-input adder trees are required provided that the throughput is 16 coefficients per clock cycle. The architecture is shown in Figure 4.13.

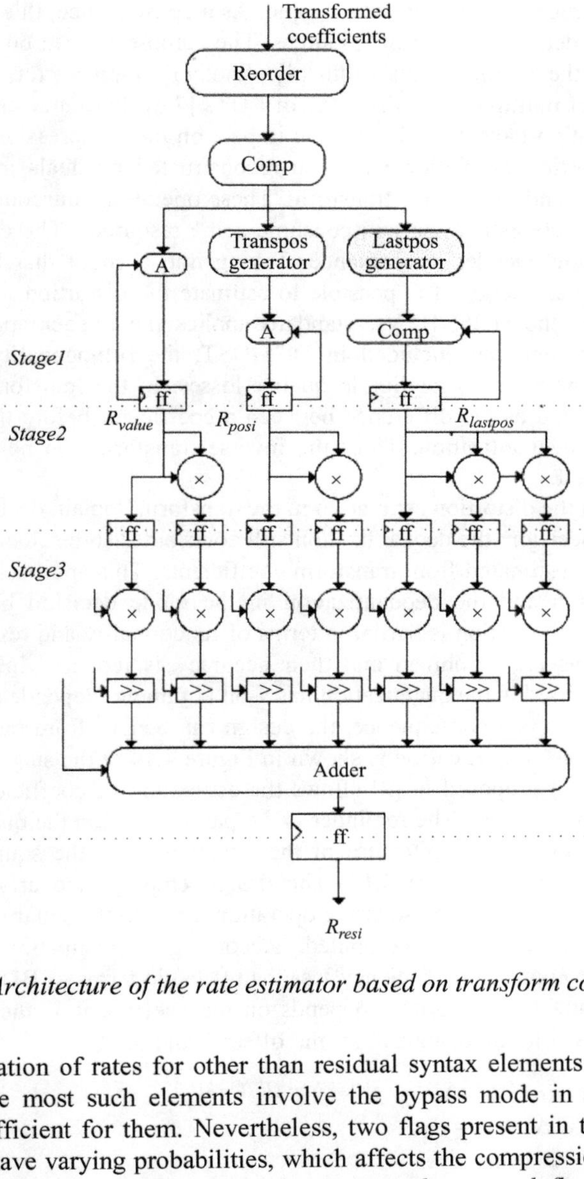

Figure 4.13 Architecture of the rate estimator based on transform coefficients [6]

The estimation of rates for other than residual syntax elements is much less complex. Since most such elements involve the bypass mode in CABAC, bin counting is sufficient for them. Nevertheless, two flags present in the prediction mode coding have varying probabilities, which affects the compression efficiency. In [6], the estimation for *cu_skip_flag* and *prev_intra_luma_pred_flag* is performed by the analysis of neighbor states. More bits are assigned when the current flag differs from those of selected for neighbors.

4.2.3 Simplified distortion estimation

The straightforward design of the distortion estimator for high-throughput architectures requires a significant number of multiplication units to compute the energy of the reconstruction error for each sample. As a consequence, this operation consumes a considerable amount of resources. The complexity can be reduced by the limitation of the multiplier bit width [4]. Another simplification consists in the replacement of multipliers by the array of LUTs [2,6]. Inaccuracies introduced by such modification have a small negative impact on the compression efficiency.

The distortion estimation based on reconstructed residuals must follow the dequantization and the inverse transform. These operations introduce a delay with respect to the rate estimation and consume some resources. The delay can negatively affect the encoder performance. Taking into account that DCT/DST preserves the signal energy, it is possible to estimate the distortion in the transform domain. Since the H.265/HEVC standard applies the integer approximation of original basis functions included in DCT/DST, the orthogonality is not ideal. Nevertheless, it involves negligible quality losses. In the transform domain, the error is computed as the difference between a coefficient before the quantization and after the dequantization. Thus, the inverse transform can be skipped in the estimation itself.

Although the distortion estimation in the transform domain decreases the delay of the mode decision, the dequantization still contributes some clock cycles. In [6], distortions are estimated from transform coefficients. This approach can eliminate the quantization and the dequantization in the mode decision based on RDO. However, the elimination is partial in terms of functionality and resources. In fact, both operations are combined and their accuracy is reduced. In particular, the multiplication used in the quantization has 5-bit argument (dependent on QP) rather than 15-bit one. As a consequence, the design can benefit from multiple constant multipliers. This simplification is shown in Figure 4.14 at the stage 1.

The method proposed in [2] allows the estimation of coefficient errors from quantization remainders. The reminder is the part rejected in the quantization after the downshifting. The architecture of the quantizer with the squared remainder output is depicted in Figure 4.15. The design employs the array of LUTs to approximate the result of the squaring operation. Prior to the squaring, the absolute value of the remainder is computed. According to evaluations, inaccuracies decrease the compression efficiency less than 0.2% in terms of BD-rate.

The quantization remainder depends on the coefficient Y, the multiplication factor MF_Q used in the quantization, the offset f, and the QP:

$$\begin{aligned}
&\Delta' = \text{fraction}\left(|Y| \cdot MF_Q \cdot 2^{\log 2(N)-(QP/6)-21}\right)\\
&\Delta = \Delta' \text{ if } \Delta' < 1 - f\\
&\Delta = 1 - \Delta' \text{ otherwise}
\end{aligned}$$

In the transform domain, the distortion depends on the transform/block size N, the quantization step size Q_s, and quantization remainders Δ computed from coefficients at block positions (i, j) as follows:

$$D = 2^{-14} \cdot N^2 \cdot Q_s^2 \cdot \sum_{i=0}^{N-1} \sum_{j=0}^{N-1} \Delta_{ij}^2$$

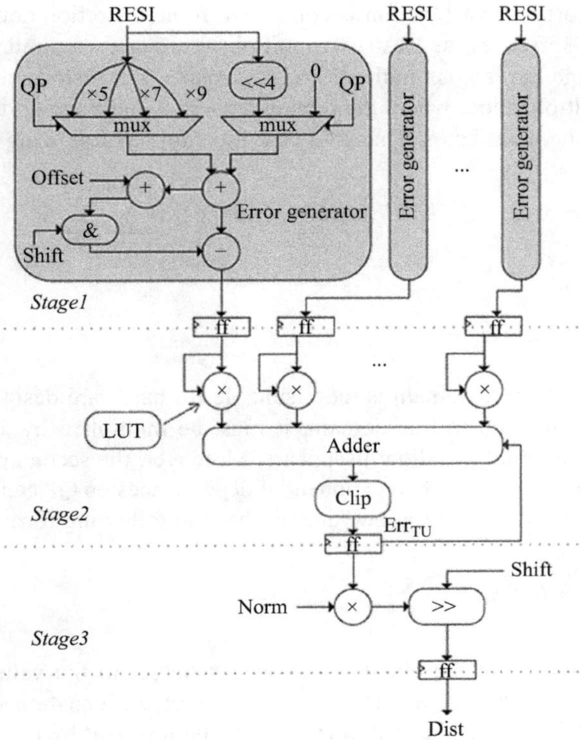

Figure 4.14 Architecture of the distortion estimator based on transform coefficients [6]

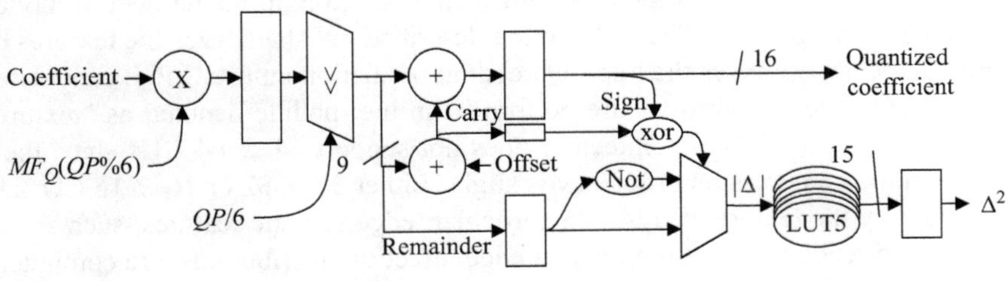

Figure 4.15 Architecture of the quantization unit with the squared remainder output [2]

The multiplication by $2^{-14} \cdot N^2$ converts the energy error from the transform domain to the residual domain. The quantization step size is computed from QP with the table of multiplication factors $MF_D()$ used in the dequantization:

$$Q_s = N^{-1} \cdot MF_D(QP\%6) \cdot 2^{1+\lfloor QP/6 \rfloor}$$

If the rate–distortion cost function is computed in the distortion domain (as in the HM reference software), at least two multiplications are required. The first one takes the rate and Lagrangian multiplier as arguments. The distortion is obtained by the second multiplication, where the remainder sum is multiplied with the selected squared multiplication factor. Thus, the cost function has following form:

$$J(R, D) = R \cdot \lambda + 2^{2\lfloor QP/6 \rfloor - 12} \cdot (MF_D(QP\%6))^2 \cdot \sum_{i=0}^{N-1} \sum_{j=0}^{N-1} \Delta_{ij}$$

where

$$\lambda = 0.57 \cdot 2^{(QP-12)/3}$$

It turns out that the rate domain is more suitable for hardware designs. To convert the above cost function to this domain, it must be multiplied by λ^{-1}. As a consequence, the first multiplication disappears. Moreover, the second part of the cost function simplifies since the two exponential dependences on QP cancel each other. Finally, the cost function in the rate domain has the following form:

$$J'(R, D) = R + Const \cdot \sum_{i=0}^{N-1} \sum_{j=0}^{N-1} \Delta_{ij}$$

The constant present in the above equation can be set to the value 11 for intra coding, which corresponds to the factor of 0.57 in the equation of Lagrangian multiplier. For this value, the multiplication can be replaced by two adders.

4.3 Intra mode decision

Due to a high complexity of the RDO, it is beneficial to limit the number of modes forwarded to this process. The architecture described in [4] analyzes the textures of source frame and removes the improper coding/prediction unit (CU/PU) candidates before RDO. These operations are performed in the module denoted as "texture-based CU/PU filter." The architecture does not support 64 $\times$ 64 CUs since their impact on the compression efficiency is slight. Either 32 $\times$ 32 or 16 $\times$ 16 CU size is selected by the texture analysis. In particular, edge statistic features, such as the main edge direction, edge strength, and edge direction distributions, are computed. According to the edge feature classification, linear relations between the pixel edge strength and its prediction error power are developed. Similar operations on 8 $\times$ 8 CU allow the decision between 4 $\times$ 4 and 8 $\times$ 8 PU sizes. The dedicated PU mode for each 8 $\times$ 8 CU is also defined during the texture preprocessing stage. The RDO process is performed in two modules working in parallel. The first computes the RD-cost of current 32 $\times$ 32 CB with the predefined 32 $\times$ 32 or 16 $\times$ 16 mode. The second obtains all RD costs of 8 $\times$ 8 partitions with the specified PU mode. The best coding tree structure is formed based on results from the two modules. In the analysis, blocks are classified according to their texture distribution features, and each class has a dedicated model. Next, the edge direction histogram is built from its edge mapping. The edge direction histogram is composed of 33 cells corresponding to 33 prediction directions. The cell with the maximum amplitude points out the main angle direction mode.

As described in Chapter 3, a number of intra-prediction modes and block dependencies limit the throughput of the video encoder. In order to improve the processing speed, the mode selection process can be performed with the use of original pixels as reference ones in the intra-prediction. In this approach, there are no block dependencies since each one can be processed almost independently from other ones. This feature allows the parallel processing at the block level and the more flexible pipelining. Results obtained in the pipeline have to be ready when the intra-prediction based on actual reference/reconstructed pixels is performed. In practice, it is sufficient to schedule the mode decision so that its result can be used in the reconstruction loop with a small delay. Longer delays require more resources to buffer mode decisions of some prediction units.

The mode decision based on original pixels is used in [7] as a so-called hybrid open–closed loop intra-prediction. In this approach, the encoder uses some pipeline stages each of which processes one macroblock. The intra mode decision based on original pixels is performed in a dedicated stage, whereas the following one employs the intra-prediction from reconstructed pixels and the reconstruction loop. The use of original pixels for the mode decision decreases the compression efficiency compared to the use of reconstructed pixels. However, it is shown that losses are slight when applied for 4×4 and 8×8 intra-predictions in the H.264/MVC encoder. Moreover, the losses are smaller since reference pixels on macroblock boundaries are taken from the reconstruction. Thus, original pixels are used as reference ones only for internal macroblock positions.

The design described in [7] takes mode decision using simplified costs computed as SATD which is less accurate compared to the RDO. As a consequence, the SATD-based mode decision introduces quality losses (about 0.35 dB/11% on average for H.264/AVC). The inaccuracy of the SATD measure partly masks that resulting from the prediction based on the original pixels. Thus, the quality losses originated from the prediction based on original pixels are more significant when the RDO is employed. This dependency is even higher in the case of the H.265/HEVC standard since there are much more angular prediction modes and prediction block sizes compared to the H.264/AVC standard.

To achieve high throughput and small quality losses, it is possible to select not one but a subset of prediction modes based on original pixels. Next, the preselected candidate modes can be processed with the intra-prediction produced from reconstructed pixels. Although the number of candidate blocks forwarded to the reconstruction loop with the RDO is limited after the preselection, the number of allowable prediction modes and sizes produced based on original pixels remains huge. This stems from the fact that the preselection should be performed by the evaluation of all prediction blocks with different sizes.

The architecture [2] performs the preselection by processing only 8×8 luma blocks for 19 out of 35 prediction modes. In particular, the analyzed modes include

DC, planar, and 17 even-numbered angular ones. Eight best prediction modes are selected based on sums of absolute differences (SAD) and forwarded to the rate–distortion analysis. Costs computed in the analysis are used to build the preselected mode subset for 8 × 8, 16 × 16, and 32 × 32 luma blocks. In particular, the rank list is built for 8 × 8 intra modes. With the observation that 16 odd-numbered modes have costs correlated with even-numbered ones, the former are added to the list if both neighbors are already included. Rank lists for larger block sizes (16 × 16 and 32 × 32) are built using the ranking of modes included in a given large block. In particular, the ranking is created by the weighted voting of ranked modes of included 8 × 8 blocks. In the case of 4 × 4 intra blocks, the preselection based on original pixels is not performed since the mode subset is fixed. Four basic intra-prediction modes (DC, planar, vertical, and horizontal) are always processed. Additional four, six, or eight modes are selected in dependence on the availability of reconstructed luma samples, which depends on timing constraints and the position of the 4 × 4 block.

The intra mode decision can be implemented without RDO using a simplified cost function. This approach allows a significant saving of logic resources at the cost of lower compression efficiency. In [8], the architecture selects the best block size and intra-prediction using sum of absolute transformed differences (SATD), as shown in Figure 4.16. To unify the processing for different block sizes, input blocks and reference samples are scaled to the 4 × 4 size. A total of 19 blocks can be processed simultaneously in separate paths, each of which includes the predictor generator, the 2D Hadamard transform, and the SAD calculator. To avoid timing constraints resulting from the delay of the reconstruction loop, the design takes early mode decision based on original pixels. When the mode and block sizes are decided, the encoder performs computations in the reconstruction loop to provide quantized coefficients and reference pixels for following coding units. When coding 1,080p video sequences, losses in compression efficiency are 12.66% in terms of BD-Rate. According to implementation results in the Xilinx Zynq FPGA technology, the design needs 83,548 LUTs and 34 DSP units. Most resources are consumed by the reconstruction loop (54,308 LUTs and 12,789 flip-flops). However, the resource consumption seems to be underestimated since only the transposition buffer of 32 × 32 samples requires 16k flip-flops. A similar underestimation occurs in the case of transposition buffers employed in Hadamard transforms. Moreover, it is not clear how the quantization and the dequantization are implemented without DSP units. Nevertheless, the incorporation of more efficient reconstruction loop following the early mode decision allows a significant reduction of resources.

Although the design complexity without the RDO is significantly reduced in [8], losses in the compression efficiency are often unacceptable. Therefore, the use of the simplified estimation of rates and distortions seems to be an attractive alternative. Architectures with simplified rate and distortion estimators are described in Sections 4.2.2 and 4.2.3, respectively. RDO simplifications in the intra encoder described in [2] lead to the average loss 2.03% in terms of BD-Rate. The architecture applies additional algorithm simplifications, which increase losses to 5.98%, on average. The simplifications include the luma-based CU-size decision, the preselection based on original samples, the skipping of residual quadtree and

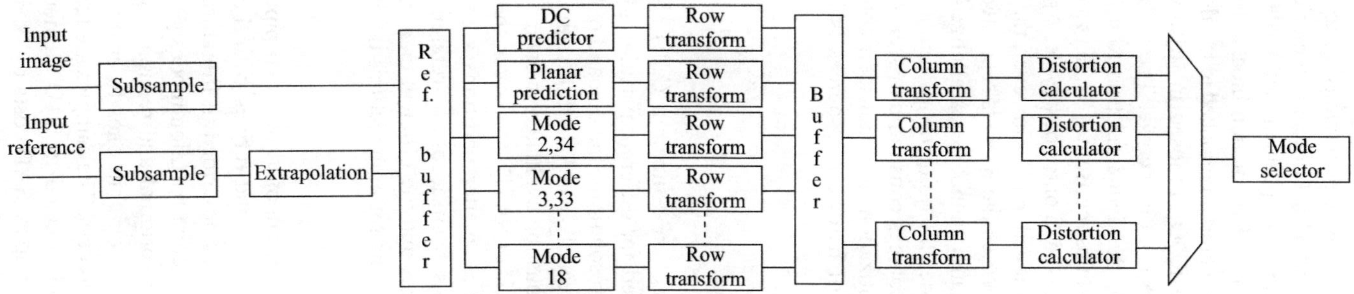

Figure 4.16 Architecture of the early mode decision in [8]

64×64 CUs, and the limited number of 4×4 prediction modes. All these modifications enable the architecture to support resolutions up to 2,160@30fps. The architecture consumes 1,080k gates and 52 kB on-chip memories. The design operates at two synchronized clock at frequencies of 200 and 400 MHz. The lower frequency is used in modules dedicated for 4×4 prediction blocks, the intra-predictor, and the entropy coder. The remaining modules are clocked at the doubled frequency.

The simplifications in the RDO proposed in [6] involve losses in the compression efficiency equal to 1.85%, on average. The architecture allows the elimination of the quantization and the dequantization to estimate mode costs in the transform domain. RDO modules consume 89.3k gates at the frequency of 330 MHz and 32 samples processed in one clock cycle. At the same parallelism, the frequency is lower than that achieved in [2]. However, the main disadvantage stems from the dataflow constraints. When the RDO is separated from the reconstruction loop, the processing in both paths cannot be parallelized and must be performed sequentially. As a consequence, the joint delay limits the encoder throughput. Furthermore, savings in the resources consumption seems to be deceptive since the quantization and the dequantization are indispensable in the reconstruction loop. Thus, the joint amount of hardware resources is larger provided the same parallelism in both paths.

4.4 Motion estimation

Architectures implementing motion estimation (ME) algorithms should provide motion vectors (MVs) estimated for different partition modes with the fractional-pel accuracy. In terms of high resolutions, the demand of high throughputs leads to complex implementations, especially when the full search method is supported. In particular, a number of MVs must be evaluated in each clock cycle, and each MV requires the computation of SAD or SATD. Moreover, the possibility to decompose pixel units into different-in-size smaller partitions significantly complicates the design. Another problem stems from access to reference samples/pixels, which should allow wide search ranges and the efficient data reuse. All these issues are addressed in the following sections for architectures described in literature.

4.4.1 Full search

Usually, ME architectures are decomposed into two pipelined modules. The first finds integer-pel MVs which the second refines with the fractional-pel accuracy. Some integer-pel ME architectures described in literature apply the full search to find MVs simultaneously for different partition modes [9,10]. Their block diagram is depicted in Figure 4.17. First, original and reference samples are fetched from external memories and written to on-chip ones to enable more flexible and low-delay access during the search process. The samples are read in the order dependent on the assumed scheduling and directed to SAD elements. To support different partition modes, the elements compute SAD costs for 4×4 subblocks. Next, such costs are aggregated with the adder tree and/or accumulators to obtain SADs for

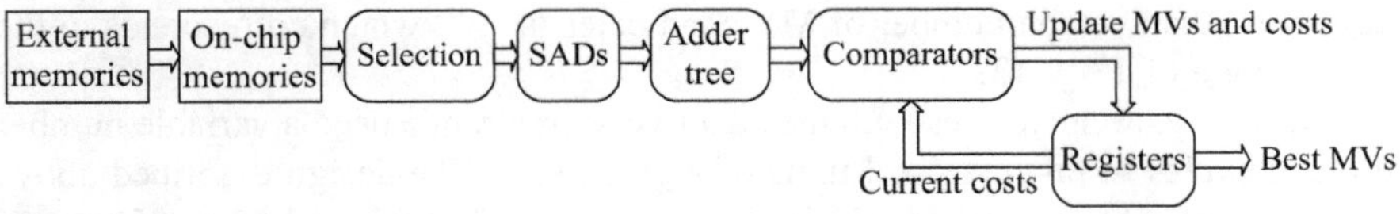

Figure 4.17 General architecture of the integer-pel ME

different partition modes. Finally, the set of comparators enable the selection of the MV with the lowest SAD cost. In particular, the best MVs and corresponding costs are saved in registers. When all MVs in the assumed search range are evaluated, the best ones are forwarded to the fractional-pel ME stage.

Although the simultaneous ME for different partition modes decreases the complexity, architectures supporting the full search still need a large amount of logic resources, especially when high-resolution videos are taken into account. For example, two H.265/HEVC ME designs described in [9,10] consume 3,560k and 2,878k gates, respectively. Such resources are sufficient to support only the 64×64 search range. Wider ranges, more suitable for UHD, would require much more logic resources. The regular dataflow applied in the two designs allows the reduced capacity of on-chip memories. In the first design, the memory size is 20 kB, whereas the second has no memories. However, the small memory size excludes the reuse of reference pixel samples between successive CTUs. Moreover, the fractional-pel ME must fetch the samples independently for each supported partition mode. As a consequence, the small memory size has a negative impact on the required bandwidth to external memories. Another disadvantage of full search architectures is the lower compression efficiency due to the unavailability of MV costs during the search process. In particular, the cost cannot be estimated for a partition since MVs of its neighbors are not known in most cases. All described disadvantages of the full-search architectures indicate that they are not suitable for high resolutions.

In video sequences, most of the content has low motion activity. Therefore, MVs with small lengths usually dominate. In order to find such MVs, the search ranges can be decreased, which reduces the computation complexity. However, high motion activity requires wider ranges, which involves more computations. The adaptation of the size of the reference area can be a solution to strike a balance between the compression efficiency and the complexity. In the architecture proposed in [11], the search range common to all partitions is selected based on the MV predictor for the current CTU. In particular, the width and the height of the reference area are increased when absolute values of two corresponding components of the MV predictor are larger. This dependence is implemented with two look-up tables separate for each dimension. However, it is not clear how the MV predictor is determined, taking into account the encoder pipeline and two (top and left) candidates. The ME design is able to change the search range from 5×5 to 65×65. As a consequence, the number of evaluated MVs can vary between 25 and 4,225 for the full search. According to the statistical analysis conducted for 4K sequences, the average number of MV candidates is 160, which corresponds to the search range of 13 × 13.

Generally, architectures with the adaptive search range need a variable number of clock cycles to process pixel units of a given size. The design described above [11] takes 495, 1,639, and 34,575 cycles for 5 × 5, 13 × 13, and 65 × 65 ranges, respectively. Although the design is claimed to support the compression of 2,160p@60fps videos with the frequency of 201 MHz, it cannot be applied in real encoder due to the strong dependency of the throughput on the motion activity. When the motion activity becomes higher at least temporally, the throughput dramatically falls. Furthermore, the declared average throughput is achieved when other stages of the encoder can process CTUs with the minimal number of clock cycles needed for the integer-pel ME (i.e., 495). In practice, it can be achieved at the unacceptable cost of resources and significant losses in the compression efficiency. Despite the reduced consumption of logic resources (274.5k gates) in the integer-pel ME, the resource consumption in the encoder would increase. Therefore, the bottom limitation on the number of clock cycles should be significantly greater, which decreases the throughput.

Apart from limitations on the throughput, the architecture proposed in [11] has other disadvantages typical for those supporting the full search. First, since MV costs cannot be used for partitions, the compression efficiency is much lower than it would stem from the limitation on the search range (BD-Rate much greater than declared 1.79%). Second, there is no reuse of reference data between successive CTUs. Although the adaptive search range reduces the bandwidth to external memories, the search path imposes the redundant fetching of reference samples even for the same CTU. Moreover, the design accesses external memories in both column wise and row wise, which is inefficient in terms of the bandwidth. The use of additional buffer could solve the problem. Another weakness consists in the assumption of low-delay access to external memories to fetch original and reference samples. In practice, the delay is much higher than assumed two cycles at the beginning of the CTU processing. Thus, the architecture needs more clock cycles for each CTU, which negatively affects the throughput.

4.4.2 Hierarchical search

The hierarchical search allows the significant reduction of the ME complexity and wider search ranges. This strategy decomposes the search into two stages assigned to the coarse and the fine ME. The first performs the full search on the subsampled reference area in the wide range, whereas the second performs the local search at the target resolution around the MV found during the coarse ME. The hierarchical search is employed in two architectures dedicated to H.264/AVC UHD encoders [12,13]. The architecture described in [12] applies the dataflow typical for the hierarchical ME, as depicted in Figure 4.18. Three pipeline stages perform the search with coarse-level, integer fine-level, and fractional accuracies of MVs. The pipeline can process three successive macroblocks. Original and reference frames are stored in separate external memories. Their data widths are 32 and 64 bits, respectively. Reference

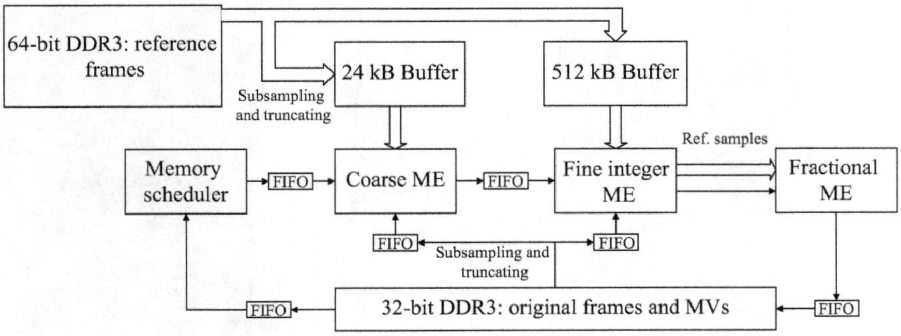

Figure 4.18 Architecture of the early mode decision in [12]

samples are buffered in on-chip memories separate for coarse and fine stages. In the case of the fractional ME, pixel data are taken from the fine search stage.

The advantage of the architecture described in [12] is the ability to support the resolution of 4,320p@30fps and the search range of 423 × 213. Such a performance and the support for the full search involve a significant amount of resources. In particular, the design consumes 2,458k gates and operates at the frequency of 210 MHz for the 40 nm technology. This resource cost is relatively high as the architecture employs the full search at both ME levels. To save computations, the shape of the search window is modified from the rectangle to the rhombus. The latter is more suitable to fit motion statistics in real video sequences, where probabilities of MV lengths in all directions are almost similar. On the other hand, the conventional rectangular shape is less efficient since its corners correspond to MVs with significantly lower probabilities of the occurrence.

The second architecture based on the hierarchical search [13] applies a modified dataflow, where the fine integer-pel and fractional-pel stages are integrated. Its block diagram is shown in Figure 4.19. Compared to the conventional dataflow, the main difference is that the interpolation of samples at fractional-pel locations is performed prior to the fine search. This order allows a low delay between the fine ME and the RDO since the integer-pel and the fractional-pel stages are integrated into one with the reconstruction loop. In particular, candidate prediction blocks can be forwarded to the reconstruction loop and RDO just after their SAD-based cost is ranked [14]. As a consequence, there is no need to separately perform the compensation, which improves the data reuse. The low delay also enables access to actual MV predictors during the fine search. As a consequence, the compression efficiency can be considerably higher compared to encoder designs with separate processing stages used for integer-pel ME, fractional-pel ME, and the RDO. First, the MV cost can be reliably estimated. Second, the prediction for the SKIP mode can be obtained and directly submitted to the reconstruction loop.

The architecture proposed in [13] embeds 128 memory modules arranged into two 8 × 8 matrices, where each module keeps every eight sample in both dimensions. One of matrices is read in each clock cycle. Read addresses are determined to get

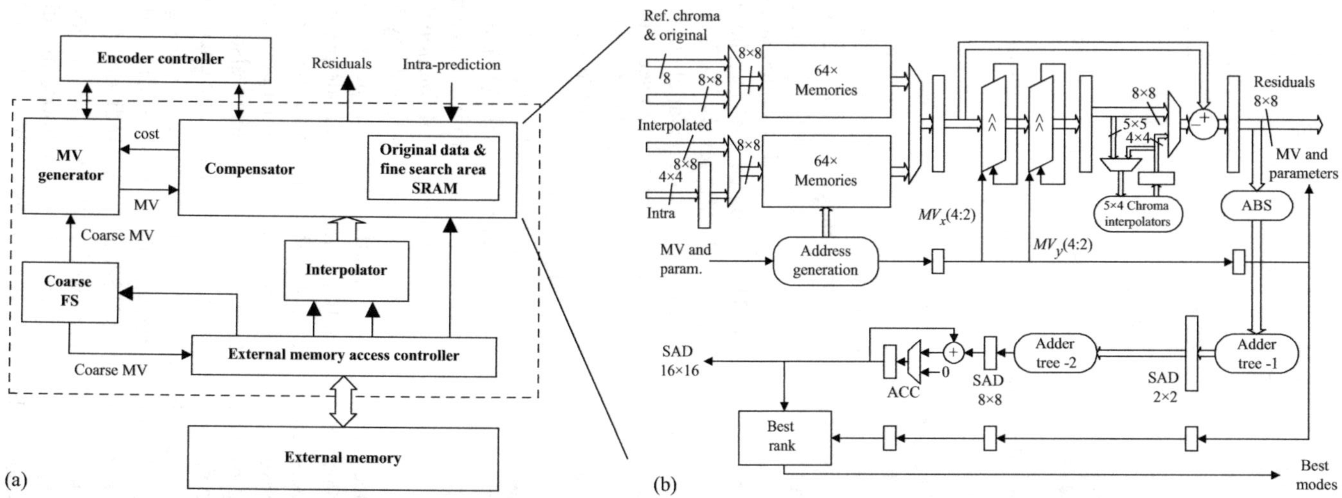

Figure 4.19 ME system architecture (a) and of the compensator architecture (b) in [13]

samples belonging to a desired block. 2D rotations restore the space orientation of samples within the block, as depicted in Figure 4.20. Addresses and numbers of rotation shifts are determined from two MV components. Since MVs generated and evaluated in each clock cycle are unrestricted by their order, the architecture can employ different fast search algorithms in the limited search range to achieve near optimal results using a small number of MVs. The search path is determined by the MV generator. The feedback from the cost computation enables the modification of the path. The fractional-pel ME is performed in the same way as the integer-pel ME. Such a unified ME facilitates the control. The design is computationally scalable, i.e., it allows the tradeoff between the number of utilized MVs and the compression efficiency.

The interpolation before the fine search increases the memory capacity 16 times. To limit the capacity, the chroma interpolation can be performed on samples read from on-chip memories. This modification introduces the two-cycle delay for chroma blocks, which has a negligible impact on the throughput. The architecture

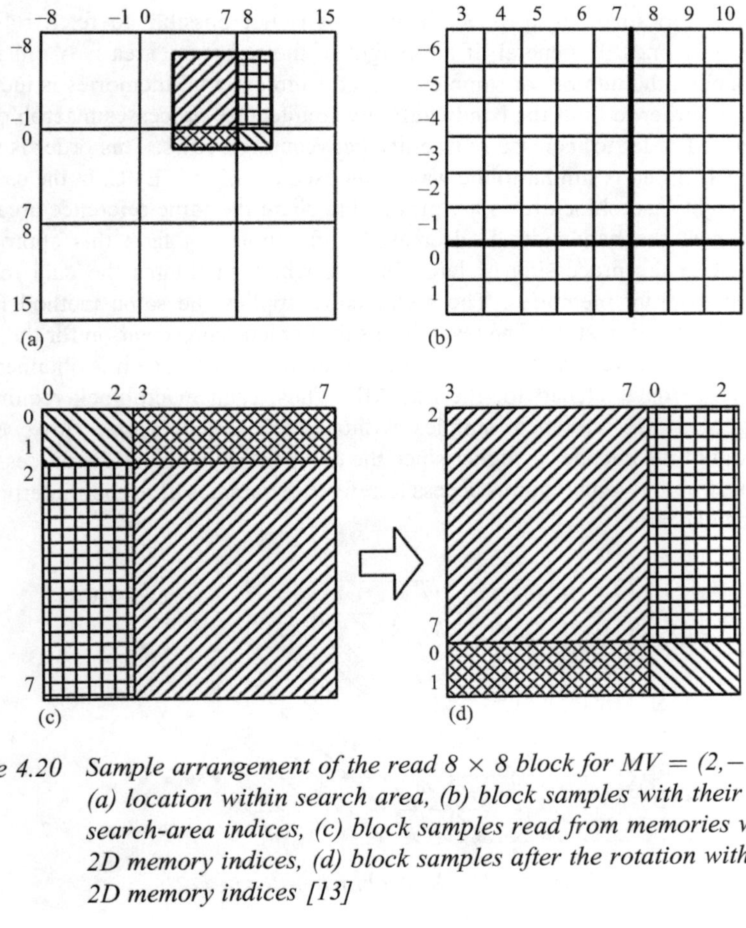

Figure 4.20 Sample arrangement of the read 8 × 8 block for MV = (2,−5): (a) location within search area, (b) block samples with their search-area indices, (c) block samples read from memories with 2D memory indices, (d) block samples after the rotation with 2D memory indices [13]

featuring the integrated ME stages and the hierarchical search [13] consumes 169k gates and 104.44 kB memories. It can support the H.264/AVC encoding for the 2,160p@30fps resolution and small losses in the compression efficiency. These results prove the usefulness of the architecture in encoders dedicated for high resolutions. Moreover, some design techniques can be applied to ME architectures developed for H.265/HEVC.

At the coarse level, both architectures employing the hierarchical search [12,13] benefits from the subsampling of the reference area four times in each dimension. As a consequence, the capacity of memories is decreased 16 times. In one design [13], coarse level memories are able to keep 16 macroblock lines, where two of them are assigned to the current frame. The mapping of whole lines removes the need to fetch the same data repeatedly. For 2,160p resolutions, the memories are 64 kB in size. The second design [12] embeds smaller memories at the coarse level. In particular, it keeps the segment of the reference area corresponding to 1,024 × 512 pixel in on-chip memories. Coarse-level samples are represented on six bits. As a consequence, the memories are 24 kB in size. Since the width of the segment is too small to keep whole lines, it is not possible to fetch reference samples only ones. In general, if the height of the reference area is *N* units (e.g., macroblocks), the number of samples fetched from external memories is increased *N* times. In order to limit the bandwidth, the architecture processes macroblocks in the modified order to increase reusability between their rows. This order is shown in Figure 4.21 and is similar to the wavefront used in H.265/HEVC. In the case of P frames, eight macroblock rows are grouped to share the same reference area. As a consequence, the bandwidth is decreased eight times. In fact, this approach is equivalent to the processing of large blocks, which facilitates the data reuse in access to external memories. The architecture applies the same method for the bandwidth reduction at the fine level. The subsampled representation for the coarse ME does not involve separate access to external memories since it is obtained from the reference data fetched for the fine ME. Thus, each macroblock requires the fetching of 1 kB reference samples without their compression. However, the bandwidth of the encoder is higher since the compensation before the processing in the reconstruction loop requires access to reference samples after their interpolation for particular modes.

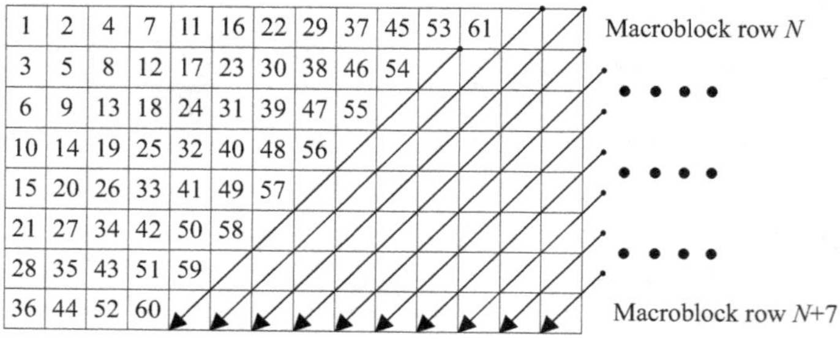

Figure 4.21 Macroblock order in [12]

The modified order of macroblocks in [12] is inconsistent with the raster scan applied in coding standards. Thus, MVs found for particular macroblocks in the group are written to memories and read in the appropriate coding order. The need for the buffering of MVs for several macroblock rows introduces a significant delay between the ME and the RDO, which excludes any feedback from the latter. On the other hand, the modified order of macroblocks enables access to MVs found for top and left neighbors for particular partitions. With additional assumptions on the prediction and partition mode decision in RDO, it is possible to estimate MV costs for the current macroblock. Although the assumptions in the ME involve slight losses in the compression efficiency, the main problem is unavailability of MV predictions for the SKIP mode. As a consequence, losses in the compression efficiency are much higher that claimed 0.08 dB (BD-Rate equal to about 2.7%).

At the fine level of the hierarchical search, the reference area fetched from external memories can be relatively narrow since MVs obtained at the coarse level should be modified with small values. The smaller search range decreases the bandwidth. In [13], the size of the fetched area is 40×40, which enables the search range of 16×16 with the MV fractional accuracy. Thus, each 16×16 macroblock requires the fetching of 256 original luma samples and 1,600 reference ones to perform the fine ME with the fractional accuracy. The architecture stores the reference samples for all fractional positions in on-chip memories. Thus, the memories at the fine level are 40 kB in size, and luma samples occupy 32 kB. This is significantly less than in the architecture storing the whole segment of $512 \times 1{,}024$ samples [12], which requires the joint capacity of 512 kB.

4.4.3 Test zone search

The test zonal (TZ) search is the ME method adopted in the HM reference software as an alternative to Full Search. Although the TZ search involves some losses in the compression efficiency, it significantly reduces the computation complexity and is suitable for wide search ranges. The algorithm combines different search methods. At the beginning, it selects the search center from two MV predictor candidates and checks some specific MVs inferred from neighboring units. Next, the zonal search is executed with either the diamond or the square pattern (diamond pattern is shown in Figure 3.9(c) in the previous chapter). The result of this step is refined with the following raster search and the star search. All these steps are difficult to map into the hardware design due to irregular access to reference samples and the variable search path.

The architecture implementing the simplified TZ search is proposed in [15]. To reduce the complexity and the bandwidth to external memories, the zonal search excludes long MVs for most directions. It turns out that the probability of direction change after the second step of the original algorithm is small. Thus, the direction is selected based on results obtained after the second step. Although such an early decision negatively affects the compression efficiency, the loss is relatively small and equal to about 1% in terms of BD-Rate. The implemented algorithm supports all CU sizes. However, only 8×8 and 16×16 CUs can be divided into symmetrical partitions. Particular PUs have different search ranges. In particular, the ranges for 16 × 16 CUs and 8 × 8 PUs are 33 × 33 and 9 × 9, respectively. PUs for the remaining sizes PU have the range of 3 × 3. The final step of the TZ search is skipped, i.e., MV is not refined with the raster search. As a consequence, the loss of about 1% is incurred. The fractional-pel ME is not performed for 8 × 8 CUs. Moreover, fractional-pel ME for larger CUs (32 × 32 and 16 × 16) determines MVs by the interpolation of costs rather than samples. The total loss in the compression efficiency is claimed to be 5.14% for all simplifications taken into account.

The loss of about 5% seems to be acceptable when the significant reduction of the computation complexity is achieved. However, the architecture in [15] imposes additional limitations on the compression efficiency, which is not taken into account. First, the MERGE mode is not supported. Second, the architecture limits access to MV predictors to the same CU size. Third, they are taken from the results of the fractional-pel ME rather than the RDO. Fourth, the CU partition mode is selected based on SATD rather than the RDO. Of particular importance is the first limitation since the removal of the MV evaluations for the MERGE mode decreases the compression efficiency of the original HM software by about 6%, on average. Furthermore, the BD-Rate is considerably higher than 5.14% when 4K sequences are coded. It stems from the fact that high-resolution videos require search ranges wider than those in the simplified algorithm.

The simplified ME algorithm described above is mapped to the architecture with three stages and two paths, as shown in Figure 4.22. The paths are assigned to small (8 × 8 and 16 × 16) and large (32 × 32 × 32) CUs. Two first stages perform integer-pel and the fractional-pel ME, whereas the last stage computes costs and selects the best mode for different CU sizes. Corresponding residuals are written to the temporary buffer of the 5.76 kB size. The architecture is designed to process 16 × 16 sample units at all coding tree levels. The size is decreased from 64 × 64 to save memory resources. In particular, buffers for original, integer-pel reference, and fractional-pel samples are reduced to 1 kB, 2 kB, and 2,752 B, respectively. Thus, the joint size of memories for the unidirectional ME is 11,580 B. The amount of logic resources is declared to be 396.6k gates.

The resource consumption of the analyzed architecture is relatively small compared to implementations based on the full search. However, the architecture is reported without a number of details. First, interfaces between particular modules are not specified. Second, the register-level dataflow of modules performing the integer-pel and fractional-pel ME is not described. Therefore, it is not possible to verify the consistency of data transfers and their processing within declared time slots. Moreover, the scheduling based on 16 × 16 units is not suitable for the ME of larger CUs. Since the MV is selected after all subblocks within a CU are processed, the fractional-pel ME for 32 × 32 and 64 × 64 CUs cannot be started after the integer-pel ME for each 16 × 16 unit. This limitation makes the architecture impractical in the form proposed in [15]. To compute the inter-prediction with the fractional-pel accuracy of reference samples in allowable time slots, the buffers should be much larger. Even for the small search range of 3 × 3, the buffer should be able to store 41 × 41 and 73 × 73 samples for 32 × 32 and 64 × 64 CUs, respectively. Thus, the 40 × 40 buffer in Figure 4.22 should be 7,010 B

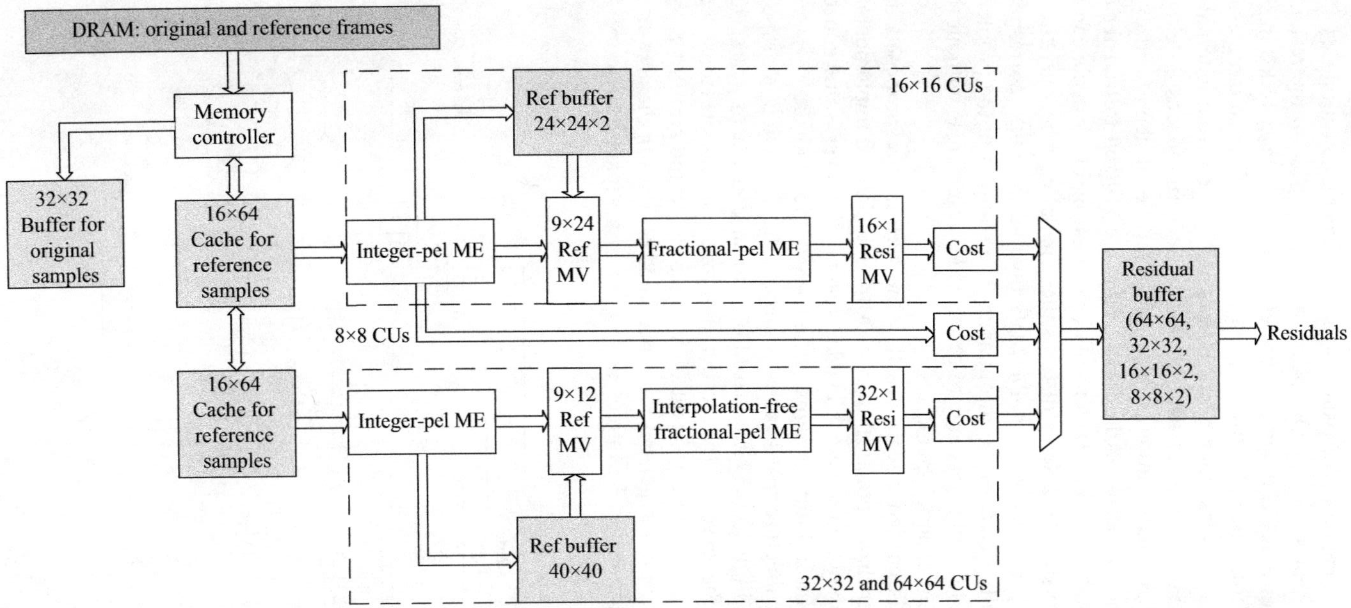

Figure 4.22 ME system architecture in [15]

rather than 1,600 B in size. Moreover, the size should be doubled to avoid the data overwrite. The limitation in the path assigned to larger CUs introduces an additional delay. To align delays in particular paths, additional space in the residual buffer is indispensable. For four CU modes, the buffer should be at least 16 kB in size.

The architecture in applies two cache buffers for reference samples. In the case of the cache miss, at least four 8 × 8 blocks are fetched from external memories. Although such cases are occasional, the occurrence introduces a delay stemming from access to external memories. The delay can be of about 100 cycles of the clock driving the memories. Additional cycles are utilized to transfer data. Since the encoder operate at 270 MHz, tens of corresponding clock cycle can be taken to complete the transfer. This delay disturbs the scheduling whenever the cache miss occurs. Although only about 2%–4% MVs involve the miss, corresponding wait states dramatically decrease the throughput. The solution to the problem could be the use of much larger buffers able to cover wide search areas without the need for quick access from external memories.

The analysis carried out proves that the architecture proposed in [15] is a conceptual rather than a practical design. Although claimed implementation results are distinctive, they cannot be achieved in reality. However, some concepts can be taken into account in future designs for high-resolution videos.

Another architecture that is able to support the TZ search is proposed in [16] and depicted in Figure 4.23. It is based on similar concept of access to reference samples as in the case of the design described in the previous section [13]. However, there are some important differences. First, the array of 8 × 8 on-chip memories stores only integer-pel reference luma samples in the area of 256 × 160 pixel. Second, the original samples are stored in a separate memory. Third, two

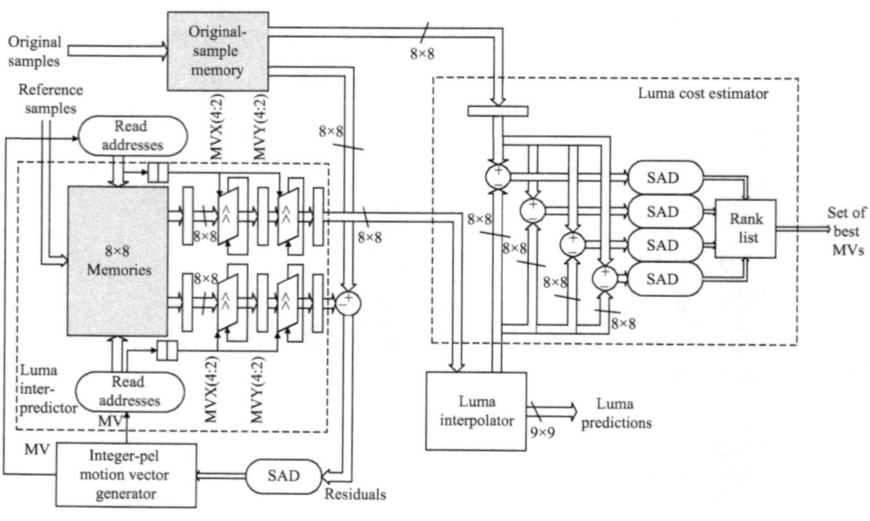

Figure 4.23 ME system architecture in [16]

parallel processing paths for the integer-pel and the fractional-pel ME are employed. Fourth, the interpolation of fractional-pel luma samples is performed on blocks read from the memory array. Fifth, the chroma interpolation for particular PU modes is computed in a separate submodule to improve the throughput.

All memories in the array utilize two ports for simultaneous read access to 8×8 blocks. Since access in the fractional-pel path is not continuous, there are some time slots available for the writing of reference samples necessary to shift the search window for the next CTU. The first path for the integer-pel ME is similar to that in [13], i.e., SAD computed for successive candidate blocks is used to make decisions after successive search steps. Final integer-pel MV can be directly used in the second path composed of the interpolator and the cost estimator with the rank list. The delay of the interpolator is minimized to enable the cooperation with the reconstruction loop and the RDO similar to that in [2]. This feature allows the support for the merge modes and the use of MV predictors in the search algorithm. The cost estimator computes SADs for four MVs in parallel. SAD results increased by the MV cost are simultaneously compared with the currently best selections for a number of PU corresponding to different partition modes. Results are ranked to allow more inter candidates for inter modes. Predictions or residuals are released as 9×9 blocks, which allows the inter-prediction for four corresponding MVs (their components differ by 1). The blocks should be buffered to select only those including candidates forwarded to the reconstruction loop.

To improve the compression efficiency, more candidates are selected for each PU in the rank list. Four candidates are selected for square PUs. The remaining (rectangular) PUs have one candidate MV, and the loss is about 0.3%. Although MVs of the candidates for adjacent partitions have small differences, the support for symmetric and asymmetric partition modes is efficient. The loss in the compression efficiency is about 0.3% compared to the case when the integer-pel ME for each PU is executed. The architecture supports the TZ search for 8×8 PUs. For larger square PU, only four MV candidates are evaluated. They are selected from the best MVs found for smaller PUs whose location coincides with the larger one. For example, 16×16 PUs has four MV candidates taken from ME results for 8×8 PUs. This simplification reduces the complexity almost three times. At least three merge mode candidates are evaluated for each square PU. Only merge modes are checked for the 64×64 CU. All the simplifications in the search algorithm lead to an average loss of 2.45% in terms of BD-Rate. However, the loss is smaller for 4K sequences and equal to 1.64%. The architecture operating at the frequency of 400 MHz is able to support the coding of 2,160p videos. It consumes 293k gates and 76 kB on-chip memories taking into account also chroma predictions. These results make the design attractive as the base to develop new architectures dedicated to higher resolutions.

4.5 Entropy coder

The support for high resolutions requires the design of high-throughput entropy coders. The main bottleneck is the arithmetic coder and decoder. For extreme conditions, they should have the throughput of above 1 Gbins/s as analyzed in

Section 3.6. This performance is achieved with either multi-symbol processing or highly optimized pipelines able to operate at high frequencies. In each pipeline stage, the multi-symbol encoding builds the chain of several units assigned to particular bin-context pairs, as shown in Figure 4.24.

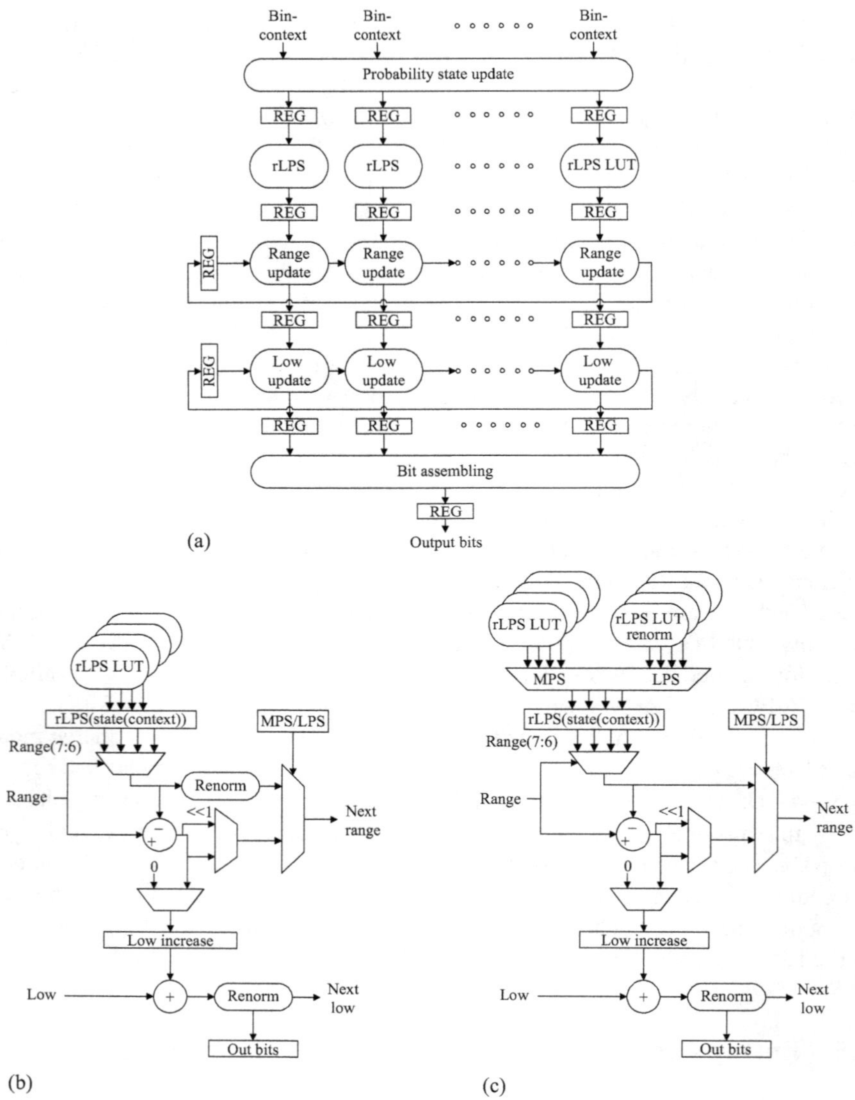

Figure 4.24 Arithmetic coder architecture (a), typical range and low update (b), and range and low update after pre-normalization (c)

In the arithmetic coder, critical paths are located in the stage updating the range register. Therefore, corresponding units should be optimized to shorten the paths. An example of such an optimization is given in [17] and depicted in Figure 4.24(c). The pre-normalization allows the displacement of the normalization operation to the preceding stage. As a consequence, the probability estimates for less probable symbols (LPS) do not need the shifter in the range update unit. The hybrid path coverage takes advantage of the simplified circuit dedicated to the less probable symbol encoding. In this method, there are four paths including units either fully featured or dedicated to less and more probable symbols. Figure 4.25 shows the diagram of units in particular paths. If less probable symbol is found in the input sequence of four bins, the coder selects one of four paths depending on the position of the symbol. Otherwise, three most probable symbols are processed. Since the simplified unit is reduced to the four-input multiplexer, a critical path is shortened. The gain in the frequency is much higher than the loss in the average number of symbols processed in one clock cycle.

The throughput of the arithmetic coder can be increased if more bypass-mode symbols are processed in parallel. Such symbols do not affect the range update. However, the range must be taken to the following stage to conditionally increase the low register. In [18], the design allows the processing of two bypass-mode symbols instead of one regular symbol in low update units. As a consequence, the architecture improves the performance by 12.76%, on average.

The design of arithmetic decoders is more challenging. Usually, architectures are designed to decode one bin in each clock cycle. The architecture shown in Figure 4.26 modifies this rule. It is able to change the number of bins. One wait

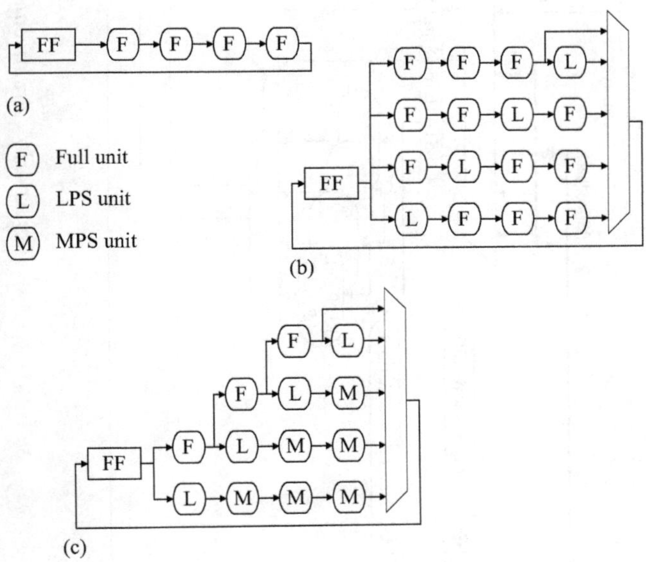

Figure 4.25 Conventional range update chain (a), basic (b), and optimized (c) hybrid path coverage organization

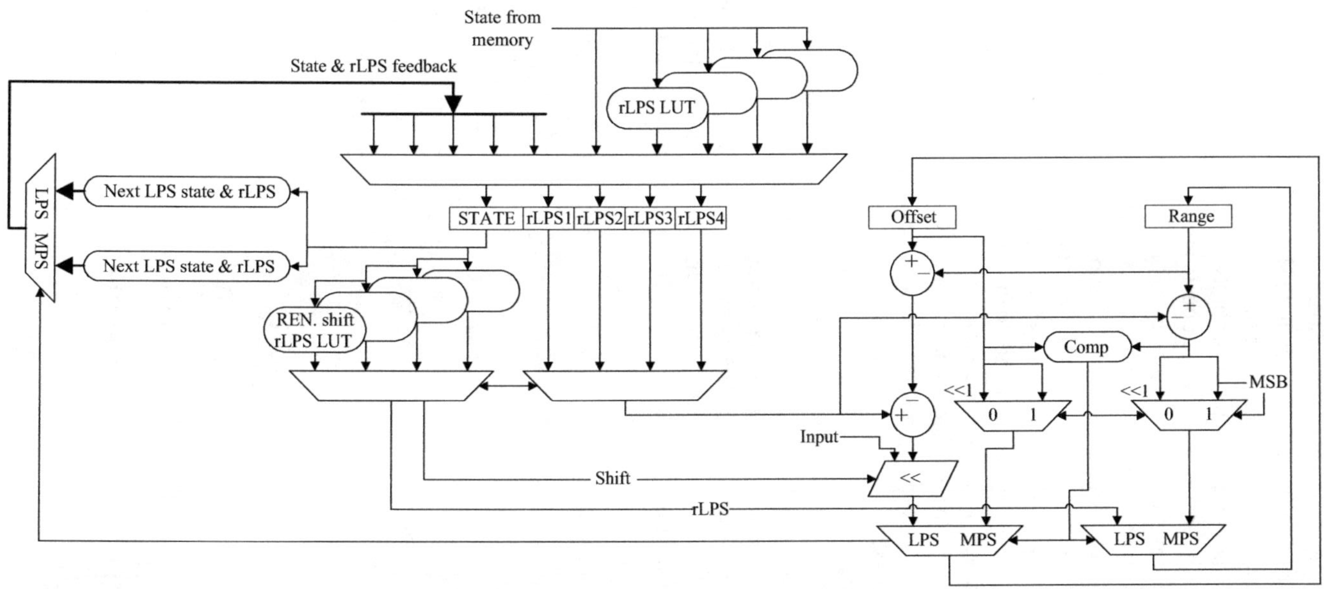

Figure 4.26 Architecture of the arithmetic decoder in [19]

cycle is inserted when selected paths have longer delay. On the other hand, two bypass-mode bins can be decoded in one clock cycle when corresponding paths have low delay. According to evaluations, the architecture has the average throughput of 0.96 bins per clock cycle. It can operate at the frequency of 1,053 MHz for the TSMC 90 nm technology.

4.6 Summary

The architectures designed to support high-resolution video coding require high clock frequencies and consume a large amount of hardware resources. The complexity reduction and the throughput improvement usually involve losses in the compression efficiency. Of importance is the consideration of all consequences of simplifications applied in implemented algorithms. In particular, mutual block dependencies must be taken into account when they are processed in parallel.

References

[1] Pastuszak G. 'Hardware Architectures for the H.265/HEVC Discrete Cosine Transform'. *IET Image Processing*, 2015, vol. 9(6), pp. 468–477.

[2] Pastuszak G., and Abramowski A. 'Algorithm and Architecture Design of the H.265/HEVC Intra Encoder'. *IEEE Transactions on Circuits and Systems for Video Technology*, 2016, vol. 26(1), pp. 210–222.

[3] Tsai S.-F., Li C.-T., Chen H.-H., Tsung P.-K., Chen K.-Y., and Chen L.-G. 'A 1062 Mpixels/s 8192 × 4320 p High Efficiency Video Coding (H.265) Encoder Chip'. *Proceedings of the Symp. On VLSI Circuits*; Jun. 2013, pp. C188–C189.

[4] Zhu J., Liu Z., Wang D., Han Q., and Song Y. 'HDTV1080p HEVC Intra Encoder with Source Texture based CU/PU Mode Pre-decision'. *Proceedings of the 19th Asia South Pacific Design Autom. Conf. (ASP-DAC)*; Jan. 2014, pp. 367–372.

[5] Liu Z., Yu X., Gao Y., Chen S., Ji X., and Wang D. 'CU Partition Mode Decision for HEVC Hardwired Intra Encoder Using Convolution Neural Network'. *IEEE Transactions on Image Processing*, 2016, vol. 25(11), pp. 5088–5103.

[6] Sun H., Zhou D., Hu L., Kimura S., and Goto S. 'Fast Algorithm and VLSI Architecture of Rate Distortion Optimization in H.265/HEVC'. *IEEE Transactions on Multimedia*, 2017, vol. 19(11), pp. 2375–2390.

[7] Ding L.-F., Chen W.-Y., Tsung P.-K., Chuang T.-D., Hsiao P.-H., Chen Y.-H., Chiu H.-K., Chien S.-Y., and Chen L.-G. 'A 212 MPixels/s 4096 × 2160p Multiview Video Encoder Chip for 3D/Quad Full HDTV Applications'. *IEEE Journal of Solid-State Circuits*, 2010, vol. 45(1), pp. 46–58.

[8] Atapattu S., Liyanage N., Menuka N., Perera I., and Pasqual A. 'Real Time All Intra HEVC HD Encoder on FPGA'. *Proceedings of IEEE 27th International Conference on Application-Specific Systems Architectures and Processors (ASAP)*; 2016, pp. 191–195.

[9] Byun J., Jung Y., and Kim J. 'Design of Integer Motion Estimator of HEVC for Asymmetric Motion-Partitioning Mode and 4K-UHD'. *Electronics Letters*, 2013, vol. 49(18), pp. 1142–1143.

[10] Vayalil N.C., and Kong Y. 'VLSI Architecture of Full-Search Variable-Block-Size Motion Estimation for HEVC Video Encoding'. *IET Circuits, Devices & Systems*, 2017, vol. 11(6), pp. 543–548.

[11] Liao T.-T., Shen C.-A., and Tseng Y.-H. 'The Algorithm and VLSI Architecture of a High Efficient Motion Estimation with Adaptive Search Range for HEVC Systems'. *Journal of Real-Time Image Processing*, 2019.

[12] Zhou D., Zhou J., He G., and Goto S. 'A 1.59 Gpixel/s Motion Estimation Processor With −211 to +211 Search Range for UHDTV Video Encoder'. *IEEE Journal of Solid-State Circuits*, 2013, vol. 49(4), pp. 827–837.

[13] Pastuszak G., and Jakubowski M. 'Optimization of the Adaptive Computationally-Scalable Motion Estimation and Compensation for the Hardware H.264/AVC Encoder'. *Journal of Signal Processing Systems*, 2016, vol. 82(3), pp. 391–402.

[14] Pastuszak G. 'Architecture Design of the H.264/AVC Encoder Based on Rate-Distortion Optimization'. *IEEE Transactions on Circuits and Systems for Video Technology*, 2015, vol. 25(11), pp. 1844–1856.

[15] Jou S.-Y., Chang S.-J., and Chang T.-S. 'Fast Motion Estimation Algorithm and Design for Real Time QFHD High Efficiency Video Coding'. *IEEE Transactions on Circuits and Systems for Video Technology*, 2015, vol. 25(9), 1533–1544.

[16] Pastuszak G., and Trochimiuk M. 'Algorithm and Architecture Design of the Motion Estimation for the H.265/HEVC 4K-UHD Encoder'. *Journal of Real-Time Image Processing*, 2016, vol. 12(2), pp. 517–529.

[17] Zhou D., Zhou J., Fei W., and Goto S. 'Ultra-high-throughput YLSI Architecture of H.265/HEYC CABAC Encoder for UHDT Applications'. *IEEE Transactions on Circuits and Systems for Video Technology*, 2015, vol. 25(3), pp. 497–507.

[18] Ramos F.L.L., Zatt B., Porto M., and Bampi S. 'High-Throughput Binary Arithmetic Encoder Using Multiple-Bypass Bins Processing for HEVC CABAC'. *Proceedings of IEEE International Symposium on Circuits and Systems (ISCAS)*; May 2018.

[19] Zhou J., Zhou D., Zhang S., Kimura S., and Goto S, 'A Variable-Clock-Cycle-Path VLSI Design of Binary Arithmetic Decoder for H.265/HEVC'. *IEEE Transactions on Circuits and Systems for Video Technology*, 2015, vol. 28(2), pp. 556–560.

Chapter 5

Low-power circuit design techniques for high-resolution video coding

Guilherme Paim[1], Leonardo B. Soares[1], Leandro M. G. Rocha[1], Brunno Abreu[1], Gustavo M. Santana[1], Cláudio M. Diniz[2], Eduardo A. C. da Costa[2] and Sergio Bampi[1]

5.1 Introduction

The pervasive use of modern portable and battery-powered devices running power-hungry applications, which demand intensive computations, poses important challenges for very-large-scale integration (VLSI) complementary metal oxide semiconductor (CMOS) circuits design. This is the case for the multicore programmable platforms, system-on-a-chip (SoC), or dedicated VLSI modules used as accelerators inside multicores or as part of SIMD (single-instruction, multiple-data) general purpose graphics processing unit (GPGPU) architecture. Multimedia services (e.g., audio, image, and video processing) are at the top of the most used applications in current years. A forecast made by [1] indicates that by 2020, the Internet Protocol video traffic will represent 82% of the global Internet traffic. This significant amount of video content flowing across the Internet demands that video-content providers, distributors, and even portable consumer cameras use the most efficient—and at the same time the least energy-consuming—video compression solutions. Saving bit-rate demand for given video quality is of utmost concern. The evolution of video compression has been tremendous, and it has been addressed in other chapters. Consider the high-efficiency video-coding (HEVC) standard, studied and profiled in [2], which became an official ITU-ISO recommendation in 2013. The Joint Video Coding Expert Group addressed higher resolution (up to 4K) videos with this launching, the state of the art around that year. The HEVC set out the goal to achieve twice the compression ratio in coding 4K videos, compared to its predecessor, also from the Joint ISO-ITU video coding team, maintaining the same perceptual video quality. With higher resolutions and in future video-coding proposals, the greatest overall bottlenecks for the mobile hardware platforms are the computational effort and the energy consumed to achieve such high compression capabilities. According to [2], the HEVC

[1]Graduate Program in Microelectronics (PGMicro), Informatics Institute, Federal University of Rio Grande do Sul, Porto Alegre, Brazil
[2]Graduate Program on Electronics and Computing, Catholic University of Pelotas, Pelotas, Brazil

Model (HM) reference software presented a computational effort up to 3.2× higher than the reference software of the previous standard H.264/AVC (advanced video coding), leading to higher energy consumption per encoded frame. Efficient HEVC-compliant encoders have to be forcefully developed, especially for the aforementioned battery-powered devices. This number indicates the challenging scenario that SoC designers have at hand to reduce power dissipation while coping with higher demanded computational effort when designing SoC systems for embedded devices.

This scenario tends to become more dramatic when considering the emerging ubiquitous concept of Internet of Things (IoT) as defined in [3,4]. The forecast presented by [5] estimated that 50 billion devices will be connected to the Internet by 2020, while more recent predictions, like that in [6], show that experts from different companies project that this number will be very high, reaching approximately 30 billion by 2020. Not all, but a portion of these devices will have video capturing and video processing and transmission requirements. The energy autonomy, in this case, will heavily depend on customizing video-processing architecture.

The main challenge resulting from this context is that these connected devices require different levels of complexity from the computing environment—often very demanding, especially for multimedia applications—which directly translates into high-power dissipation and energy consumption. It becomes a serious issue when battery-powered devices, e.g., smartphones, tablets, smartwatches, camcorders, have to support either recording or playback of high-definition videos. For these platforms, energy-aware hardware design is necessary to make better use of the battery resources while achieving the computational demands of modern multimedia services. Thus, developing hardware accelerators for video coding–decoding modules is of utmost importance to reduce the power/energy requirements of HEVC-compliant devices. This chapter addresses dedicated hardware architecture for video encoding, which will be crucial to provide energy efficiency for IoT smallest edge devices or general mobile platforms used by billions of consumers worldwide.

In HEVC, motion estimation (ME) is used during inter-prediction to exploit temporal redundancies in video sequences, which are abundant due to the high frame rates. The first step of this process, called integer ME (IME), consists in performing several computations to search in previously encoded frames, which are stored in memory, for blocks similar to the one being currently encoded. This first step generates an integer motion vector pointing to the best block found in this search. The second step, named fractional ME (FME), refines the IME by searching through a set of interpolated candidates with sub-pixel offset, producing a fractional motion vector [7]. Three metrics are used to estimate the distortion between two video blocks in mode decision: sum of absolute differences (SAD), sum of absolute transformed differences (SATD), and sum of squared errors. The SAD metric is used during IME as a similarity criterion, but it can also be used in FME, replacing the SATD metric [8,9], and in intra-prediction, to figure out which intra mode is the best [10]. Since ME must be applied several times to encode each frame, SAD becomes one of the most time-consuming operations of the HEVC encoder. Figure 5.1 shows profiling of the HM reference software [11] using GProf [12] to measure encoding time, presenting a breakdown of the time spent in each of several modules of the video-encoding process. For this

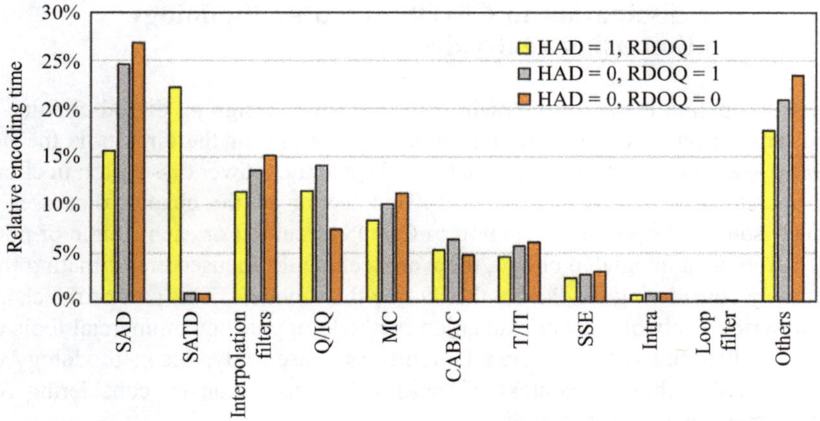

Figure 5.1 Encoding time breakdown obtained from encoding 1 s (50 frames) of the BasketballDrive sequence (1,920 × 1,080 pixel)

analysis, the encoder was run for different combinations of the Hadamard (HAD) transform and the rate-distortion optimized quantization (RDOQ) parameters. The HAD parameter defines whether SAD or SATD will be used during the FME stage (a HAD value of 0 indicates that SAD metric will be executed in the FME), and RDOQ is responsible for deciding whether the encoder takes the quantization error into account when deciding the best partitioning scheme. The computationally demanding characteristic of the aforementioned metrics along with the battery constraints of portable devices require that both performance-driven and energy-efficient approaches are considered in all abstraction levels when design CMOS digital circuits for video-coding applications. Notably, this chapter will give particular attention to three blocks of HEVC video coding—SAD, SATD, and interpolation filters—that are responsible for a large portion of the encoding time, as presented in Figure 5.1.

The sources of power dissipation in CMOS circuits, as well as a methodology for accurate power dissipation estimation using real-video data, are presented in Section 5.2. Section 5.3 presents state-of-the-art low-power techniques used in dedicated hardware accelerator designs. One of the methods introduced in this section shows the hybrid encoding of arithmetic operators and presents a new hybrid-encoded adder operator. The power-efficient hybrid encoding representation groups *m* bit and uses gray encoding to potentially reduce circuit switching activity, both internally and at the inputs of the arithmetic operators. This section also discusses the exploration of different adder compressor structures in SAD operation and filter interpolation hardware architecture. Adder compressor architecture performs the simultaneous addition of *N* operands. Combinations of 3-2, 4-2, 5-2, and 7-2 adder compressors are discussed. Finally, an approximate computing technique is presented in Section 5.3.3 which is based on the coefficient pruning for SATD hardware architecture. The application of the low-power techniques in SAD, SATD, and interpolation filters is presented in detail in Section 5.4.

5.2 Power dissipation in CMOS and a methodology estimation using real-video data

Power dissipation is one of the main concerns when designing digital circuits. An important aspect when considering reliability analysis of the circuits is the peak power. However, the most critical factor is the average power dissipation in circuits that operate over a time interval [13]. This section of the chapter discusses the primary sources of power dissipation in CMOS circuits. For each source of power dissipation in an integrated circuit, there are techniques focused on estimating them accurately considering the factors that define their models. This part of the chapter will describe a reliable power evaluation methodology using commercial tools that may be applied to any type of digital circuit. As a case study, this methodology will be illustrated within the context of multimedia application by considering real-video sequences as input stimuli.

5.2.1 Sources of power dissipation

There are three primary sources of power dissipation in CMOS circuits under normal operation [14]: (i) power dissipation caused by leakage currents and sub-threshold currents, (ii) short-circuit power dissipation, which occurs due to the direct current flow from the power supply to the ground during the switching process in a transistor gate, and (iii) dynamic power dissipation, which is the result of loading and unloading the capacitances of the circuits. The main aspects that contribute to power dissipation in CMOS circuits are presented below.

5.2.1.1 Static power

Semiconductor devices inherently have reverse currents due to their *pn* junctions, which appear even when the circuit is in an idle state. Additionally, these devices present subthreshold currents that appear due to the inherent non-idealities of the input-signal voltage. These two effects are the source of the static power dissipation in CMOS circuits.

The leakage currents occur when one transistor is off and the other active transistor charges (up/down) the drain to the substrate potential [13]. Figure 5.2 shows the representation of the reverse leakage current.

This current is given by the characteristic equation of a reverse bias diode as shown in (5.1) (visualized in Figure 5.2), where I_S is the reverse saturation current, V is the diode voltage, and $V_T = KT/q$ is the thermal voltage:

$$I_{leakage} = I_s(e^{V/V_T} - 1) \tag{5.1}$$

According to [13], power dissipation due to the subthreshold current occurs as a result of the diffusion of charge carriers between the source and the drain when the gate–source voltage (V_{gs}) exceeds the weak inversion point but still presents a lower value than the threshold voltage (V_t). In this manner, the metal-oxide-semiconductor transistor behaves similarly to a bipolar transistor and the sublimate current is exponentially dependent on V_{gs} and V_{ds}—drain–source voltage—as shown in (5.2),

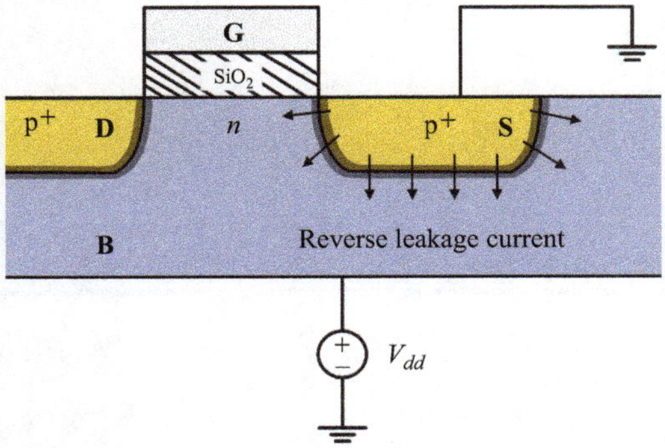

Figure 5.2 Reverse leakage current characterization

where K is a technology-dependent parameter, $n = 1 + \Omega t_{ox}/D$, t_{ox} is the thickness of the oxide of the gate, D the width of the channel depletion, and $\Omega = \epsilon_{si}/\epsilon_{ox}$:

$$I_{ds} = Ke^{V_{gs}-V_t/nV_T}(1 - e^{-(V_{ds}/V_T)}) \tag{5.2}$$

5.2.1.2 Short-circuit power

Short-circuit power dissipation occurs when current flows through a direct path from the power supply to ground. It happens when a static CMOS circuit is switched on by an input signal with rising and fall times different than zero, with the p and n-channel Metal Oxide Semiconductor Field Effect transistors being active simultaneously for a short period. Figure 5.3 shows the behavior of an inverter that exemplifies the short-circuit activity, assuming that there is no charge in the output. Note in Figure 5.3 that in a rise transition, at the input of the inverter, the NMOS transistor starts to drive when the input voltage (V_{in}) is equal to the threshold voltage (V_{tn}) of the NMOS transistor. The PMOS transistor will stop conducting when V_{in} is the difference between the power supply (V_{dd}) and the threshold voltage of the PMOS transistor (V_{tp}), i.e., $V_{in} = V_{dd} - |V_{tp}|$. In all further examples, we assume $V_t = V_{tn} = |V_{tp}|$ unless specified.

The NMOS transistor will be in saturation in the time interval τ_1 to τ_2 since $V_{gs} > V_t$ and $V_{ds} > V_{gs} - V_t$. For this condition, the current for the NMOS transistor is given by (5.3) for the situation $0 < I < I_{\max}$, where β is the effective transconductance parameter of the NMOS transistor (expressed in A/V^2):

$$I = \frac{\beta}{2}(V_{in} - V_t)^2 \tag{5.3}$$

The average current I_{mean} is used to determine the average short-circuit current that is drained from the power supply. An approximation for this average current is given by [15] and is represented by the following equation:

$$I_{mean} = 2\frac{2}{T_{clk}}\int_{\tau_1}^{\tau_2} I(t)dt = \frac{4}{T_{clk}}\int_{\tau_1}^{\tau_2} \frac{\beta}{2}(V_{in}(t) - V_t)^2 dt \tag{5.4}$$

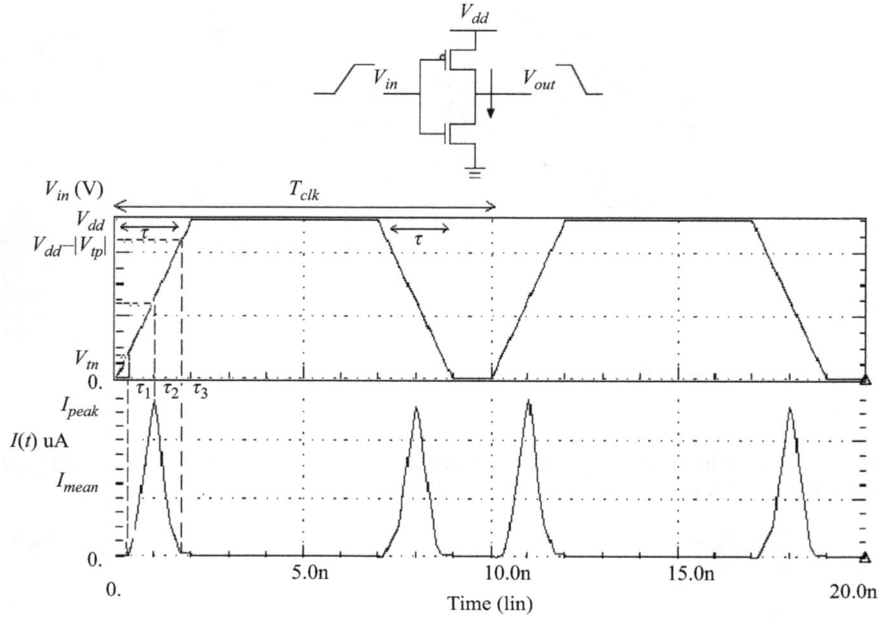

Figure 5.3 Short-circuit current for an inverter with no charge in the output

The term V_{in} in (5.4) represents the voltage at the inverter input and it is given by (5.5). τ_1 and τ_2 are, respectively, the rise and fall times of the short-circuit current, shown in Figure 5.3, whose values are given by (5.6) and (5.7). For all cases, τ denotes the time period between the instants when $V_{in} = V_t$ and $V_{in} = V_{dd} - V_t$:

$$V_{in} = V_{dd}\frac{t}{\tau} \tag{5.5}$$

$$\tau_1 = V_t\frac{\tau}{V_{dd}} \tag{5.6}$$

$$\tau_2 = \frac{\tau}{2} \tag{5.7}$$

Rewriting (5.4) with the terms described in (5.5)–(5.7), we obtain the value of the mean short-circuit current according to the following equation:

$$I_{mean} = \frac{\beta}{12V_{dd}}(V_{dd} - 2V_t)^3\frac{\tau}{T_{clk}} \tag{5.8}$$

Finally, (5.9) expresses the short-circuit power, P_{sc}, that is calculated as the product of the power supply by the average current:

$$P_{sc} = V_{dd}I_{mean} = \frac{\beta}{12}(V_{dd} - 2V_t)^3\frac{\tau}{T_{clk}} \tag{5.9}$$

As Equation 5.9 shows, the P_{sc} has a linear increase with the slew-time (τ) of the input transition. Costa [16] have demonstrated in their paper that P_{sc} according to Equation 5.9 is just an approximation: in fact, P_{sc} has a smaller polynomial dependence on VDD, and P_{sc} is not zero for τ tending to zero (sharp input transitions).

5.2.1.3 Switching power (dynamic power)

Dynamic power dissipation is the result of charging and discharging the capacitances of the circuit due to the switching activity which is intrinsic to the operation of CMOS circuits. Assuming a logic gate G connected to a load capacitance C_L, the changing of the gate output value will cause a current flow that will either charge or discharge C_L. This cycle will dissipate power according to (5.10), where A is the output node activity measured in signal rising edges per clock cycle for a complete charge/discharge, and f is the frequency of operation:

$$P_{dyn} = C_L A V_{dd}^2 = \alpha_{0\to 1} f C_L V_{dd}^2 \tag{5.10}$$

For a digital circuit operating at frequency f, most of the circuit nodes will not change at every clock cycle. Conversely, it is most likely that these nodes will change their values with a probability $\alpha_{0\to 1}$ which represents the node rising edge activity factor. Considering a circuit with more than one node, the value of α is computed as a function of input statistics and logical models that describe the device.

The energy drained from the power supply for a $0 \to V_{dd}$ transition at the output of a CMOS gate is given by $C_L V_{dd}^2$. Figure 5.4 shows a circuit model highlighting the switching power component [13].

It is noticeable that approximately half of the energy drained from the power supply is stored in the load capacitor, and the other half is dissipated in the PMOS network as heat. For a $V_{dd} \to 0$ transition at the output, there is no charge drained from the power supply, and the energy stored in the capacitor ($C_L V_{dd}^2/2$) is dissipated in the NMOS pull-down network.

If a simple signal rising edge transition is performed at each clock cycle at the rate f_{clk} (clock-signal frequency), power is given by ($C_L V_{dd}^2 f_{clk}$). However, there are cases in which the signal transition occurs at different frequency rates, and we have to consider the value of the number of signal rising edge transitions per clock cycle or the activity factor of rising edge transitions of nodes. In this case, (5.10) represents the average power, which corresponds to the average number of switching rising edge transitions in a period ($\alpha_{0\to 1}$). Since the internal nodes of a circuit may switch their values, the transition activity has to be calculated for all nodes of

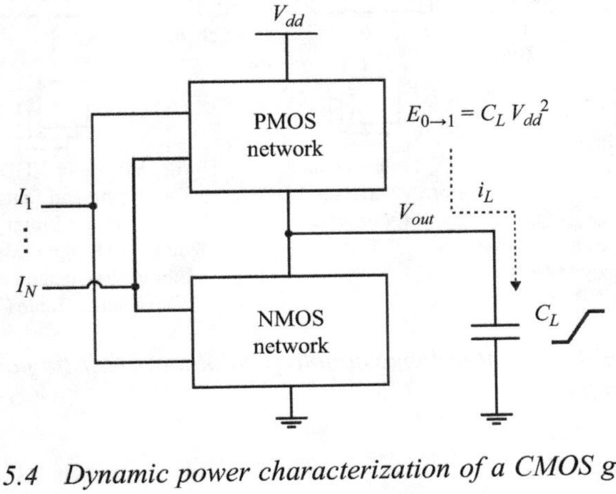

Figure 5.4 Dynamic power characterization of a CMOS gate [13]

the circuit, and the total power of the circuit is given according to the following equation:

$$P_{dyn(total)} = \sum_{i=1}^{\#\,nodes} \alpha_{0\rightarrow 1i} C_i V_{dd}^2 f_{clk} \tag{5.11}$$

During the transition 0→1, half of the energy is stored on the output capacitance. On the discharge transition, the consumed energy comes only from the capacitor. Hence, if we consider both output transitions from 0→1 and 1→0, the transition count would double while the energy drawn stays the same. Therefore, the average power equations (5.10 and 5.11) would need to be divided by two.

5.2.2 Power estimation in the current industrial ASIC design flow

Signal transitions cause dynamic power dissipation in digital integrated circuits. Commercial synthesis tools generate, by default, probabilistic input values to estimate the power dissipated by the circuit. This approach, however, is often pessimistic and rarely represents the actual input behavior of the application for which the circuit was conceived. Therefore, an accurate power estimation methodology is necessary to both represent the actual application behavior and to provide the designer with a reliable quality of results that will enable the optimization of critical power-hungry blocks.

Figure 5.5 shows the proposed methodology for power dissipation extraction. First, the register-transfer level (RTL) description is synthesized using a commercial synthesis tool—such as the Cadence Genus Synthesis Solution™ and Synopsys Design

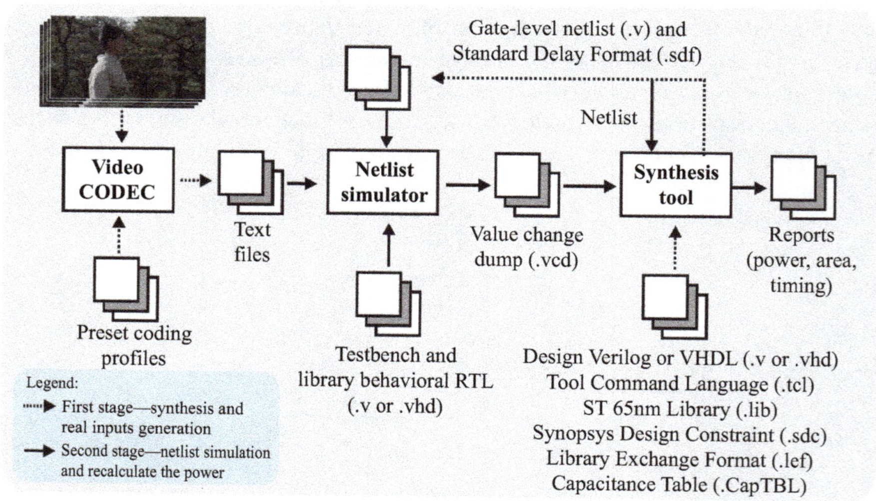

Figure 5.5 The ASIC synthesis methodology to obtain realistic power-extraction results

Compiler—for a given technology. As a by-product, the tool generates the circuit netlist in Verilog, the Standard Delay Format (SDF) file that annotates the specific delays for gates and nets and the design reports of area, power dissipation, and critical path. Then, the simulation tool—like Cadence Incisive™ and Synopsys VCS—simulates the generated netlist with the testbench files—written in very high speed integrated circuits hardware description language (VHDL)/Verilog/SystemVerilog—feeding the circuit with input data from standard benchmark images or videos to generate the dump file that is in either value change dump or toggle count format file formats.

The input data used as stimuli are extracted from a video codec software and stored into text files. Usually, the video sequences processed by the codec software come from the common test conditions (CTCs) that are specific to each video-coding standard. With these text files, the simulation tool receives these values and executes the testbench with the gate-level netlist files and SDF delay files to generate accurate switching activity. At the end of the simulation, the dump file is generated containing the switching activity of all nodes within the circuit netlist. Lastly, the synthesis tool is executed a second time with the same parameters except that, in this turn, the dump file is fed to the software to generate accurate power dissipation reports.

Modern synthesis tools support the estimation of the gate interconnection penalties in area, power, and timing. Cadence logic synthesis tool [17] disposes of the physically aware layout estimation mode, while the Synopsys Design Compiler tool offers the topographical mode to analyze the impact of interconnections. These tools estimate the length of the nets and take into account the load capacitance effects in the power dissipation, considering a relatively pessimistic layout routing estimation. Such an analysis requires the inclusion of the Library Exchange Format (LEF) files, which mostly contain the physical layout information of the library. The LEF macro includes the inner library cell's capacitance, and the tech LEF comprises the process metal capacitance for the interconnection capacitance estimation [18]. Additionally, standard-cell libraries usually offer an additional file, namely, capacitance table, which describes the technology capacitances in a more precise and fine-grained way since it considers the process variations [18].

5.3 Compute-intensive video-coding blocks overview

Several metrics can be applied to determine the degree of similarity between two blocks. They differ from each other in ease of implementation, efficiency, and result accuracy, i.e., how precisely they can define the similarity between the blocks. The two main metrics used to estimate block similarity in video codecs are shown in the next subsections.

5.3.1 Sum of absolute differences

The most used metric in IME process, to determine the best match between two video blocks, is the SAD. Optionally, this metric specifies the best fractional offset in FME and also decides the best intra-prediction mode for each block.

SAD metric estimates, in ME, the distortion between the block to be encoded (the original block), in the current picture, and the candidate blocks in the reference pictures (reference blocks). Equation (5.12) shows the way of representing the SAD metric, where O is the original block, R is the reference block, i and j are the pixel positions, and m and n are the dimensions of the block in samples:

$$SAD = \sum_{i=0}^{m-1} \sum_{j=0}^{n-1} \left| O_{i,j} - R_{i,j} \right| \tag{5.12}$$

Figure 5.6 shows a diagram of the SAD hardware architecture. It is composed of eight subtractors, to subtract eight samples of the original block (O in Figure 5.6) with eight samples of the reference block (R in Figure 5.6). This operation generates the remaining samples, and eight registers store them. Eight absolute operators receive as input the signed residue and perform the absolute operation. An adder tree accumulates the eight absolute values and generates a partial SAD value. At each clock cycle, a register stores a partial SAD value. Thus, for an 8×8 block, the final SAD value is calculated after eight clock cycles. This architecture uses binary adders (macro-function adder implemented by the synthesis tool) and binary encoding for the input operands and the output SAD value.

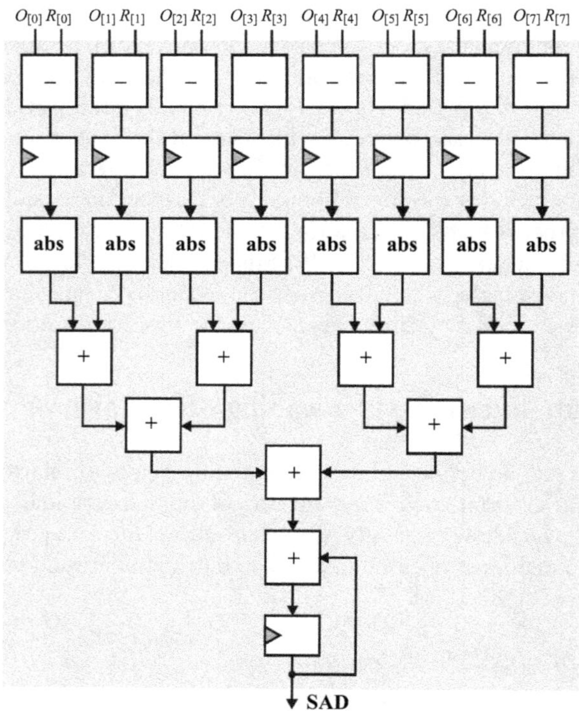

Figure 5.6 Original SAD architecture (Equation (5.12))

5.3.2 Sum of absolute transformed differences

SATD is a metric that measures distortion between two blocks in video coding. This metric is used to provide better overall video compression results when compared to the SAD at the expense of larger amount of arithmetic operations as well as higher energy consumption. Block matching algorithms employ these distortion metrics to find similarities between pixel blocks during ME aiming to benefit from these temporal redundancies and, consequently, reduce the bitstream size. For the SATD computation, the differences between two blocks are transformed by the 2-D forward HAD transform (HT) as shown in the following equation:

$$W = H \cdot Y \cdot H^T \tag{5.13}$$

where W and Y denote the transformed coefficients and the input matrix of differences, respectively. The H and H^T refer to the transform matrix and its transposed form, respectively. Equation (5.14) shows a HAD matrix H for a block size of 4×4 pixel:

$$H = \frac{1}{2}\begin{bmatrix} 1 & 1 & 1 & 1 \\ 1 & -1 & 1 & -1 \\ 1 & 1 & -1 & -1 \\ 1 & -1 & -1 & 1 \end{bmatrix} \tag{5.14}$$

Therefore, the SATD is calculated with the aforementioned HT as shown in (5.15), where $w_{i,j}$ denotes the HT coefficient from the ith row and jth column:

$$SATD = \sum_{i,j} |w_{ij}| \tag{5.15}$$

Figure 5.7 shows an example of the fully parallel 4×4 SATD implementation. The second and fourth steps are the horizontal and vertical 1D HT. Figure 5.8 shows the detailed schematic that computes both the vertical and horizontal transforms.

5.3.3 Interpolation filter

Figure 5.9 shows a diagram of the original interpolation filter architecture [19]. It receives as input a motion vector and a block index that points to the integer-pel block to be interpolated. In the output, it produces the fractional-pel block. It contains two engines to accelerate luma and chroma interpolation filters, buffers to store integer-pel and fractional-pel and scheduling (control) module to feed the filtering engines. Each acceleration engine (gray-colored blocks in Figure 5.9) is composed by 12 interpolation datapaths (ID) in parallel to meet the throughput requirement of HEVC interpolation filter [19]. The focus is the improvement of the interpolation datapaths by employing efficient adder compressors. Hence, there are details about the internal architecture of the interpolation datapaths of the original architecture. Other modules of the architecture (e.g., buffers, address generation unit, reference picture memory, and predicted unit buffer) are not the focus to be presented here.

Figures 5.10 and 5.11 depict the detailed diagrams of the interpolation filter datapaths for luma and chroma in the original architecture, respectively. Luma filter

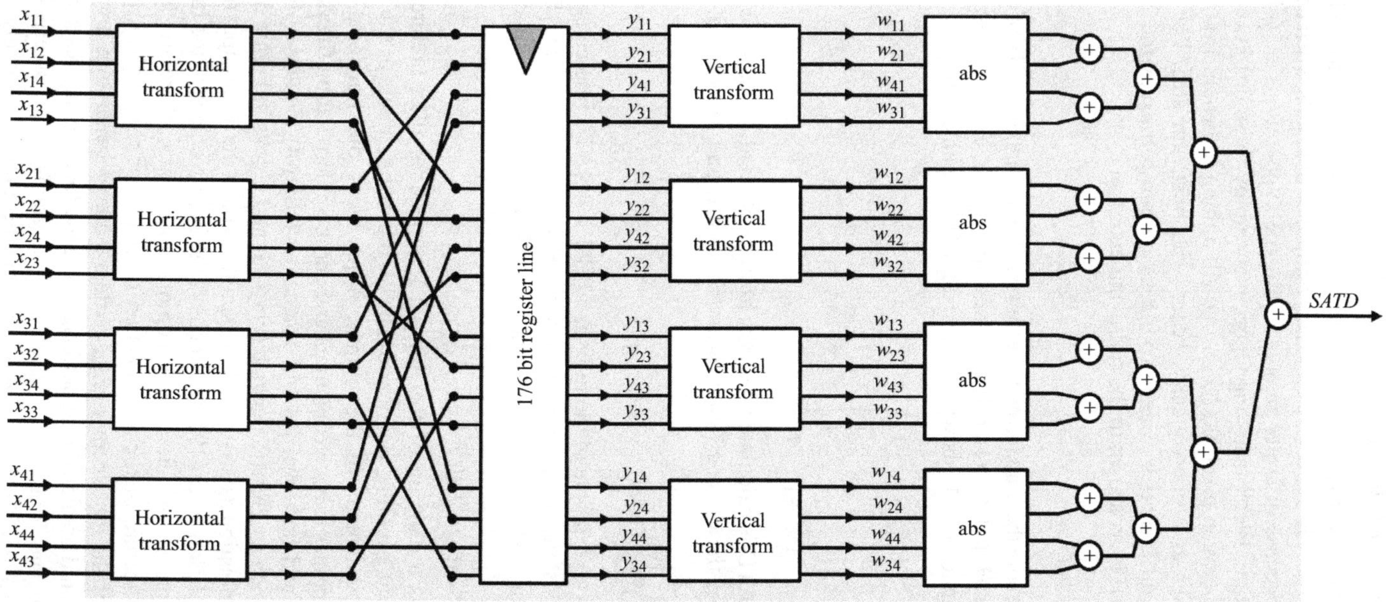

Figure 5.7 Fully parallel 4 × 4 SATD implementation

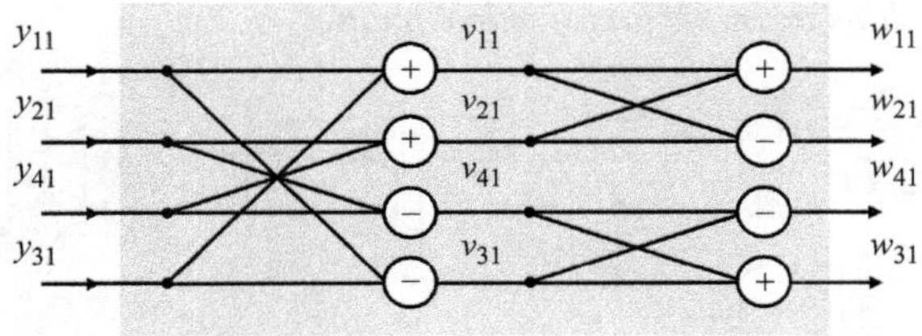

Figure 5.8 Internal structure of horizontal and vertical transforms

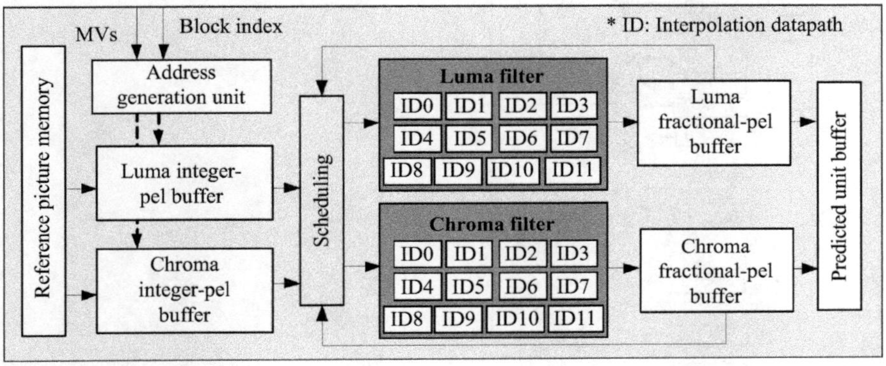

Figure 5.9 Overview of the [18] interpolation filter hardware architecture

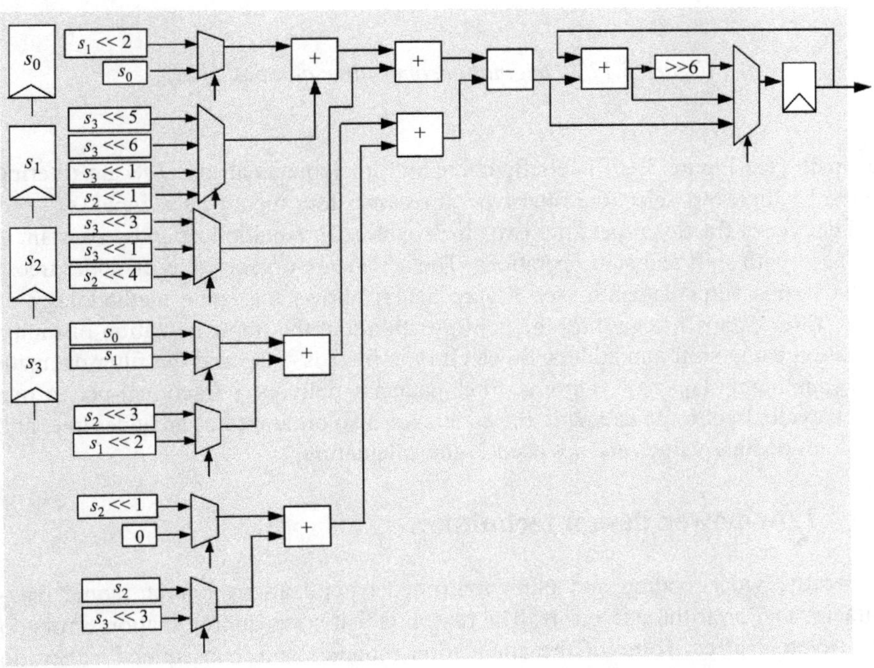

Figure 5.10 Original luma filter datapath [18]

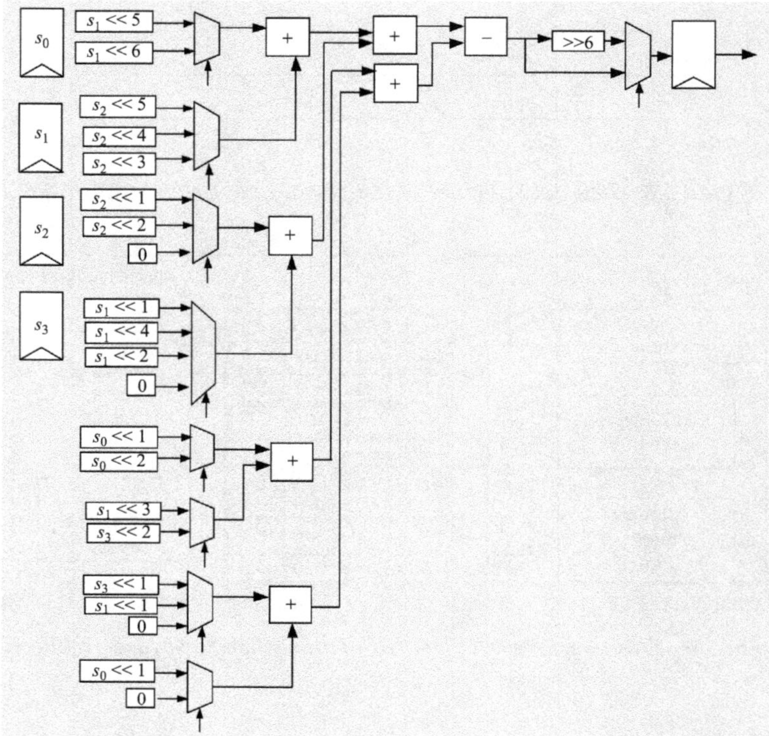

Figure 5.11 Original chroma filter datapath [18]

datapath (see Figure 5.10) is configurable and implements all the 7-/8-tap filtering types. Multiplexers select the filter type. It receives four input pels as input ($s_0 - s_3$) and delivers a fractional-pel after two clock cycles. The multiplication by constant is replaced with shift and sum operations. The adders are organized as an adder tree.

Chroma filter datapath (see Figure 5.11) follows the same methodology of luma filter datapath regarding the configuration and the implementation of multiplication using shift and adders. Since chroma filter is 4-tap and the filter datapath has four inputs ($s_0 - s_3$), chroma filter datapath delivers a fractional-pel in one clock cycle. In chroma datapath, the adders are also organized as an adder tree, and the intermediate values are not used in the calculation.

5.4 Low-power design techniques

Generally, video coding and other multimedia applications tolerate some inaccuracies and/or arithmetic errors. The reason is that some approximation errors do not severely affect some of the application modules and, consequently, they do not produce a significant loss in the subjective quality of this type of digital signal.

For instance, human vision has less sensitivity to high-frequency components in digital images and videos. Therefore, errors that occur in these components are inherently difficult to visualize due to limited visual perception in humans [20]. Thus, the concept of imprecise computation or approximate arithmetic arises, whose main idea is to use techniques that propose a simpler hardware with relaxed accuracy requirements aiming at reducing energy consumption or increasing the computational performance of digital VLSI systems. According to the accuracy required by the application, each module can be approximated to a certain degree, resulting in controlled error rates on the output. This section of the chapter focuses on the specific themes within the area of design of SoCs with low-power strategies such as adder compressors, operand encoding, and imprecise adders.

5.4.1 Compression-based operators

Adder compressors appear in literature as efficient addition structures since they can simultaneously add many multi-bit inputs with minimal carry propagation. The first compressor was presented in [21], and it featured an arrangement to add four 1-bit inputs simultaneously. This structure was named 4-2 Carry Save module, and it is composed of a combination of full-adder cells in truncated connection so that a fast compression is possible.

5.4.1.1 Internal structures of adder compressors

Figure 5.12(a) presents the internal structure of the 3-2 adder compressor. Two EXOR gates form the critical path of the adder compressor. Equation (5.16) gives the final sum S of the 3-2 adder compressor ($S = A + B + C$). The 3-2 adder compressor can also be used as a full-adder (mux-based full-adder) when using the input C as a carry input.

$$S = Sum + 2Carry \tag{5.16}$$

Figure 5.12(b) shows the internal structure of the 4-2 adder compressor [22]. It has a reduced critical path compared to a construction using only full adders since it has only three EXOR gates instead of four gates on the critical path. The 4-2 compressor has five inputs (A, B, C, D, C_{in}), where C_{in} is the input carry, and three outputs $(Sum, Carry, C_{out})$. In this adder compressor, the carry output C_{out} is independent of the input carry (C_{in}). Removing the carry signal dependence from the critical path makes it possible for this structure to have better performance when compared to traditional adder structures in multi-bit operands due to the limited carry propagation. Equation (5.17) gives the final sum S result of the 4-2 adder compressor ($S = A + B + C + D + C_{in}$).

$$S = Sum + 2(C_{out} + Carry) \tag{5.17}$$

Figure 5.12(c) and (d) shows the internal structures of 5-2 [23,24] and 7-2 adder compressors, respectively. The 7-2 adder compressor structure [25] has a critical path composed of 6 EXOR gates. This compressor has seven primary inputs, two carry inputs (C_{in}), two carry outputs (C_{out}), *Sum* and *Carry* output terms. It is noticeable

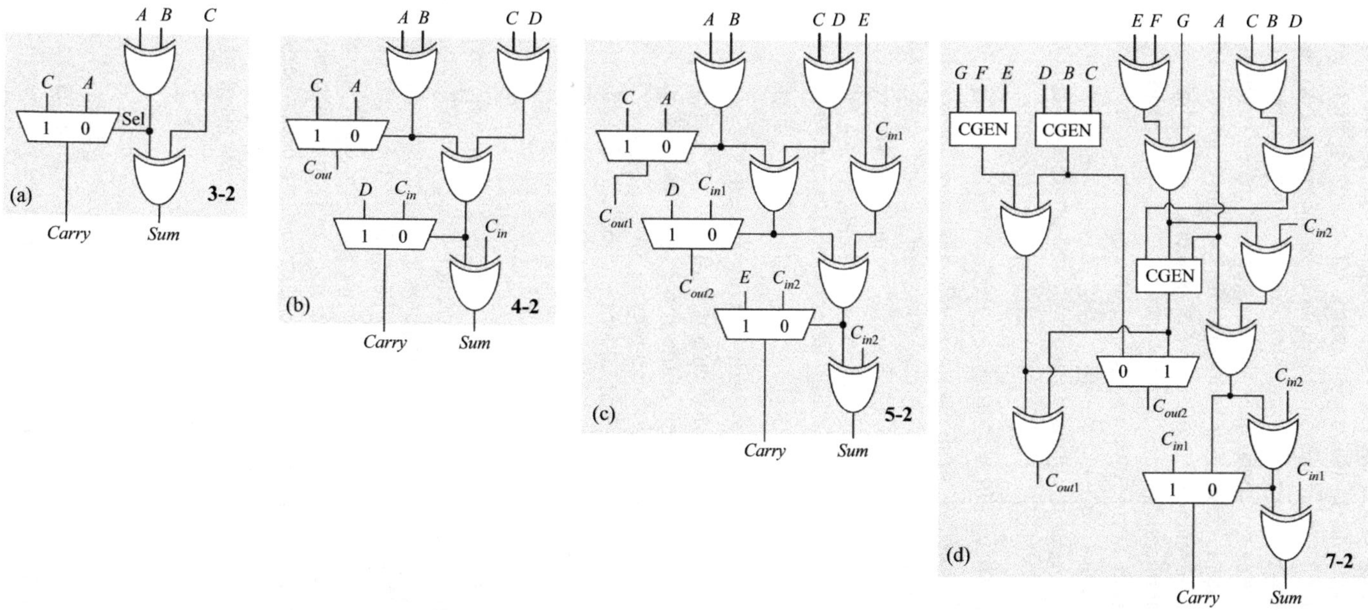

Figure 5.12 Single-bit adder compressors internal structures: (a) 3-2, (b) 4-2, (c) 5-2, (d) 7-2

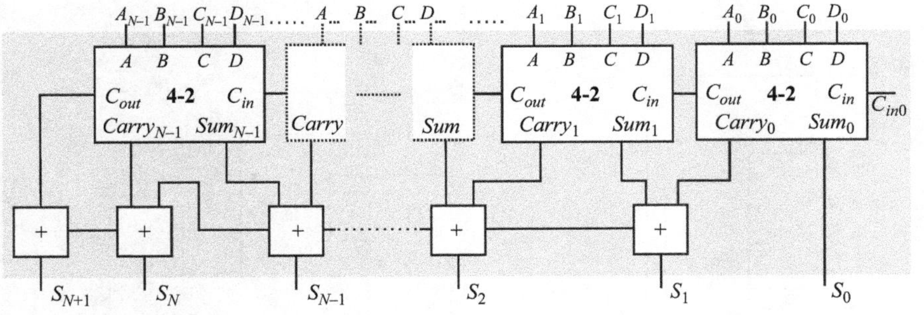

Figure 5.13 N-bit addition example using 4-2 adder compressors

that, besides the EXOR gates and MUX, the 7-2 compressor also uses a carry generator module (CGEN). The CGEN module is a majority gate to generate carry signals according to the input values.

The implementation of an *N*-bit adder compressor requires a final recombining adder to sum the partial *Carry* and *Sum* values. Figure 5.13 presents a cascade of half and full adder circuit using a ripple carry-based approach to compose the recombination adder for an 8-bit 4-2 compressor.

5.4.1.2 Hierarchical adder compressors

The hierarchical adder compressors use the smaller adder compressors presented in the previous subsection as building blocks to produce high-order compressors. Figure 5.14 shows several structures of an 8-2 adder compressor using different adder compressors units, enabling the simultaneous addition of eight numbers.

The 8-2 adder compressor comprises 13 inputs, in which eight of them are primary inputs (A, B, C, D, E, F, G, H) and five are carry inputs—C_{in} $(C_{in0}, C_{in1}, C_{in2}, C_{in3}, C_{in4})$. Further, this circuit has seven outputs, where two of them are primary outputs (*Sum* and *Carry*) and the others are carry outputs—C_{out} $(C_{out0}, C_{out1}, C_{out2}, C_{out3}, C_{out4})$. The 8-2 adder compressor also has the same weight for all inputs and the *Sum* output. The remaining outputs are of one-bit order higher. Equation (5.18) shows the final sum result for the 8-2 adder compressor.

$$S = Sum + 2(C_{out0} + C_{out1} + C_{out2} + C_{out3} + C_{out4} + Carry) \tag{5.18}$$

Recent works [26–28] exploit the different internal structures for the hierarchical 8-2 adder compressor using combinations of primary 3-2, 4-2, 5-2, and 7-2 adder compressors, as presented in Figure 5.14(a)–(d).

5.4.2 Hybrid encoding

One of the most promising encodings that can be used to reduce switching activity is the gray code since only one-bit change between consecutive values is tolerated. Therefore, for highly correlated signals, the switching activity can be significantly reduced when gray-based encoding is employed [14].

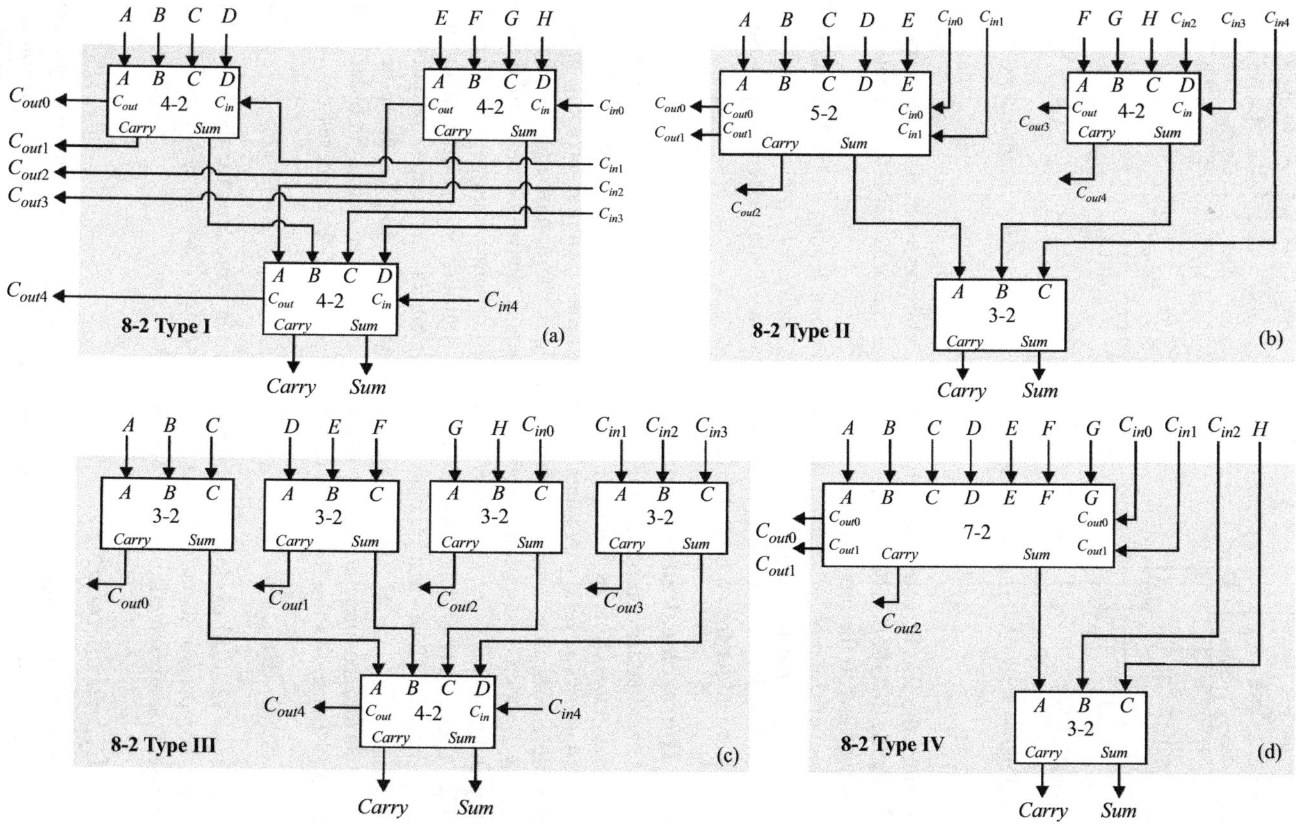

Figure 5.14 *8-2 Adder compressor implementation based on (a) Type I with 4-2 compressors, (b) Type II with 5-2, 4-2, and 3-2 compressors, (c) Type III with 4-2 and 3-2 compressors, and (d) Type IV with 7-2 and 3-2 compressors*

The process of implementing arithmetic operators that operate directly upon gray code inputs was investigated in [29]. However, it was observed that the combinational logic required by the arithmetic modules is considerably large. Furthermore, it was not possible to develop a regular structure for the operators such that could accommodate different word sizes. Hybrid operand encoding was proposed in [30] and it is a compromise between the minimal input dependency presented by the binary encoding and the low-switching characteristic of the gray encoding.

5.4.2.1 Hybrid code definition

The idea of the hybrid code is to split the operands into groups of m bit, encode each group using gray encoding, and use the binary approach to propagate the carry signals between groups. In this manner, it is possible to minimize the number of transitions inside each group, and it is also possible to easily build arithmetic operators with a regular structure. Table 5.1 exemplifies the binary, gray, and hybrid encoding for 2s complement 4-bit numbers and radix-4 ($m = 2$).

Equations (5.19)–(5.21) calculate the number of transitions, for n bit and a group of m bit, to count sequences such as that of Table 5.1 for the binary, gray, and hybrid codes, respectively.

$$\#trans.(Bin) = 2^{n+1} - 2 \tag{5.19}$$

$$\#trans.(Gray) = 2^{n} \tag{5.20}$$

$$\#trans.(Hyb) = 2^{m}\frac{2^{n} - 1}{2^{m} - 1} \tag{5.21}$$

Table 5.1 2s complement binary, gray, and hybrid (m = 2) code representations

Decimal Number	Binary code	Hybrid code (m = 2)	Gray code
0	0000	0000	0000
1	0001	0001	0001
2	0010	0011	0011
3	0011	0010	0010
4	0100	0100	0110
5	0101	0101	0111
6	0110	0111	0101
7	0111	0110	0100
−8	1000	1100	1100
−7	1001	1101	1101
−6	1010	1111	1111
−5	1011	1110	1110
−4	1100	1000	1010
−3	1101	1001	1011
−2	1110	1011	1001
−1	1111	1010	1000

An additional feature of the hybrid code is the easy way of converting to and from binary, as indicated in Figure 5.15 (for $m = 2$ example). Translating words of W bit between binary and hybrid, in any direction, requires only $W/2$ EXOR gates. Therefore, the overhead regarding transcoders is low, making this encoding very desirable for the arithmetic operators.

5.4.2.2 Hybrid adder

The work in [31] discusses that the implementation of adder circuits operating directly on hybrid code results in regular structures that are more complex than those used on binary adders. Hence, it proposes a hybrid adder whose structure is illustrated in Figure 5.16. This adder is basically a binary adder for 2-bit operands

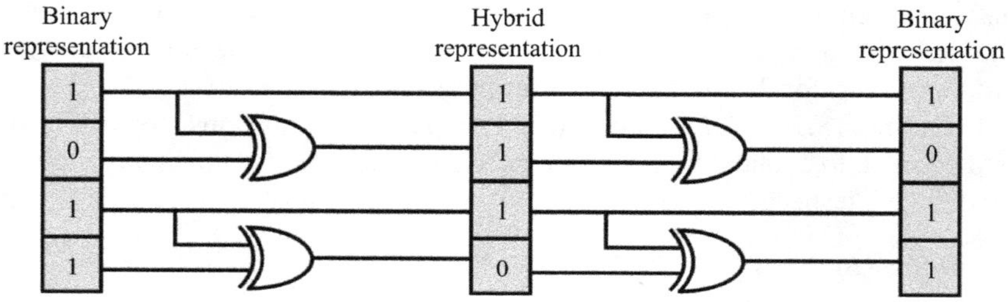

Figure 5.15 Conversion between binary and hybrid (m = 2) codes

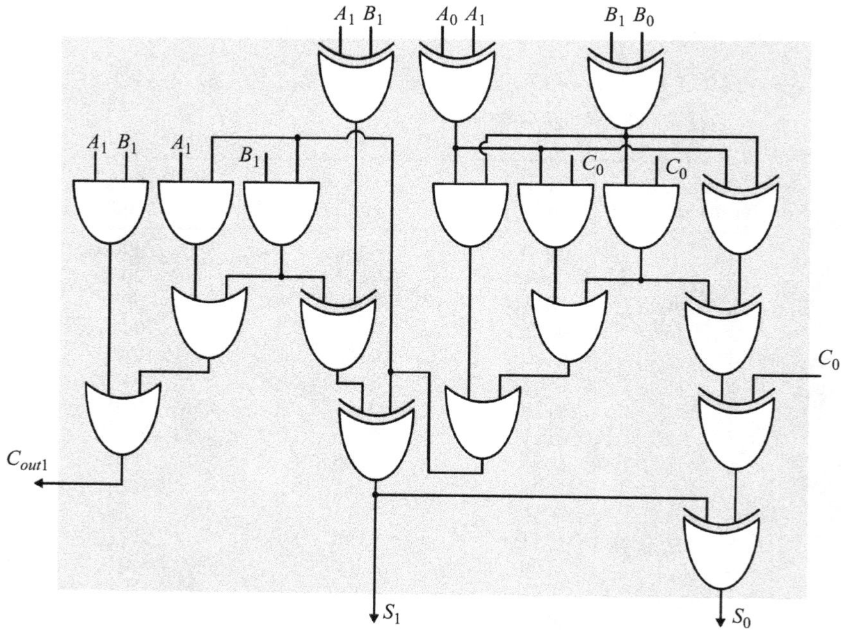

Figure 5.16 Two-bit hybrid adder scheme of [31]

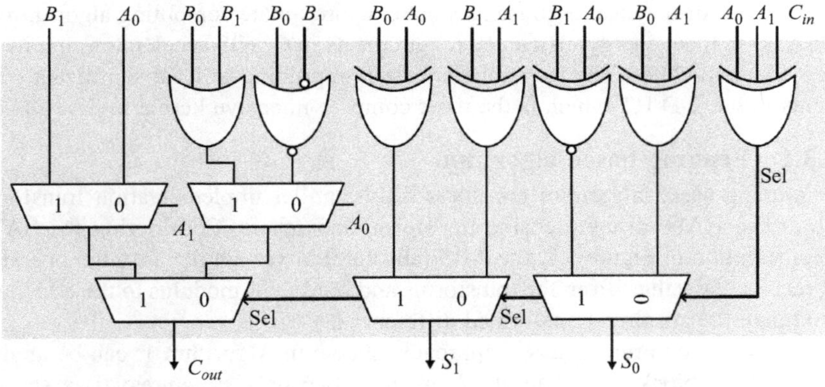

Figure 5.17 Two-bit hybrid adder scheme of [32]

($A_1, \ldots, A_0$ and $B_1, \ldots, B_0$) with a binary-to-hybrid converter at the inputs and a hybrid-to-binary converter at the outputs. The addition of two 2-bit operands requires seven XOR gates, six AND gates, and four OR gates, resulting in a total of seventeen.

Instead of using hybrid transcoders on inputs and outputs, [32] proposed an improved adder that operated directly on hybrid operands without the need for any extra conversion. This novel adder is based on multiplexers and performs an addition of two 2-bit operands directly on hybrid encoding. The novel hybrid adder is composed of five EXOR gates, five multiplexers, one AND gate, one OR gate and two inverters, as shown in Figure 5.17. It has fewer gates and a shorter critical path compared with the hybrid adder proposed in [31] (see Figure 5.16).

5.4.3 Approximate computing using pruning for SATD

Among the different energy-efficient techniques, approximate computing has emerged as one of the most important techniques used to increase performance and simultaneously reduce power dissipation [33]. The key in this approach is to reduce the accuracy constraints required by the application and design simpler circuits to speed up the critical path timing and to consume less energy. Approximate computing techniques take advantage of error-tolerant applications which do not need high accuracy all the time, but only good enough results or output quality. For example, highly popular multimedia applications in portable devices such as video coding, audio filtering, and image processing are intrinsically related to human senses. Since human sensors process analog information and have difficulty to realize digital approximations [34], multimedia signals are error-tolerant applications. In other words, it is possible to adopt approximate computing techniques to improve energy efficiency in multimedia applications by adequately exploiting the user experience at different levels of quality.

This part of the chapter presents a novel approximate computing algorithm for the SATD to meet energy-efficiency requirements in CMOS accelerator circuits for video codec modules. The technique is based on pruning of least significant coefficients in the 2-D HT which is the most compute-intensive kernel in SATD.

5.4.3.1 Pruning-based algorithm

The pruning-based algorithm considers fully parallel implementation transforms such as the HAD, discrete cosine transform, and others. Considering the SATD implementation in Figure 5.7, the ABS (absolute) boxes are the absolute operators that receive the values from the transforms and send their modulus to the adder tree to compute the absolute transformed differences.

The proposed pruning-based approach, shown in Algorithm 1, can be applied to any SATD block size. The algorithm uses two data structures: (i) a stack S containing the HT absolute coefficients w_{ij} sorted by a given policy and (ii) a tree R of dependencies among the prior terms used in the HT to compute each coefficient $w_{i,j}$. The stack is necessary to decide the order of the coefficients being discarded from the matrix H shown in (5.14). The tree is used by the proposed algorithm to solve dependencies among the discarded coefficients $w_{i,j}$ and prior terms that can be pruned from the architecture because the remaining HT coefficients do not use them. This method also takes as input the number of coefficients n to be discarded. The output is the approximate SATD RTL design generated according to the remaining terms in the pruned tree R.

Since the proposed method uses a tree of dependencies, an example is shown in Figure 5.18 to clarify the reason why this is the central structure of the algorithm. This example refers to the sub-tree related to the vertical transform presented in Figure 5.8. All the coefficients $w_{i,j}$ are leaves in the tree, and they are computed by the prior terms denoted as father nodes. For instance, the prior terms v_{31} and v_{41} are used only for the computation of the w_{31} and w_{41} coefficients. Hence, if those coefficients are discarded, the proposed algorithm must prune from the tree v_{31}, v_{41}, and $y_{11} - y_{31} + y_{41} - y_{21}$ terms. One can observe that all the remaining terms from the tree must be preserved since the coefficients w_{11} and w_{21} were not discarded.

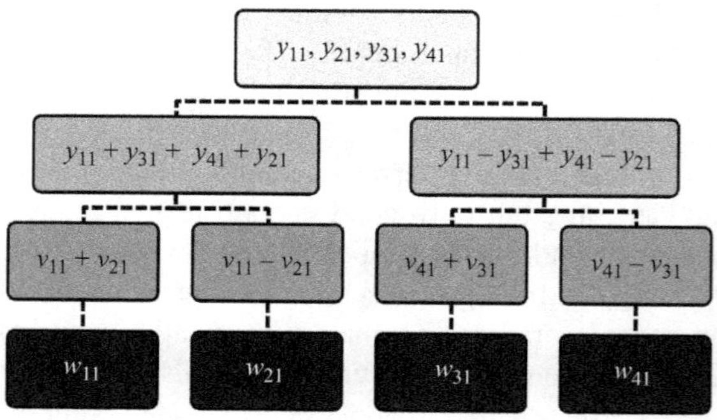

Figure 5.18 Example of the tree of dependencies for the vertical transform

Algorithm 1 The pruning-based algorithm for SATD

```
Input: S – a stack of coefficients sorted by any policy
       R – a tree of dependencies among HT terms or any transform
       n – the number of coefficients to be discarded
Output: Q – the approximate RTL netlist
1: while n > 0 do
2:     pruning := TRUE
3:     s := pop(S)
4:     l := search_leaf(s,R)
5:     while pruning do
6:         k := father(l,R)
7:         remove(l,R)
8:         if has_child(k,R) then
9:             pruning := FALSE
10:        else
11:            if is_root(k,R) then
12:                remove(k,R)
13:                pruning := FALSE
14:            else
15:                l := father(k,R)
16:                remove(k,R)
17:                if has_child(l,R) then
18:                    pruning := FALSE
19:                end if
20:            end if
21:        end if
22:    end while
23:    n := n -1
24: end while
25: Q := generate_netlist(R)
```

The pruning algorithm has an external loop that is performed *n* times to remove the next coefficient and its exclusive associated prior terms at each iteration. For instance, when considering the 4 × 4 SATD, the *n* parameter can range from 1 to 16 (i.e., no use of HT). The Boolean *pruning* variable is used to control the pruning iteration in line 5 in Algorithm 1.

First, the pop function extracts a given coefficient from the top of the coefficient stack previously sorted. Next, the algorithm searches the tree R for the extracted coefficient. Once the coefficient is found, it is stored in the variable *l* and the pruning process begins. The next step is to visit its father node and store its location in variable *k*. Since the location of the coefficient to be discarded is stored in variable *l*, this information is used to remove this coefficient from the tree effectively.

Then, it evaluates if the current father node has a remaining child. If this verification is true, the pruning must stop, since some coefficients or terms still depend on the current father node. Otherwise, the algorithm has to check whether the father node is also the root node. If this condition is true, the tree does not have any remaining child, and all the coefficients were pruned and, consequently, the pruning must stop. If the father node stored in k is not the root, it may be removed since it does not have any child. It is essential to store the location of the father of k in l before removing k from the tree. Then, the algorithm checks if l has any child. When the outcome is false, the pruning iteration must continue to remove the l node and, possibly, more prior terms. When true, the pruning must stop because l still has a child and cannot be pruned from the tree.

5.5 Low-power techniques for video-coding blocks—applications and results

Video data are dominating internet traffic and mobile computing applications. In recent years, the increase of video resolution in both displays and cameras present in mobile devices is demanding higher computational effort to improve video compression. The HEVC standard [35] emerged to double the compression capability when compared to the previous H.264/AVC [36] for the same perceptual video quality [3]. As a consequence, the computational effort can substantially increase reaching up to 3.2× more computation required when compared to the previous standard in order to sustain this compression improvement [3]. Therefore, additional low-power techniques are necessary to alleviate the intrinsic power-hungry characteristic of state-of-the-art video coders. This part of the chapter discusses the use of the previously presented low-power techniques to be applied to video-coding blocks. First, we present a novel adder based on hybrid encoding and a SAD hardware architecture implemented with the new adder. The SAD block is also exploited by using adder compressors. Next we show the design space exploration of power-efficient SAD architecture using different combinations of adder compressors. Interpolation filter architecture also explore the use of adder compressors. Since a reasonable number of additions compose these filters, efficient adder compressors are used to reduce both area and power of the filter datapaths. A video is intrinsically a dynamic media, since at each frame, the content may vary entirely or partially. Depending on the type of the video, the ME can be more or less accurate without significant degradation. Therefore, we propose an algorithm to prune the SATD computation, which can be further explored and configured in run time.

5.5.1 SAD block with adders compressors and hybrid adder

SAD calculation is one of the most time-consuming operations of video encoders compatible with the HEVC standard. SAD hardware architecture employ an adder tree to accumulate the coefficients from the absolute difference between the two video blocks. This part of the chapter exploits both different adder compressors structures and hybrid adders into the SAD hardware architecture.

5.5.1.1 SAD with adders compressors

Observing the SAD architecture structure (Figure 5.6), the intermediate values on the adder tree are not used in the calculation. Hence, the adder tree can be replaced by efficient adder compressors, as can be observed in Figure 5.19.

The use of 8-2 compressors in the adder tree becomes interesting since it allows the realization of simultaneous addition of eight operands in an efficient way, which has the potential to reduce power dissipation in SAD architecture. As shown in Figure 5.13, adder circuits are required for the partial sums recombination (recombination line).

The decision to evaluate different adders in the recombination line is given by the fact that this line represents the critical path of the circuit, thus requiring efficient adders for its implementation. The focus is an architectural exploration of SAD architecture using four different 8-2 adder compressor structures. The architecture also exploits five different adders in the recombination line (see Figure 5.13). Thus, the architecture versions differ with each other on how the 8-2 adder compressor was

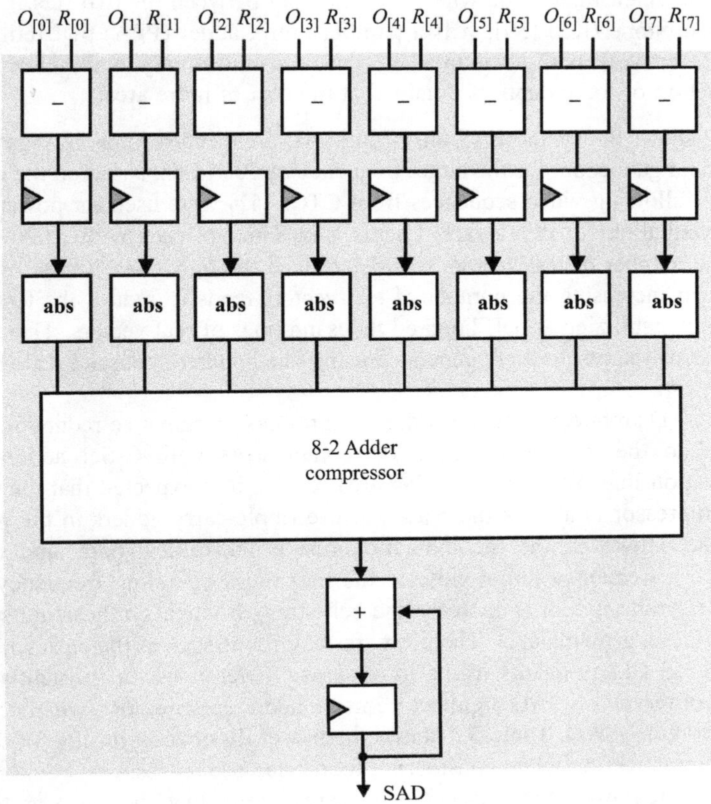

Figure 5.19 Diagram of SAD architecture using 8-2 adder compressor

developed and what type of adder was used in the recombination line of the partial sums. The adders exploited in the recombination line of the 8-2 adder compressors are presented in the following:

- Ripple carry with full-adder (RC-F) [37]: This adder is composed of a cascade of full-adder blocks. Although the RC-F is composed by a simple structure that enables a straightforward implementation, it presents a poor performance, because the carry is propagated into each addition block.
- Ripple carry with mux-based full-adder (RC-M): This adder is the same as RC-F with the difference that this one is composed by full-adder based in mux.
- Carry lookahead (CLA) [37]: This addition scheme generates all incoming carries in parallel and avoids waiting until the correct carry propagates from the stage of the adder where it has been generated.
- Carry-select (C-Sel) [38]: It is an adder that has a high fan out. In this scheme, each group generates two sets of bits sums and an outgoing carry. One set assumes that the incoming carry into the group is 0, and the other assumes that it is 1. When the incoming carry into the group is assigned, a multiplexer selects which final value will be selected out between the two sets.
- Kogge–Stone (KS) [39]: It is a parallel prefix adder (PPA) that introduces a prefix network with a minimal fan out of a unit at each node. However, the reduction of logic depth is obtained at the cost of more area.

Table 5.2 shows the synthesis results of the SAD architecture versions regarding the cell area and gate count for the target frequency of 300 MHz, based on the results in [40]. The following video sequences from CTCs [41] were used for power estimation: BasketballDrive, BQTerrace, Cactus, and Kimono, each with 1,920 × 1,080 pixel. The number of test vectors was determined through a sensitivity test. It was noticed that increasing the number of test vectors tends to reduce the total power results until stabilizing at half hundred thousand lines of real vectors. Therefore, all power dissipation results were generated using one hundred thousand real vectors as input per test.

Few SAD architecture using adder compressors present area reduction slightly compared to the tool adder, e.g., 4-2 compressors with C-Sel adders in the recombination line. When considering area results, it is expected that the smallest adder compressor is always the ones that use ripple-carry adders in the recombination line. However, the recombination line is the critical path, and we have considered a frequency target close to the maximum operating frequency, which forces the synthesis tool to increase the cell strength when synthesizing slow circuits like these, growing area. Therefore, faster circuits such as the ones with C-Sel, CLA and KS adders results in the lower area. Nevertheless, a substantial power saving is observed on SAD architecture using adder compressors, when compared with the original SAD. Table 5.3 shows the power dissipation results for the SAD evaluated architecture when considering four video sequences.

The main power dissipation results demonstrate that the SAD architecture when using 8-2 adder compressors composed by 4-2 adder compressor with KS adders in the recombination line saves about 25.5% power dissipation compared

Table 5.2 Circuit area and gate count results for SAD architecture using adder compressors @ 300 MHz

Recombination adder	Adder tree combination								Tool	
	3-2, 4-2		4-2		5-2, 4-2, 3-2		7-2, 3-2			
	Gate count	Cell area (μm²)	Gate count	Cell area (μm²)	Gate count	Cell area (μm²)	Gate count	Cell area (μm²)	Gates count	Cell area (μm²)
Ripple-carry mux-based FA (RC-M)	1,875	1,500	1,844	1,474	1,885	1,508	1,903	1,522	1,833	1,466
Ripple-carry with FA (RC-F)	1,858	1,487	1,839	1,471	1,897	1,517	1,914	1,531		
Carry-select (C-Sel)	1,908	1,526	1,815	1,452	1,907	1,525	1,823	1,458		
Carry lookahead (CLA)	1,919	1,535	1,847	1,477	1,909	1,527	1,858	1,486		
Kogge–Stone (KS)	1,963	1,570	1,838	1,470	1,963	1,570	1,828	1,462		

Table 5.3 Power dissipation (μW) results for SAD architecture using adder compressors @ 300 MHz

		Adder tree combination												Tool (+)		
		4-2, 3-2			4-2			5-2, 4-2, 3-2			7-2, 3-2					
Video	Recombination	Leak.	Dyn.	Total	Leak.	Dyn.	Total	Leak.	Dyn.	Total	Leak.	Dyn.	Total	Leak.	Dyn.	Total
BD	RC-F	13.3	934.4	947.7	12.8	919.8	932.6	14.3	1,193.4	1,207.7	14.8	1,311.2	1,326.0	12.9	1,126.3	1,139.1
	RC-M	13.8	1,007.9	1,021.8	13.0	950.7	963.7	14.3	1,166.9	1,181.2	14.8	1,307.9	1,322.7			
	C-Sel	14.2	1,103.4	117.6	12.3	833.1	845.4	14.9	1,177.0	1,191.9	12.8	1,062.6	1,075.4			
	CLA	14.6	1,075.7	1,090.3	12.9	945.6	958.5	14.7	1,193.5	1,208.2	13.5	1,138.3	1,151.8			
	KS	14.9	1,188.3	1,203.2	12.2	831.4	843.6	15.6	1,325.6	1,341.2	12.3	1,071.6	1,083.9			
BT	RC-F	13.3	1,219.2	1,232.5	12.8	1,185.8	1,198.6	14.3	1,505.2	1,519.5	14.8	1,641.9	1,656.7	12.9	1,469.5	1,482.4
	RC-M	13.8	1,221.5	1,335.3	13.1	1,235.1	1,248.1	14.3	1,485.3	1,499.7	14.8	1,629.9	1,644.7			
	C-Sel	14.3	1,408.2	1,422.5	12.4	1,115.5	1,127.9	14.9	1,524.9	1,539.9	12.9	1,398.8	1,411.7			
	CLA	14.6	1,435.5	1,450.1	12.9	1,248.2	1,261.1	14.8	1,544.8	1,559.6	13.5	1,444.4	1,457.9			
	KS	14.9	1,625.9	1,640.8	12.3	1,081.3	1,093.5	15.6	1,736.5	1,752.1	12.4	1,375.2	1,387.6			
CT	RC-F	13.3	922.5	935.9	12.8	905.9	918.8	14.4	1,148.1	1,162.5	14.8	1,254.4	1,293.3	12.9	1,131.9	1,144.8
	RC-M	13.8	980.7	994.5	13.1	928.0	941.1	14.4	1,125.5	1,139.9	14.8	1,259.0	1,273.8			
	C-Sel	14.3	1,072.6	1,086.9	12.3	860.2	872.5	14.9	1,162.2	1,180.1	12.8	1,058.6	1,071.4			
	CLA	14.6	1,065.9	1,080.5	12.9	934.0	946.8	14.8	1,190.3	1,205.1	13.5	1,097.6	111.4			
	KS	15.0	1,188.0	1,202.9	12.2	846.7	858.9	15.6	1,326.8	1,342.4	12.4	1,075.9	1,088.3			
KI	RC-F	13.3	1,303.6	1,316.9	12.8	1,275.5	1,288.3	14.3	1,609.4	1,623.8	14.8	1,749.8	1,764.6	12.7	1,547.5	1,560.2
	RC-M	13.8	1,382.3	1,396.1	13.1	1,300.8	1,313.8	14.4	1,575.9	1,590.2	14.8	1,754.8	1,768.9			
	C-Sel	14.3	1,493.4	1,507.7	12.3	1,211.0	1,224.0	14.9	1,638.5	1,653.4	12.8	1,489.7	1,502.5			
	CLA	14.6	1,488.5	1,503.1	12.9	1,313.2	1,326.1	14.8	1,677.0	1,691.8	13.5	1,531.3	1,544.8			
	KS	14.9	1,657.5	1,672.4	12.2	1,175.4	1,187.6	15.6	1,887.3	1,902.9	12.4	1,505.0	1,517.3			

with the tool's version (+). The 8-2 compressor provides this significant power saving because it adds eight simultaneous values with higher parallelism and reduced critical path compared with the adder selected by the synthesis tool. The SAD architecture versions regarding energy per SAD operation is also compared. The calculation of energy per SAD operation is given in (5.22), where C is the number of clock cycles of the architecture, T is the clock period, and P_{TOTAL} is the total power dissipated. The architecture work in one clock cycle to generate SAD for eight samples. They operate in the same clock period of 3.33 ns (target timing for 300 MHz).

$$E_{SAD} = C \times T \times P_{TOTAL} \tag{5.22}$$

Table 5.4 shows results in energy per 1 × 8 SAD operation for the evaluated architecture for each video sequence. Figure 5.20 summarizes the percentage energy savings, which also correspond to the power savings for proposed SAD architecture based on each evaluated 8-2 adder compressor topology, comparing with the adder from the tool (+). Among the evaluated 8-2 adder compressor structures, only the one that employs a combination of 3-2, 4-2, and 5-2 compressors is less

Table 5.4 Energy (PJ) per 8 × 1 *operation for SAD architecture using adder compressors*

		Adder tree combination				
1,080p video sequence	**Recombination adder**	**3-2, 4-2**	**4-2**	**5-2, 4-2, 3-2**	**7-2, 3-2**	**Tool**
BasketballDrive	RC-F	3.16	3.11	4.03	4.42	3.80
	RC-M	3.41	3.21	3.94	4.41	
	C-Sel	3.73	2.82	3.97	3.58	
	CLA	3.63	3.19	4.03	3.84	
	KS	4.01	2.81	4.47	3.61	
Cactus	RC-F	4.11	3.99	5.07	5.52	4.94
	RC-M	4.45	4.16	5.00	5.48	
	C-Sel	4.74	3.76	5.13	4.71	
	CLA	4.83	4.20	5.20	4.86	
	KS	5.47	3.65	5.84	4.63	
BQTerrace	RC-F	3.12	3.06	3.87	4.23	3.82
	RC-M	3.32	3.14	3.80	4.25	
	C-Sel	3.62	2.91	3.93	3.57	
	CLA	3.60	3.16	4.02	3.70	
	KS	4.01	2.86	4.47	3.63	
Kimono	RC-F	4.39	4.29	5.41	5.88	5.20
	RC-M	4.65	4.38	5.30	5.90	
	C-Sel	5.03	4.08	5.51	5.01	
	CLA	5.01	4.42	5.64	5.15	
	KS	5.57	3.96	6.34	5.06	

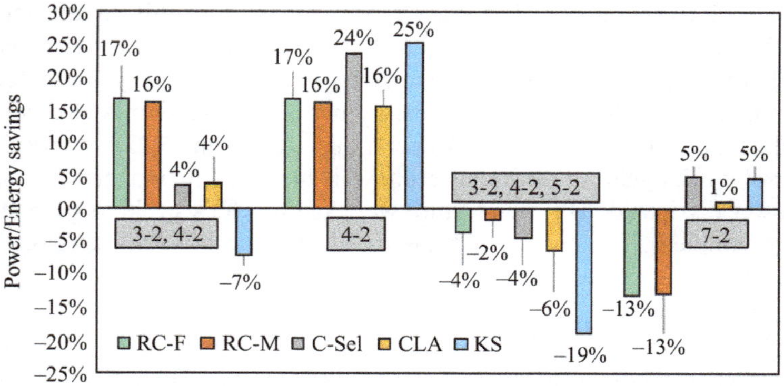

Figure 5.20 Power/Energy savings (average) for SAD using adder compressors compared with the baseline adder tree automatically selected by the synthesis tool ("+" in the RTL)

power-efficient compared with the tool's adder (+). This inefficiency is due to the time path imbalance of 3-2, 4-2, and 5-2 structures, which increases the power caused by glitches. The remaining adder compressor configurations present power/energy reduction compared to the tool's adder, depending on the recombination line method. The SAD architecture using the 4-2 adder compressor provides the best power/energy results.

5.5.1.2 SAD with hybrid adder

Figure 5.21 shows the SAD hardware architecture employing hybrid adders. The input samples use binary encoding. The conversion from binary-to-hybrid encoding (*to Hyb* modules in Figure 5.21), for all the residue, occurs on the inputs of the adder tree. The new hybrid adder implements the adder circuits of the adder tree and also the accumulator. There is also an evaluation of the implementation of hybrid adder proposed by [31]. The final SAD value is converted back to binary encoding (*to Bin* module in Figure 5.21) for output. Figure 5.15 showed a diagram of hybrid to binary and binary to hybrid conversion.

The synthesis of SAD architecture with hybrid adder was targeted to 400 MHz with 1.1 V power supply. Table 5.5 shows the leakage, dynamic, and total power of SAD architecture using hybrid adder. By the way of comparison, the adders of SAD architecture were also implemented with macro-function (using "+" in VHDL), leaving the synthesis tool to generate the adder (it is referred as "Tool" in Table 5.5), and with the hybrid adder proposed in [32]. Power results are shown in Table 5.5 for the six video sequences we have evaluated. SAD architecture using our proposed hybrid adder reduces power (on average of the six video sequences) in 7.42% and 12.97% compared to the tool adder and hybrid adder in [32], respectively. Those power savings are obtained by the use of our proposed hybrid adder in SAD architecture, that reduces switching activity of the circuit, reducing dynamic power dissipation, with the cost of a small increase in area (see Table 5.6) due to the extra XOR gates.

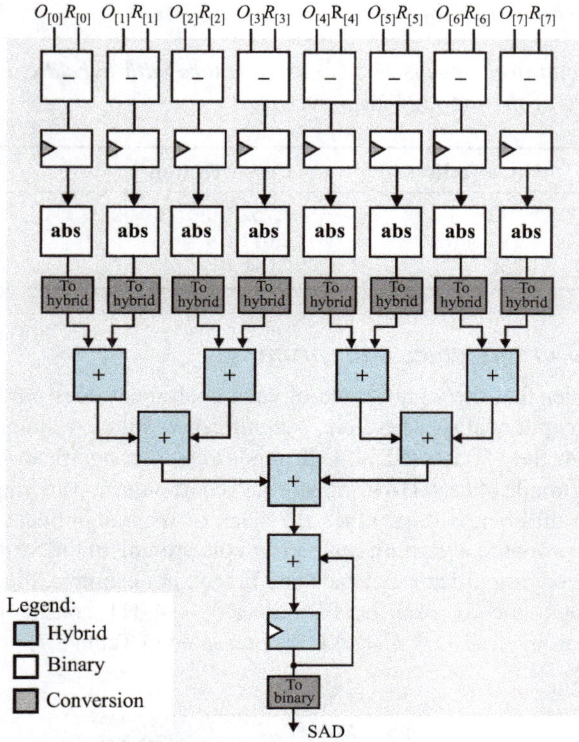

Figure 5.21 SAD architecture with the hybrid adders

Table 5.5 Power dissipation results (μW) of hybrid encoded SAD architecture and the state-of-the-art [32] power savings

Video	**Adder**	**Leakage**	**Dynamic**	**Total**	**Savings (%)**
Blowing bubbles	Tool	14.36	410.76	425.14	7.32
	[31]	17.13	428.40	445.54	11.57
	[32]	16.42	377.55	393.98	–
RaceHorses	Tool	14.81	412.58	427.40	7.70
	[31]	17.29	446.29	463.59	13.22
	[32]	16.87	378.55	395.69	–
BQSquare	Tool	14.57	412.58	389.88	6.95
	[31]	17.38	446.29	416.68	14.98
	[32]	16.60	378.81	362.76	–
Steam locomotive train	Tool	14.41	417.39	431.80	7.46
	[31]	17.19	445.75	462.94	13.69
	[32]	16.45	383.09	399.55	–
BQTerrace	Tool	14.36	410.83	425.19	7.42
	[31]	17.57	427.35	444.92	11.06
	[32]	16.42	377.55	393.98	–
Nebuta Festival	Tool	14.64	419.97	434.61	7.69
	[31]	17.44	445.44	462.87	13.32
	[32]	16.70	384.47	401.17	–

Table 5.6 Circuit area results of SAD architecture with hybrid encoding and the state-of-the-art [32] area overhead

Adder	No. of cells	Cell area (μm^2)	Overhead (%)
Tool	913	1,752.0	+17.35
[31]	1,191	2,011.0	+2.24
[32]	1,238	2,056.0	–

5.5.2 SATD architecture with pruning

One may consider that the significance of each coefficient does not have the same weight for the output quality. Therefore, a significance-driven technique to prune the coefficients from the HT is used [42]. In this scope, the significance is defined as the average magnitude of each HAD transformed coefficient considering a set of video sequences from different classes. Once the stack of most significant coefficients is determined, the proposed algorithm can perform the pruning of the SATD architecture based on the average significance of each coefficient. For example, Figure 5.22 shows the average magnitude for each coefficient in 4 × 4 HT considering four video sequences (i.e., from class *A*, *B*, *C*, and *D*, as presented in Table 5.7).

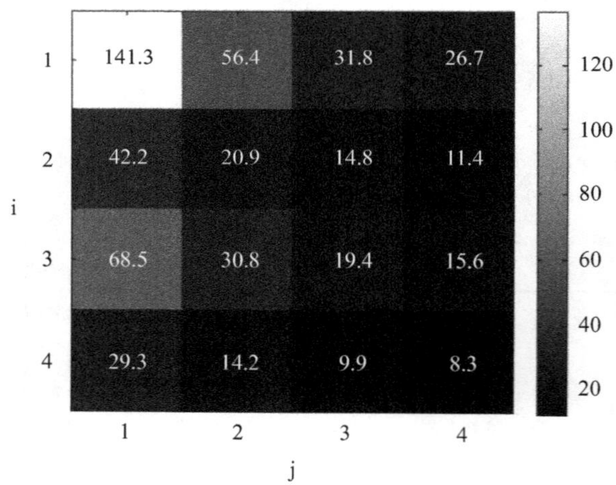

Figure 5.22 The average magnitude of each coefficient in 4 × 4 HT

For instance, to discard the ten least significant coefficients from the precise example in Figure 5.7, the following coefficients will be removed: w_{44}, w_{43}, w_{24}, w_{42}, w_{42}, w_{23}, w_{34}, w_{33}, w_{22}, w_{14}, and w_{41}. One can observe that, based on the proposed algorithm, the tree of dependencies is pruned to reduce the number of cells in the circuit. Figure 5.23 shows the resulting approximate SATD architecture, where the approximate architecture presents pruned horizontal or vertical transforms.

Table 5.7 Video sequences specification for SATD comparisons

Class	Video sequence name	Resolution (pixel)
A	Traffic	2,560 × 1,600
B	BQTerrace	1,920 × 1,080
C	RaceHorses	832 × 480
D	BlowingBubbles	416 × 240

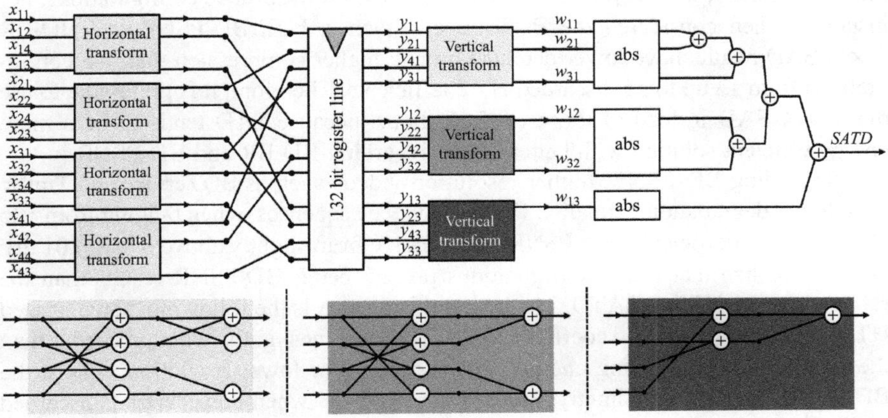

Figure 5.23 Approximate SATD architecture with ten discarded coefficients

The pruned horizontal and vertical transform blocks have fewer adders than the complete ones. The reduction in the number of adders for the approximate SATD architecture in comparison with the precise one is 37.9%. In conclusion, the additional cost of the approximate SATD architecture shown in Figure 5.23 is 1.7× when compared to the 4 × 4 SAD architecture. The precise 4 × 4 SATD has an additional cost of 3.1×. When compared with the ten pruned coefficients approach, the precise architecture has 1.8× more computational cost.

One can notice in Figure 5.23 that some registers can be discarded when compared with the precise architecture. The fundamental solution is to provide a configurable scheme where these registers are clock gated instead of being discarded. Based on that, dynamic power is reduced in favor of accuracy-configurable energy-efficient design. To select different levels of accuracy for the configurable approach, Table 5.7 shows the four videos considered to evaluate the output quality and bitrate per number of discarded coefficients. In this scope, 64 frames from these videos are encoded by considering the HM 16.0 software and the standard test conditions presented in [41]. The random access configuration was selected instead of all intra or low delay. It is because it is the most complex and most generic configuration for picture ordering. Each video sequence was run for the following quantization parameters: 22, 27, 32, and 37. It is because those points are integrated to evaluate the Bjøntegaard-delta bit-rate (BD-BR) and PSNR (BD-PSNR). For BD-BR and BD-PSNR metrics, the precise 4 × 4 SATD is the reference. For this analysis, all default block sizes regarding SATD computation are allowed.

Video quality (BD-PSNR) and video compression (BD-BR) results can be seen in Figure 5.24(a) and (b), respectively. In the graphs, discrete points for each configuration of discarded coefficients are shown, and the dotted red line represents the 4 × 4 SAD (i.e., when 16 coefficients are discarded in the 4 × 4 HT). In fact, this line represents a boundary to evaluate if configurations are ranging from 1 to 15 discarded coefficients that have worse results than using the 4 × 4 SAD. Based on that, for all the cases, the 4 × 4 SAD presents better results than some pruned configurations. For instance, when considered the BQTerrace sequence, both BD-PSNR and BD-BR 4 × 4 SAD results have lower degradation and higher compression than the configurations from 13 up to 15 discarded HT coefficients. Therefore, it is preferable to use the 4 × 4 SAD instead of those configurations, because SAD tends to be a more energy-efficient solution with better results regarding BD-BR and BD-PSNR.

Regarding BD-PSNR, higher resolution videos such as BQTerrace and Traffic have lower degradation in quality. For those video sequences, when ranging from 1 to 10 discarded coefficients, BD-PSNR results range from slightly above 0 to −0.01 dB. One can realize that some configurations present better BD-PSNR results than the reference precise 4 × 4 SATD. The hypothesis for that is the following: some pruned HT (e.g., the two discarded coefficient HT) could present slightly better properties than the complete HT regarding energy concentration. For lower resolution videos, the BD-PSNR ranges from slightly above 0 to −0.05 dB when considered 1 discarded coefficient HT and the 4 × 4 SAD, respectively.

The BD-BR results are just a complimentary evaluation of the BD-PSNR. For higher resolution videos (i.e., Figure 5.24(a)(I) and (IV)) and considering the range from 1 discarded coefficient up to the 4 × 4 SAD, the BD-BR configuration increases up to 0.4% and 0.5%, respectively. For lower resolution videos in Figure 5.24(b)(II) and (III), the increase in BD-PSNR is not higher than 1%.

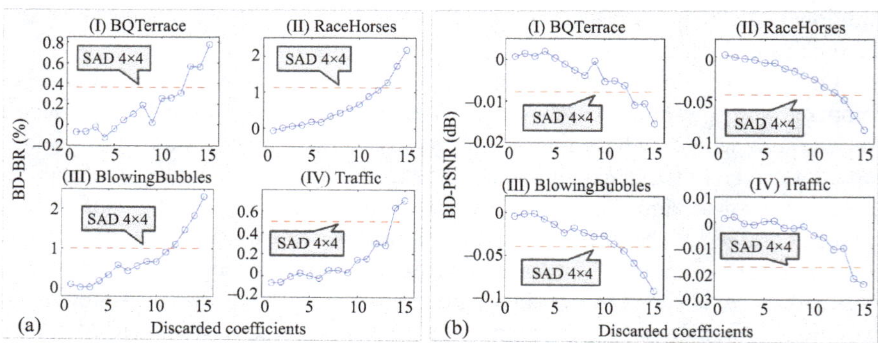

Figure 5.24 Compression-efficiency results (a) BD-BR results for approximate SATD architecture and (b) BD-PSNR results for approximate pruned SATD architecture

5.5.2.1 Results and discussions of the SATD pruning techniques

This section shows results for the case study on configurable 4 × 4 SATD. The experimental setup is presented followed by the results regarding video quality, compression, and energy efficiency.

As shown in Table 5.7, four video sequences are considered from the HEVC benchmark to evaluate the video quality and compression analysis. For each video sequence, 64 frames are encoded considering random access configuration due to its higher complexity when compared to low delay. All these videos are encoded with four different quantization parameter (QP) values (22, 27, 32, and 37) as recommended in CTC by [41]. Differently from the analysis previously mentioned, a more constrained quality and compression evaluation are implemented to validate the approximate architecture.

Since only the 4 × 4 SATD is implemented, the original version of the HM-16.0 software was modified to use only this block-size for SATD operation. It is performed to evaluate the precise baseline 4 × 4 SATD response without considering the implications of any other block sizes. The pruning algorithm was also implemented inside the HM source code to evaluate the video quality and compression when the pruned versions are running.

The approximate and the baseline 4 × 4 SATD architecture were described in VHDL followed by synthesis. The synthesis was performed by the RTL compiler tool, where the designs were mapped onto the 45 nm Nangate FreePDK, and the netlists were generated. The netlists were simulated to extract switching activity. The input stimuli were extracted from the HM-16.0 and represent 40,000 4 × 4 residual blocks from different video sequences. After switching activity is annotated, it is used to estimate the power dissipation in the circuits.

Table 5.8 shows the mean energy per operation (MEOp) and area results (in terms of number of cells and number of NAND2 X1 equivalent gate count—EG) of approximate 4 × 4 SATD architecture compared to SAD and baseline (precise) 4 × 4 architecture. It shows five pruning options to design approximate SATD architecture according to the algorithm proposed in [42]: with 2, 4, 6, 8, and 10 discarded coefficients from the 4 × 4 HT.

Table 5.8 Energy efficiency and area results of approximate SATD architecture and comparison with SAD and baseline SATD architecture

SATD configuration (no. of discarded coefficients)	MEOp (PJ) @ 650 MHz	Area (EG)	No. of cells
SAD	2.5	3,847	2,171
10	7	7,553	3,305
8	9.9	9,426	4,299
6	14.7	11,863	5,632
4	16.2	12,813	6,177
2	20.8	14,930	7,339
Baseline 4 × 4 architecture (precise)	23.9	15,986	8,082

Based on these results, one can conclude that the pruning-based algorithm brings substantial energy reduction for the approximate 4 × 4 SATD architecture with small rate-distortion degradation. Mean energy per operation is reduced from 23.9 PJ (precise SATD architecture) to only 7 PJ when 10 coefficients are discarded, which represents a reduction of 70.7%, while area is reduced by 52.75%.

5.5.3 Interpolation filter with adder compressors

The new interpolation filter that generates half- and quarter-pixel (pel) for FME and motion compensation is one of the most complex coding tools introduced in HEVC. There are two interpolation filters: a 7-/8-tap interpolation filter for luminance (luma) pels and 4-tap filter for chrominance (chroma) pels. The whole HEVC interpolation filter consumes 20%–30% and 20%–40% of the execution time of encoder and decoder software, respectively [19]. Therefore, to achieve real-time encoding and decoding, it is essential to develop efficient hardware architecture for the HEVC interpolation filter, since it is one of the most complex coding tools in HEVC.

5.5.3.1 Interpolation filters architecture using adder compressors

With the observation that luma and chroma interpolation filter datapaths have adder trees, and one does not use the intermediate values in the calculation, we can replace the adder trees with efficient adder compressors. Luma datapath has seven values (output from multiplexers) that need to be added. Hence, we developed a modified luma interpolation filter datapath employing a 7-2 adder compressor (see Section 5.4.1.1, Figure 5.12(d)), as shown in Figure 5.25. In the original architecture, there is a subtraction operator at the end of the adder tree. One can solve it by complementing the inputs of the four multiplexers at the bottom of Figure 5.25.

Chroma datapath has eight values (output from multiplexers) to be added. A modified chroma interpolation filter datapath employing an 8-2 adder compressor was developed, as shown in Figure 5.26. In this datapath, the last four inputs are complemented because of the subtraction operation. Four different internal structures of the hierarchical 8-2 adder compressor were exploited (see Section 5.4.1.1, Figure 5.12).

5.5.3.2 Interpolation filter architecture results

The interpolation filter datapaths using adder compressors were implemented in hardware description language (HDL) and synthesized to 45 nm Nangate Open Cell Library using Cadence Encounter RTL Compiler version 8.10 synthesis tool [17]. The timing constraint for clock period was set to 1.5 ns. Table 5.9 shows the synthesis results of luma and chroma interpolation datapaths regarding cell usage, area, and power. Power results are generated using a simulation of 10,000 random test vectors as input in the architecture.

Table 5.9 shows the results of different internal structures of adder compressors. Two versions are presented for luma filter datapath: (i) the original luma filter datapath, where adders are implemented using adder operation in HDL, leaving the synthesis tool to perform the adders and (ii) luma filter datapath using efficient 7-2 adder compressor [25]. Five versions are exploited for chroma filter

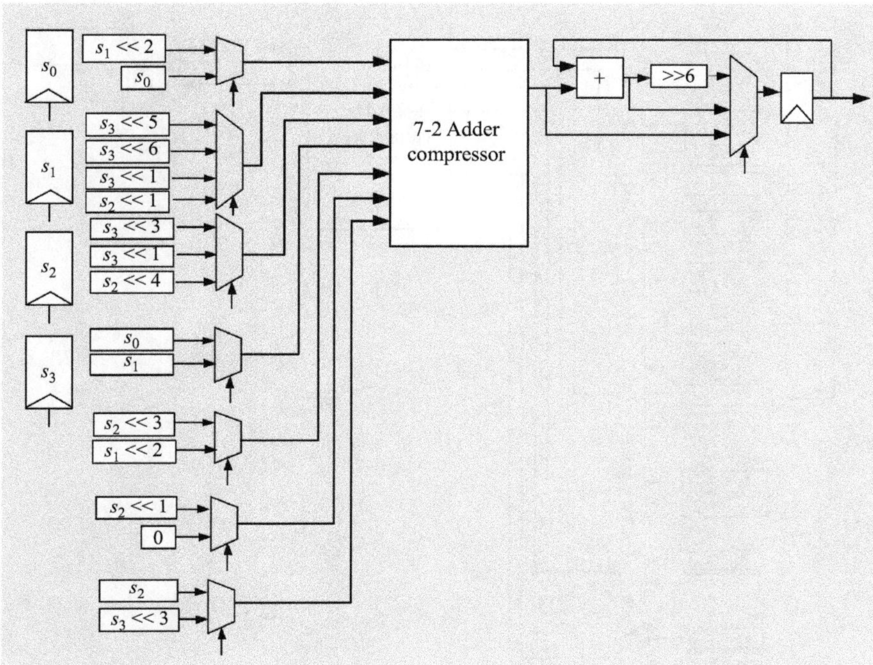

Figure 5.25 Modified luma filter datapath using 7-2 adder compressor

datapath: (i) the original chroma filter datapath, where adders are implemented using adder operation in HDL, leaving the synthesis tool to implement the adders; (ii) 8-2 adder compressor using 4-2 adder compressors; (iii) 8-2 adder compressor using 3-2 and 4-2 adder compressors; (iv) 8-2 adder compressor using 5-2, 4-2, and 3-2 adder compressors; and (v) 8-2 adder compressor using 7-2 and 3-2 adder compressors.

Luma datapath using a 7-2 adder compressor reduces the area in 13.94% and power-delay product (PDP) in 25% when compared with the original architecture implemented with the ripple-carry adder. The most power-efficient version of the hierarchical 8-2 adder compressor Type III applied to chroma datapath is the version that employs only 4-2 and 3-2 adder compressors (see Figure 5.14(c)). However, the most area-efficient version of hierarchical 8-2 adder compressor is the one that uses 3-2 and 7-2 adder compressors (see Figure 5.14(d)). It leaves the option of the designer if the architecture is focused more on the area or power reduction.

It is important to note that comparing the results of our modified datapaths with the original datapaths using adder operations (leaving the synthesis tool to map to efficient adders in the standard cell library), the comparison is not very fair, favoring better circuit optimizations for the synthesis tool (the original datapaths). It is because the tool uses very efficient adders with optimizations in logic, circuit, and layout levels for the desired frequency target. However, these circuits are black boxes for the RTL designers. Nevertheless, the 7-2 and 8-2 adder compressors achieve small area and power savings when compared with the synthesis tool.

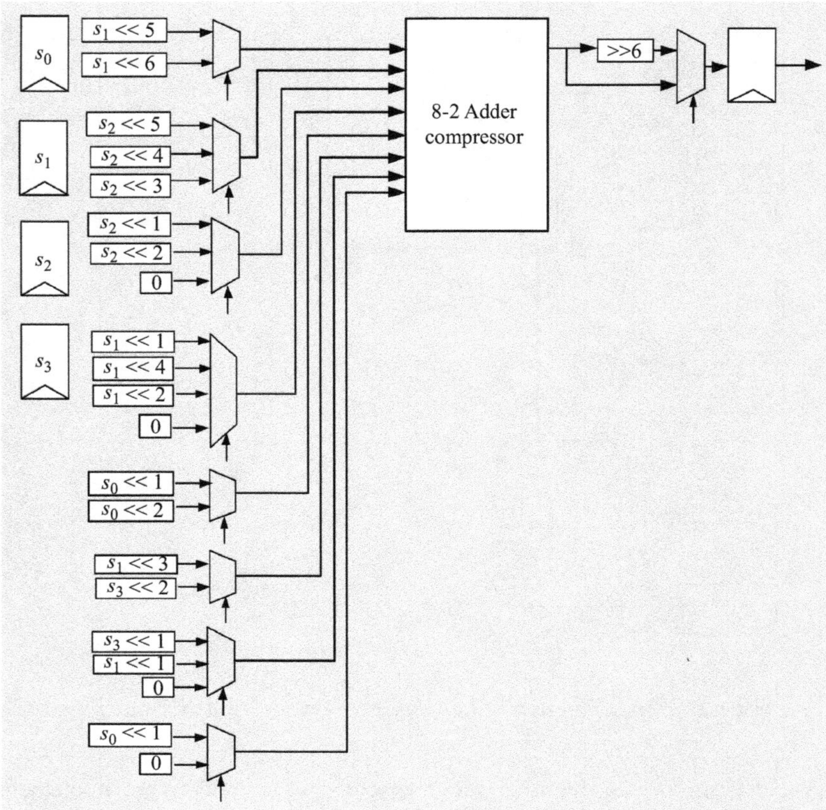

Figure 5.26 Modified chroma filter datapath using 8-2 adder compressor

5.6 Conclusions

This chapter presented new architecture design solutions for critical modules of the dedicated video compressor hardware. The low-power design techniques used in the design of these architecture proved to save a significant amount of power, and consequently energy, in the execution of key computation kernels used in video coding. Especially significant are the savings for higher resolution video coding, mainly for Full HD and 4K resolutions. The general concepts regarding power dissipation in CMOS VLSI circuits were first reviewed, giving to the reader a more detailed background on static, short-circuit, and dynamic power. Evaluating power consumption in video encoders demands a realistic profiling and power-estimation method. In this chapter, a methodology focused on obtaining accurate power dissipation results using real-video inputs was also presented. Most hardware design reports in the literature, found so far, just reported power estimates coming from heuristics of logic-synthesis tools. These estimates are known to be inaccurate, and the proposed method addressed

Table 5.9 Luma and chroma interpolation datapaths synthesis results

	Cells	**Area (μm^2)**	**Delay (ns)**	**Leakage Power (μW)**	**Dynamic Power (μW)**	**Total Power (μW)**	**PDP ($\times 10^{-13}$)**	**PDP reduction (%)**
Luma datapath								
Original with RCA	384	1,692	1.872	26.0	415.9	441.9	8.27	-
i) 7-2	337	1,443	1.570	24.4	370.8	395.2	6.20	25
ii) 8-2 (w/ 4-2)	392	1,589	1.478	25.6	367.7	393.3	5.81	30
iii) 8-2 (w/ 3-2 and 4-2)	395	2,446	1.604	25.2	353.1	378.3	6.06	27
iv) 8-2 (w/ 5-2, 4-2 and 3-2)	423	1,702	1.628	25.8	388.6	414.4	6.74	18
v) 8-2 (w/ 7-2 and 3-2)	329	1,404	1.549	24.1	365.5	389.6	6.03	27
Chroma datapath								
Original with RCA	435	1,761	1.612	27.2	396.8	424.0	6.83	-
i) 8-2 (w/ 4-2)	430	1,643	1.466	26.0	355.4	381.5	5.59	18
ii) 8-2 (w/ 3-2 and 4-2)	445	1,660	1.563	25.6	337.6	363.2	5.68	17
iii) 8-2 (w/ 5-2, 4-2 and 3-2)	449	1,721	1.598	26.1	360.3	386.5	6.18	10
iv) 8-2 (w/ 7-2 and 3-2)	396	1,546	1.499	24.7	354.1	378.9	5.68	17

in this chapter is very accurate for the use cases of interest to the portable video devices. This book chapter presents several techniques for low-power design, such as the use of adder compressors, which reduce the critical path when compared to normal adders, and the use of hybrid encoding, which presents a smaller bit transitions comparing with the standard binary representation, leading to smaller dynamic power results. Finally, a background on some video-coding concepts, such as the SAD and SATD metrics, were presented, in order to introduce low-power techniques which are relevant for the video-coding accelerators addressed for filtering, SAD, SATD, and HT.

References

[1] CISCO. Cisco Visual Networking Index: Forecast and Methodology, 2015–2020, 2016. Found in: http://www.cisco.com/c/en/us/solutions/collateral/service-provider/visual-networking-index-vni/complete-white-paper-c11-481360.html. Accessed: 30 Jan. 2017.

[2] Vanne, J.; Viitanen, M.; Hämäläinen, Timo D. and Hallapuro, A. Comparative Rate-Distortion-Complexity Analysis of HEVC and AVC Video Codecs. IEEE Transactions on Circuits and Systems for Video Technology, vol. 22, no. 12, pp. 1885–1898, 2012, [s. l.].

[3] Atzori, L.; Iera, A. and Morabito, G. The Internet of Things: A Survey. Computer Networks, vol. 54, no. 15, pp. 2787–2805, 2010, [s. l.].

[4] Kamilaris, A. and Pitsillides, A. Mobile Phone Computing and the Internet of Things: A Survey. IEEE Internet of Things Journal, vol. 3, no. 6, pp. 885–898, 2016, [s. l.].

[5] CISCO. The Internet of Things: How the Next Evolution of the Internet Is Changing Everything, 2011. Found in: https://www.cisco.com/c/dam/en_us/about/ac79/docs/innov/IoT_IBSG_0411FINAL.pdf. Accessed: 30 Jan. 2017.

[6] Nordrum, A. Popular Internet of Things Forecast of 50 Billion Devices by 2020 Is Outdated, 2016. Found in: http://spectrum.ieee.org/tech-talk/telecom/internet/popular-internet-of-things-forecast-of-50-billion-devices-by-2020-is-outdated. Accessed: 30 Jan. 2017.

[7] Penny, W.; Paim, G.; Proto, M.; Agostini, L. and Zatt, B. Real-Time Architecture for HEVC Motion Compensation Sample Interpolator for UHD Videos. In *2015 28th Symposium on Integrated Circuits and Systems Design (SBCCI)*, 2015, pp. 1–6.

[8] Chang, C.-H.; Gu, J. and Zhang, M. Hardware Implementation for the HEVC Fractional Motion Estimation Targeting Real-Time and Low Energy. Journal of Integrated Circuits and Systems, vol. 11, no. 2, pp. 106–120, 2016.

[9] Pastuszak, G. and Trochimiuk, M. Algorithm and Architecture Design of the Motion Estimation for the H.265/HEVC 4K-UHD Encoder. Journal of Real-Time Image Processing, vol. 12, no. 2, pp. 517–529, 2016.

[10] Corrêa, M.; Porto, M.; Zatt, B. and Agostini, L. A Low-Area and High Throughput Intra Prediction Architecture for a Multi-Standard HEVC and H.264/AVC Video Encoder. In *2015 28th Symposium on Integrated Circuits and Systems Design (SBCCI)*, 2015, pp. 1–6.

[11] HEVC Test Model (HM) v. 16.7, http://HEVC.hhi.fraunhofer.de.
[12] GProf Profiler. https://ftp.gnu.org/old-gnu/Manuals/gprof-2.9.1/html_mono/gprof.html.
[13] Chandrakasan, A. and Brodersen, R. Low Power Digital CMOS Design, New York: Kluwer Academic Publishers, 1995.
[14] Weste, N. and Eshraghian, K. Principles of CMOS VLSI Design, 2nd edition. Boston: Addison-Wesley, 1994.
[15] Veendrick, H. Short-Circuit Dissipation of Static CMOS Circuitry and Its Impact on the Design of Buff Circuits. IEEE Journal of Solid-State Circuits, vol. 19, pp. 468–473, 1984.
[16] Costa, E. A. C.; Cortes, F. P.; Cardoso, R.; Carro, L. and Bampi, S. Modeling of short circuit power consumption using timing-only logic cell macromodels. In: Proceedings 13th Symposium on Integrated Circuits and Systems Design (SBCCI), pp. 222–227, 2000.
[17] CADENCE EDA TOOLS, http://www.cadence.com.
[18] MANUAL: Encounter RTL Compiler Synthesis Flows, http://www.cadence.com.
[19] Diniz, C.; Shafique, M.; Bampi, S. and Henkel, J. High-Throughput Interpolation Hardware Architecture with Coarse-Grained Reconfigurable Datapaths for HEVC. In *IEEE International Conference on Image Processing (ICIP)*, 2013, pp. 2091–2095.
[20] Park, J.; Choi, J. and Roy, K. Dynamic Bit-Width Adaptation in DCT: An Approach to Trade Off Image Quality and Computation Energy. IEEE Transactions on Very Large Scale Integration (VLSI) Systems, vol. 18, no. 5, pp. 787–793, 2010.
[21] Weinberger, A. 4-2 Carry-Save Adder Module, IBM Technical Disclosure Bulletin, 1981.
[22] Oklobdzija, V.; Villeger, D. and Liu, S. A Method for Speed Optimized Partial Product Reduction and Generation of fast parallel multipliers using an algorithmic approach. IEEE Transaction on Computers, vol. 45, pp. 294–306, 1996.
[23] Chang, C-H.; Gu, J. and Zhang, M. Ultra Low-Voltage Low-Power CMOS 4-2 and 5-2 Compressors for Fast Arithmetic Circuits. IEEE Transactions on Circuits and Systems – I, vol. 51, pp. 1985–1997, 2004.
[24] Prasad, K and Parhi, K. Low-Power 4-2 and 5-2 Compressors. In *Proceedings of the 35th Asilomar Conference on Signals, Systems and Computers*, vol. 1, 2001, pp. 129–133.
[25] Rouholamini, R.; Kavehie, O.; Mirbaha, A.; Jasbi, S. and Navi, K. A New Design for 7:2 Compressors. In *IEEE/ACS International Conference on Computer Systems and Applications (AICCSA)*, 2007, pp. 474–478.
[26] Altermann, J.; Costa, E. and Bampi, S. Fast Forward and Inverse Transforms for the H.264/AVC Standard Using Hierarchical Adder Compressors. In *IEEE/IFIP International Conference on VLSI and System-on-Chip (VLSI-SoC)*, 2010, pp. 310–315.
[27] Diniz, C.; Fonseca, M.; Costa, E. and Bampi, S. Enhancing a HEVC Interpolation Filter Hardware Architecture with Efficient Adder Compressors.

In *IEEE International New Circuits and Systems Conference (NEWCAS)*, 2015, pp. 1–4.

[28] Diniz, C.; Fonseca, M.; Costa, E. and Bampi, S. Evaluating the Use of Adder Compressors for Power-efficient HEVC Interpolation Filter Architecture. Analog Integrated Circuits and Signal Processing, vol. 89, no. 1, pp. 111–120, 2016.

[29] Costa, E.; Monteiro, J. and Bampi S. Power Efficient Arithmetic Operand Encoding. In *14th Symposium on Integrated Circuits and Systems Design*, 2001, pp. 201–206.

[30] Costa, E.; Monteiro, J. and Bampi S. A New Architecture for 2's Complement Gray Encoded Array Multiplier. In *15th Symposium on Integrated Circuits and Systems Design*, 2002, pp. 14–19.

[31] Costa, E.; Monteiro, J. and Bampi, S. A new array architecture for signed multiplication using Gray encoded radix-2^m operands, Integration, the VLSI Journal, vol. 40, no. 2, pp. 118–132, 2007.

[32] Ferreira, R.; Silveira, B.; Fonseca, M.; Diniz, C. and Costa, E. Low Power Sum of Absolute Differences Architecture Using Novel Hybrid Adder. In *IEEE 8th Latin American Symposium on Circuits & Systems*, 2017, pp. 213–216.

[33] Han, J.; and Orshansky, M. Approximate Computing: An Emerging Paradigm for Energy-Efficient Design. In *Test Symposium 9ETS*, 2013, pp. 1–6.

[34] Zhu, N.; Goh, W. L.; Zhang, W.; Yeo, K. S. and Kong, Z. H. Design of Low-Power High-Speed Truncation-Error-Tolerant Adder and Its Application in Digital Signal Processing. IEEE Transactions on Very Large Scale Integration (VLSI) Systems, vol. 18, no. 8, pp. 1225–1229, 2010.

[35] ITU-T; ISO/IEC. High-Efficiency Video Coding, ITU-T Recommendation H.265 and ISO/IEC 23008-2, 2013.

[36] ITU-T; ISO/IEC. Advanced Video Coding, ITU-T Recommendation H.264 and ISO/IEC 14496-10 (MPEG-4 AVC), 2011.

[37] Parhami, B. Computer Arithmetic: Algorithms and Hardware Designs, New York: Oxford University Press, 2000.

[38] Bedrij, O. Carry-Select Adder. IRE Transactions on Electronic Computers, vol. EC-11, no. 3, pp. 340–346, 1962.

[39] Kogge, P. and Stone, H. A Parallel Algorithm for the Efficient Solution of a General Class of Recurrence Equations. IEEE Transactions on Computers, vol. C-22, no. 8, pp. 786–793, 1973.

[40] Silveira, B.; Paim, G.; Abreu, B.; *et al.* Power-Efficient Sum of Absolute Differences Hardware Architecture Using Adder Compressors for Integer Motion Estimation Design. *IEEE Transactions on Circuits and Systems I (TCAS-I)*, vol. 1, p. 1–12, 2017.

[41] Bossen, F. Common Test Conditions and Software Configurations. In: JCT-VC document no. L1100, Jan 2013.

[42] Soares, L. B.; Diniz, C. M.; da Costa, E. A. C. and Bampi, S. A Novel Pruned-Based Algorithm for Energy-Efficient SATD Operation in the HEVC Coding. In *29th Symposium on Integrated Circuits and Systems Design*, IEEE Press, 2016, p. 13.

Chapter 6
Real-time architectures for 3D video coding

Vladimir Afonso[1], Mario Saldanha[1], Ruhan Conceição[1], Murilo Perleberg[1], Marcelo Porto[1], Bruno Zatt[1], Altamiro Susin[2] and Luciano Agostini[1]

The popularization of devices capable of dealing with digital videos, especially handheld devices, keeps growing pushed by the interest of the users in high-quality and immersive video-related experiences, and by the platforms with support for video sharing such social networks (Facebook, Instagram, etc.), messaging platforms (WhatsApp, Facebook Messenger, Viber, etc.), and video-streaming services (Netflix, YouTube, Amazon Prime Video, etc.).

In 2016, the data traffic related to mobile video on the internet accounted for 60% of total mobile data traffic [1]. CISCO forecasts [1] that more than three-fourths of the total mobile data traffic will be video by 2021, accounting for 78% of total mobile data traffic. This way, the increasing interest of the users for videos with improved image quality is collaborating for the advancement of handheld devices, as smartphones and tablets, and for the emerging of new high-quality and immersive video technologies. In this context, it is expected that these emerging video technologies are increasingly integrated with the handheld devices, which face severe constraints in terms of performance, storage, bandwidth, and, mainly, energy consumption since these devices are battery powered. In order to provide high-throughput and energy efficiency, dedicated very large scale integration (VLSI) design focusing on video coding is mandatory but becomes more challenging when considering handheld devices performing real-time encoding along with emerging video technologies.

Emerging tridimensional video technologies attract crescent attention from the entertainment industry, leading to the investment in three-dimensional (3D)-movies, living games, virtual reality, augmented reality, mixed reality, among other technologies. The popularization of handheld devices capable of dealing with 3D content is expected to be driven by emerging acquisition and displaying technologies like Google Glass [2], Microsoft HoloLens [3], Microsoft RealSense 3D [4], Structure Sensor [5], and Stereolabs ZED [6].

[1]Federal University of Pelotas (UFPel), Graduate Program in Computation (PPGC), Video Technology Research Group (ViTech), Pelotas, Brazil
[2]Federal University of Rio Grande do Sul (UFRGS), Institute of Informatics (INF), Graduate Program in Microelectronics (PGMicro), Porto Alegre, Brazil

Tridimensional videos require much more data to be represented than two-dimensional (2D) videos since 3D videos are constructed using multiple views, commonly captured from multiple cameras, which represent the same scene at the same time at different viewpoints. Thus, a set of videos are encoded and transmitted simultaneously resulting in a considerable increase in the computational effort of the 3D encoder and the bandwidth for 3D video transmission, when compared to 2D videos. Therefore, new approaches to surpass this drawback and enable real-time 3D-video processing have been investigated in the last years.

The first intuitive solution to encode tridimensional videos is a simulcast coding scheme [7], where the frames are predicted from frames belonging to the same view, i.e., each view is independently encoded. Although simulcast being less complex than other encoding models that exploit inter-view prediction, such a coding scheme is much less efficient regarding compression when compared with other approaches that are able to explore the similarities among neighbor views.

However, the multiple views of the same scene (multiview (MV)) allow applications like stereoscopic and autostereoscopic displaying, 3DTV, and free viewpoint television that may require a large set of views. Such applications become unfeasible with simulcast approach because using this scheme, the 3D encoded video information tend to increase proportionally to the number of views used in the scenario.

In order to efficiently exploit the coding process of tridimensional videos, the MV video coding (MVC) [8] was first standardized as an extension of the H.264/advanced video coding (AVC) [9]. The main idea of MVC is to exploit the inter-view redundancies among neighbor views reducing the encoded data. The MVC uses the H.264/AVC tools to encode these multiple videos and explore the information redundancies between neighbor views in a new process called inter-view prediction [8]. The main tool used in the inter-view prediction is the disparity estimation (DE), which is based on the motion estimation (ME) used to encode 2D videos [10]. For each block of the current frame in a dependent view (DV), the DE searches for a similar block in a neighbor views, i.e., high similarity between the current block and the blocks of the neighbor views frames at the same instant of time [10]. DE application increases the coding efficiency in the MV context, and this was the first tool included in video-coding standards to explore the specificities of 3D information. When compared to Simulcast approach using H.264/AVC, the MVC is able to reduce the encoded data volume in about 20%–50% [11].

Some works targeting the hardware design for MVC tools were published in the literature. The authors in [12] presented a ME and DE architecture for the MVC focusing on battery-powered devices in which the SW (Search Window) is dynamically expanded depending upon the search trajectories of the neighboring macroblocks. This way, on-chip and off-chip memory energy consumptions are reduced due to the off-chip memory accesses decrease and the data prefetching application. The work presented in [13] presents an MVC encoder hardware using a view-parallel interleaved scheduling to process full HD videos in real-time. The work [14] proposed a frame waiting-time aware scheduling and a hardware resource aware scheduling scheme for MVC. The first one reallocated the ME/DE operation to other ME/DE hardware in the same or different view aiming a more balanced encoding time in each view. The last one decides the schedule of each frame, considering the waiting time in on-chip static random access memory (SRAM). In [15] the authors proposed a new ME/DE scheme for MVC, where the scheme is composed of a fast ME/DE algorithm and a multi-level pipelined parallel hardware architecture. The proposed algorithm exploits the correlation available in 3D videos (i.e., spatial, temporal, and view) to avoid the search step of these modules. The hardware architecture used parallel ME/DE modules to improve the throughput and a shared array of the sum of absolute differences (SAD) to exploit the four levels of parallelism available: view, frame, reference frame, and block partitioning.

The high-efficiency video coding (HEVC) standard also defined an extension for MV tridimensional videos named as MV-HEVC [7] that follows a similar encoding structure of H.264/AVC MVC but using the new HEVC tools. Then, the MV-HEVC also is able to explore inter-view redundancies, increasing the coding efficiency. The MV-HEVC requires a higher computational effort when compared to MVC because the HEVC encoding tools are much more complex (and efficient) than the H.264/AVC encoding tools.

Only one work was found in the literature focused on hardware design targeting the MV-HEVC. The work [16] presents a hardware design for the MV-HEVC focusing on the real-time decoding of MV videos in mobile devices.

The state-of-the-art 3D video-coding standard uses the concept of MV plus depth (MVD) [17], where a depth map is associated with each texture frame. The depth map is a grayscale image that represents the distance between the scene objects and the camera where dark tones represent objects far from the camera and light tones are used for objects near the camera. The MVD is an important evolution for 3D videos representation and encoding. The use of MVD allows reducing the number of captured and encoded texture views, since the texture views together with its depth maps can be used at the decoder side to generate synthesized (or virtual) views, as will be explained in the next section.

The MVD motivated the definition of a second HEVC extension targeting 3D videos. This extension was called 3D-HEVC [7]. This extension was published in 2015 and is the state of the art for 3D video-coding standards. Since the 3D-HEVC must consider the depth maps together with texture, novel encoding tools were also defined to efficiently deal with this type of information. These tools will be detailed in the next section including the new encoding modes for intra-frame prediction (depth modeling mode (DMM-1 and DMM-4) and depth intra skip (DIS)) and an alternative residual flow (segment-wise DC coding (SDC)) [7]. All the original HEVC tools were also inherited and are available to be used in the 3D-HEVC.

The new coding tools introduced by 3D-HEVC to efficiently encode MVD content result in a higher complexity than HEVC. Therefore, the development of real-time 3D systems is a very challenging task, demanding dedicated energy-aware hardware solutions to enable real-time encoding, especially for handled devices. This way, the relevance of this research topic is evident due to the high 3D-HEVC complexity and its features.

Since the 3D-HEVC was recently released, there are a few published papers targeting efficient hardware designs for the new proposed encoding tools or specifically focusing on the 3D-HEVC challenges. However, since the HEVC tools are also required in the 3D-HEVC implementations, previously published works targeting the HEVC can also be used in 3D-HEVC implementations with some adaptations.

All prediction modes available at the HEVC intra-frame prediction are also used in the 3D-HEVC. Then, the hardware designed targeting the HEVC intra-frame prediction can be used in 3D-HEVC implementations. Then, works like [18–21] should be considered when designing dedicated hardware for a complete 3D-HEVC codec. These architecture can be adapted to include the new prediction modes defined by the 3D-HEVC, as will be detailed in the next section.

The hardware designed targeting the HEVC inter-frame prediction (ME and motion compensation) can be directly used in the 3D-HEVC inter-frame prediction. Since the inter-frame and inter-view predictions have very similar behavior, solutions designed targeting the HEVC inter-frame prediction can be replicated to work in the 3D-HEVC inter-view prediction with minor adaptations. Then, works [22–25] can be used to design a 3D-HEVC codec.

The hardware designed targeting other HEVC encoder modules, forward and inverse transforms, forward and inverse quantization, entropy coding, and in-loop filters, also can be directly reused in the 3D-HEVC context with minor adaptations, including the works [26–29]. The HEVC entropy coding, due to its natural control flow behavior, is harder to be parallelized, then designs of high-throughput hardware are not easy to be found in the literature, but works [30–32] can be used in this scenario. Finally, the published hardware designs targeting the HEVC in-loop filter also can be reused in 3D-HEVC implementations, including works [33–35].

Since the hardware designs targeting the HEVC are focused in other chapters of this book, this chapter will not bring additional information about these works. Moreover, such solutions do not feature optimizations considering the runtime 3D-HEVC behavior and processing/memory requirements. Therefore, the focus here will be those designs targeting the 3D-HEVC extension. Among the main works related with the 3D-HEVC, it is possible to highlight [36–43]. The work [36] presents a motion and DE system with memory hierarchy and some algorithmic simplifications to allow the real-time processing of high-resolution 3D-HEVC videos with low power consumption. Efficient and scalable hardware design for the new 3D-HEVC DMM-1 and DMM-4 intra-frame prediction tools is presented in [37]. The work [38] presents a low-power and high-throughput design targeting DIS, which is also a new intra-frame prediction tool defined by the 3D-HEVC. A DMM-1 decoder implemented in hardware and supporting the processing of high-resolution videos in real time is presented in [39]. The work [40] presents an architecture for the Simplified Edge Detector (SED) algorithm employed in 3D-HEVC depth maps intra-prediction. The work presented in [41] presents power efficient DE architecture with high-throughput targeting the 3D-HEVC. The work [42] presented a complexity and a hardware architecture analysis for the implementation of synthesized view-distortion estimation. Finally, a hardware architecture composed of the new tools for depth maps intra-prediction is presented in [43]. Three of these works were selected to be detailed in this chapter, considering the main contributions of the authors in this research field. These works cover the main innovations and challenges of the 3D-HEVC encoders.

The rest of this chapter is organized as follows. Section 6.1 introduces important concepts about the 3D-HEVC and highlights the main challenges related to 3D-HEVC real-time video coding designs. Section 6.2 presents details about a hardware design targeting the DMM-1 and DMM-4 tools. A hardware solution targeting the DIS encoding tool is presented in Section 6.3. Section 6.4 discusses a hardware solution targeting the 3D-HEVC inter-frame and inter-view predictions. Finally, Section 6.5 renders the conclusions of this chapter.

6.1 3D-high efficiency video coding: a 3D-video-coding standard supporting the multiview plus depth format

3D-HEVC is the state of the art for 3D-video encoding as it proposes novel-encoding tools along with an MVD encoding format. 3D-HEVC deals with each point of view through two channels, one related to the texture (regular video content) and another related to the depth (always associated with the texture for the same view). Although texture pictures are the ones showed to the spectators, usually composed of pixels obtained from three color matrices of samples, the depth maps are used to represent the distance between the objects of a scene and the camera through only one matrix of samples. In depth maps, the objects near to the camera are represented by light gray tones and the objects far from the camera are represented by dark gray tones. Figure 6.1 shows an example of a texture picture (here also in grayscale) of the *Shark 3D*-video [44] sequence and its correspondent depth map.

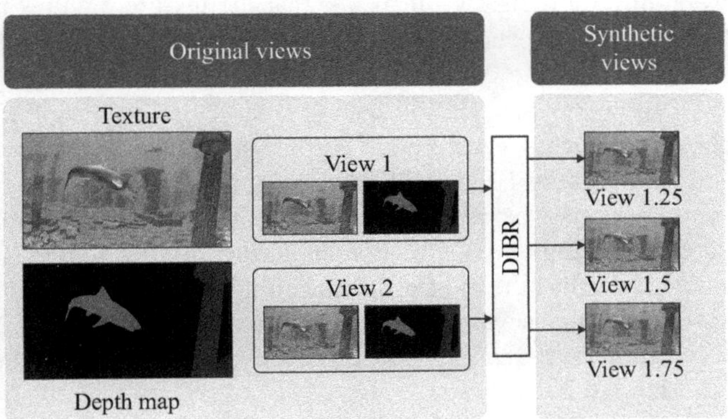

Figure 6.1 Example of texture pictures and their respective depth maps considering a system with two original and three synthetic views

By adopting the MVD format, the 3D-HEVC enables the generation of intermediate synthetic views through a view-synthesis process called depth-image-based rendering (DIBR) [17] at the receiver side. This way, the encoder transmits a reduced number of original texture pictures and their respective depth maps, and the receiver generates synthetic texture pictures at different viewpoints using the DIBR process. With MVD data format, as many synthesized views as needed can be generated at the decoder side. Figure 6.1 shows an example of a system with two original views (two textures plus two depth maps) and three synthesized views.

As it is possible to observe in Figure 6.1, depth maps present two main characteristics: (i) the presence of sharp edges and (ii) the presence of large homogeneous regions [7]. To efficiently deal with these characteristics, the 3D-HEVC implements new encoding tools for the depth-map encoding when compared with the ones used in the HEVC standard, since these aspects have an important role in the view synthesis process. It is important to notice that an accentuated distortion in depth maps can result in poor image quality of synthetic views. Therefore, these new tools introduced in 3D-HEVC are very important in the coding process, and they result in additional complexity and energy requirements when compared with a 2D approach.

3D-HEVC encoding structure is based on access units (AUs), where each AU contains all the pictures (textures+depth maps) referred to the same time instant. Each AU contains one base view (BV) and one or more DVs. BV does not depend on other views as reference and DVs are views that depend on other views as a reference in the encoding process. In the conventional 3D-HEVC coding process, the texture frames are encoded before depth maps for all encoded views. However, 3D-HEVC also allows the flexible coding order (FCO) [45], where texture frames are encoded before depth maps in BVs, and depth maps are encoded before texture frames in the DVs. Thus, the FCO approach enables the 3D-HEVC to take different advantage of the encoding order aiming to improve the encoding performance since depth map information can be used to accelerate texture frames encoding.

Both non-FCO and FCO approaches can be found in the literature. By default, the 3D-HEVC follows a non-FCO. It means that the texture pictures are encoded before the depth maps independently of which the view is being encoded [7].

Besides, similarly to the 2D video approaches, 3D-HEVC employs a block-based hybrid video-coding scheme following the structure presented in Figure 6.2. Before the encoding process, each picture is split into square-shaped blocks (called coding tree units (CTUs) in HEVC). By default, the CTUs have 64×64 pixel in 3D-HEVC, and they may be recursively partitioned into smaller square-shaped blocks named CUs (coding units). CUs may be further divided into PUs (prediction units) to evaluate and select the most efficient coding tool (following rate-distortion (RD) criteria) among the several tools available for each block. Among these evaluated tools, there are the tools in intra-prediction and inter-prediction steps, as can be seen in Figure 6.2.

The main innovations of the 3D-HEVC considering the intra-frame prediction are in the tools specially developed to deal with depth maps, as the DMM modes and the DIS [7]. The original HEVC intra-coding tools, as Planar, DC, and angular

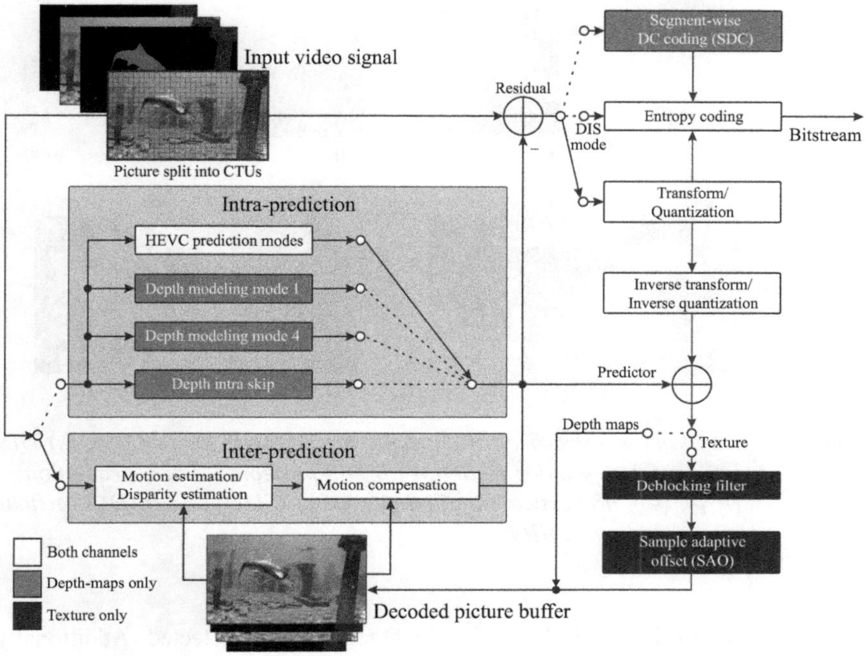

Figure 6.2 Simplified 3D-HEVC encoding model

modes [46], are applied to texture pictures and the depth maps. Regarding the inter-prediction, the inter-view redundancy [7] (using DE) is exploited along with the well-known ME used in 2D approaches for the inter-frame prediction [46].

The DMM modes (DMM-1 and DMM-4) were included in the 3D-HEVC intra-frame prediction intending to increase the coding efficiency of depth maps by preserving the sharp edges during the encoding process at the cost of a high computational effort and high use of storage in memory [7]. DMMs are applied for square-blocks with size ranging from 4×4 to 32×32 samples. These modes can be considered as two additional intra-frame prediction modes when comparing with the original HEVC since the original HEVC intra-coding tools (Planar, DC, and angular modes [46]) are also available to be applied to texture pictures and depth maps.

Figure 6.3 illustrates the 3D-HEVC depth maps intra-prediction tools DMM-1, DMM-4 and DIS. DMM-1 implements a partitioning strategy to efficiently represent edges information. Each block is divided by a wedgelet partition, which separates the encoding block into two regions using a straight line, as demonstrated in Figure 6.3(a). Each region is represented by the average value of each region, known as constant partition value (CPV) [7]. To obtain the best wedgelet partition, DMM-1 performs a search process with a refinement step. This process is applied over a predefined set of wedgelet partitions and the wedgelet with the lowest

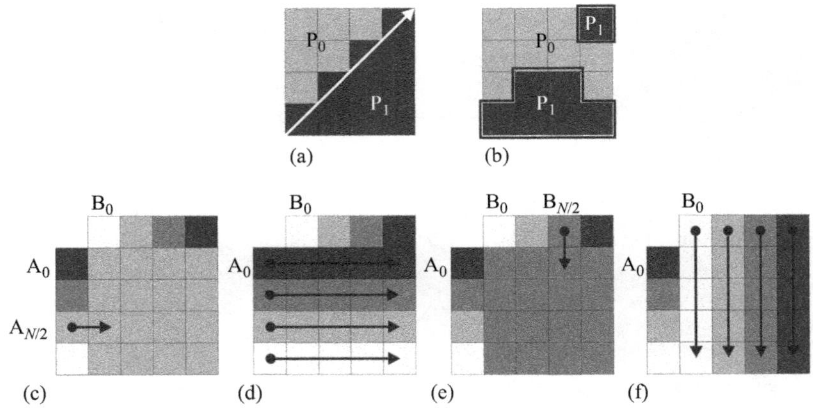

Figure 6.3 Depth maps intra-prediction tools: (a) DMM-1, (b) DMM-4, (c) DIS horizontal single depth (SD_H), (d) DIS horizontal intra–prediction (IP_H), (e) DIS vertical single depth value (SD_V), and (f) DIS vertical intra–prediction (IP_V)

distortion between the original and the predicted signal is selected. Additionally, the refinement step is performed and a subset of eight wedgelets is evaluated, which are neighbors of the wedgelet selected in the previous step. Finally, the wedgelet with minimum distortion obtained in the refinement step is added to the RD list and evaluated in terms of RD-cost.

Figure 6.3(b) presents the DMM-4 encoding tool, which also divides the block to be predicted into two regions and uses a CPV to represent each region. The DMM-4 uses the collocated texture block to calculate a threshold value and estimate the depth block partitioning. The DMM-4 threshold value is obtained from the average of four corner reference samples of the collocated texture block associated with the depth-map block [47]. Furthermore, differently from DMM-1, discontinuous regions may belong to each region, as can be seen in Figure 6.3(b).

DIS encoding tool was introduced to deal with the large homogeneous regions presented in the depth maps, avoiding the encoding process of the residues for these regions. DIS is applied at the level of CUs (i.e., square-blocks with size ranging from 8×8 to 64×64), whereas the other intra-prediction encoding tools are generally applied at the level of PUs [7]. Basically, the DIS-encoding tool implements four prediction modes: (i) horizontal intra-prediction (IP_H), (ii) horizontal single depth (SD_H), (iii) vertical intra-prediction (IP_V), and (iv) vertical single depth (SD_V) [38]. These modes are exemplified in Figure 6.3(c)–(f), where IP_H and IP_V predict the block copying the reference samples into horizontal and vertical directions for entire depth block, respectively, whereas SD_H and SD_V copy the reference sample in middle of the left column (vertical direction) and in the middle of the above column (horizontal direction) for the entire depth block, respectively.

3D-HEVC also introduces SDC tool for depth maps as an alternative to the conventional residual coding with transform/quantization (TQ) [46] used in HEVC. SDC can be applied in intra-prediction and inter-prediction in order to achieve further bitrate gains in smooth regions. Instead of using the residual data of depth maps samples, SDC considers only a single value (known as DC value) to represent the residue information for HEVC prediction modes and for inter-prediction, and two DC values for DMM-1 and DMM-4 (one for each region). The DC value for HEVC intra-prediction is obtained from the average of corner samples in the depth map block, and for DMMs, the CPV value is used. More details of DC calculation can be consulted in [48].

In the inter-frame context, the ME and the DE exploit the temporal and inter-view redundancies, respectively, and they also take a prominent position in 3D-HEVC due to their high efficiency regarding compression results and intense processing and memory communication [36]. Figure 6.4 presents an example of inter-frame and inter-view predictions, where 3D-HEVC uses the ME process to explore the temporal redundancies by exploiting the similarities presented at different time instants within the same view using a block matching algorithm (BMA) that compares the blocks being encoded with the ones from reference frames. Likewise, 3D-HEVC uses the DE process to exploit the inter-view redundancies, i.e., it also uses a BMA (commonly the same used in ME process) to compare the blocks being encoded with reference blocks, but in this case, the frames used as reference belong to the same time instant of neighboring views.

After the prediction steps, the 3D-HEVC coding process is similar to the HEVC, i.e., the residues pass by TQ steps before following to the Entropy Coding. Also, after the TQ steps, the decoding process (composed of inverse transform and inverse quantization steps, deblocking filter (DF) and the sample adaptive offset (SAO) filter) are performed at the encoder side in order to obtain a reconstructed picture to be stored in the Decoded Picture Buffer and used as reference to encode the next pictures. Whereas most prediction modes can apply the residues to the TQ

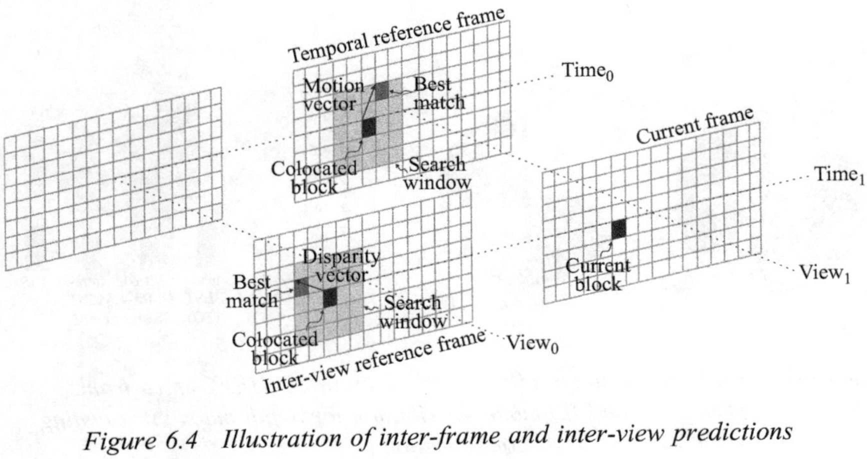

Figure 6.4 Illustration of inter-frame and inter-view predictions

steps (or adopted an SDC in the case of depth maps [49]), DIS mode avoids these steps in the depth-map coding, as presented in Figure 6.2. Moreover, depth maps encoding flow does not perform DF and SAO to preserve edge information, as demonstrated in Figure 6.2.

The next sections present memory, processing, and complexity-related challenges posed by the 3D-HEVC encoding tools that hinder the development of 3D-HEVC real-time video-coding architecture.

6.1.1 Memory, processing, and complexity challenges on 3D-HEVC intra-prediction

Figure 6.5 presents an analysis of 3D-HEVC intra-prediction regarding the time profiling of 3D-HEVC intra-coding and the mode distribution for depth maps intra-coding. These results were obtained through evaluations using the 3D-HTM reference software version 16.0 [47] and respecting the common test conditions (CTC) [44] for 3D-HEVC under all-intra (AI) encoder configuration. It is important to emphasize that, when running in AI configuration, only intra-frame prediction tools are available to encode texture and depth maps.

Analyzing the time profiling in the left of Figure 6.5(a), it is possible to note that the texture coding represents only 13.7% of the encoder computational effort, whereas depth maps coding represents 86.3% when considering the AI scenario. Detailed profiling of depth coding time is also presented in Figure 6.5(a), where DMM-1 is the most time-consuming encoding tool being responsible for 27.2% of the depth maps intra-computational effort, followed by residual encoding tools with 26.6% and 25.3% for TQ and SDC, respectively. The "intra-HEVC" considers the evaluation of the prediction modes inherited from HEVC texture coding and represents 12.1% of the time spent in depth maps intra-coding. The remaining encoding modes together, DMM-4 and DIS, represent less than 9%. Even though these modes do not represent high encoding time when compared to other tools, they can provide significant improvements in 3D-HEVC encoding efficiency [49].

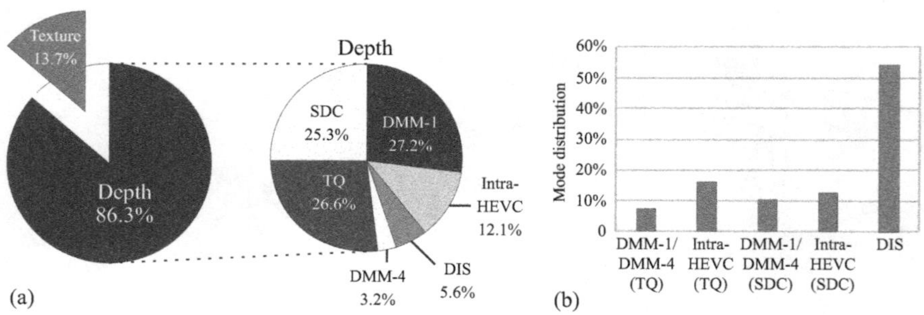

Figure 6.5 (a) Computational effort distribution in 3D-HEVC intra-frame prediction and (b) mode distribution for depth maps intra-coding when running in AI configuration

Figure 6.5(b) shows a second experiment where the encoding flows were evaluated together with the mode selected to encode the depth maps. Then, the DMMs and the intra-HEVC were evaluated when processed by TQ and by SDC residual encoding flows. The DIS was also evaluated and, as explained before, it does not use the residual encoding flows (TQ or SDC), since no residues are transmitted to reduce the bitrate of the encoded video. One can notice that DIS is selected as the best encoding mode for more than 50% of encoded blocks, while any of the remaining flows reached 20% of the encoded blocks. These results can be explained since depth maps are composed of large homogenous regions and the DIS typically fits better with these kinds of regions; however, depth maps also present edge and gradient regions, where DMMs and intra-HEVC are crucial to provide high encoding efficiency. The results showed a similar mode-selection distribution considering DMMs and intra-HEVC encoding flows.

Therefore, according to the data presented in Figure 6.5, one can conclude that depth maps intra-coding is a highly costly computational task for the 3D-HEVC encoder. The highest computational effort is mainly concentrated in DMM-1 and residual encoding flows. In order to reduce this computational effort, some works proposed to avoid the entire DMM-1 evaluation [50] and decrease the number of modes evaluated in residual encoding flow [51].

However, when hardware architecture able to achieve real-time processing with low power dissipation are considered, new challenges can be noticed. For instance, in the residual coding flow, limited parallelism can be exploited since these steps share data with entropy encoding that works with a sequential approach. Besides, in DMM-1, the number of wedgelets to be evaluated depends on the block size being predicted, as presented in Table 6.1, and these wedgelets must be stored in memory to be evaluated.

One can notice in Table 6.1, the large number of wedgelets patterns allowed for both DMM-1 encoding and decoding process. This storage of all wedgelets patterns is undesirable, mainly considering 3D-HEVC systems implemented in dedicated hardware. DMM-1 requires 183,264 bit to store all wedgelets patterns of all block sizes supported by the encoding tool, according to the 3D-HTM implementation [47] (i.e., squared-block sizes ranging from 4×4 to 32×32). Also, the DMM-1 algorithm requires continuous access to the wedgelets patterns on the memory and intensive processing, which demands high energy consumption. Thus, to obtain performance and energy-consumption compatible with real-time

Table 6.1 Number of possible wedgelets in DMM-1

Block size	**Total no. of possible wedgelets**
4×4	86
8×8	802
16×16	510
32×32	510

processing of 3D-HEVC high definition videos, it is necessary to reduce memory size, memory accesses, and processing effort related to DMM-1 calculations.

6.1.2 Memory, processing, and complexity challenges on 3D-HEVC inter-frame and inter-view predictions

ME and DE are among the costliest steps of video encoders regarding energy and memory issues [36] while exploiting temporal and disparity redundancies, respectively. ME compares the current block (the block being encoded) with reference blocks of previously processed frames to find the most similar one. ME considers reference frames that belong to the same view but are at different video-scene time instants. This process is done under a similarity criterion following a search pattern determined by a BMA typically constrained within an SW over the reference frames. DE consists in a process very similar to the ME. However, DE compares the current block with reference blocks that belong to different views but represent the same time instant. Considering the similar behavior of ME and DE in the inter-frame and inter-view predictions, these two encoding tools will be analyzed together in this section.

The test zone search (TZS) BMA algorithm and the SAD similarity criterion are among the most widely used approaches in the latest encoder implementations, besides appearing as the BMA and similarity criterion adopted in the 3D-HTM by default [47]. Both ME and DE encoding tools are evaluated in the encoder at the level of PU blocks, and the 3D-HEVC allows ME/DE to be performed with 24 block sizes, ranging from 4×8 up to 64×64 pixel in order to improve coding efficiency. In the case that a full RD optimization process [47] is adopted, i.e., all possible combinations considering the 24 PU sizes are evaluated in the encoding process, the decision on which PU size must be used occurs after the complete evaluation of the possibilities through the encoding loop showed in Figure 6.2. In other words, a costly evaluation process is required to achieve the best trade-off between compression and distortion. This way, ME/DE demands a huge number of block-matching operations leading to intense processing, memory communication, and, consequently, high energy consumption.

Memory access is one of the main components regarding energy consumption and performance considering ME/DE hardware designs due to the number of candidate blocks to be compared by the BMA. Data-reuse schemes as the Level-C [52] have been used in dedicated ME and ME/DE solutions to reduce the memory bandwidth. The Level-C scheme allows the reuse of overlapping regions between neighboring SWs since neighboring blocks share much of the SW. This way, these data may be kept in the on-chip memory avoiding external memory retransmission.

The work [36] presents an evaluation of the external memory traffic considering 3D-HEVC encoding of three-view (three texture pictures plus their respective depth maps) HD 1,080p MVD videos using hardware-oriented TZS (HOTZS) static scheduling and horizontal disparity search (HDS) fast block-matching algorithms (this work is better detailed in Section 6.4). In this work, search patterns, much less complex than the original TZS algorithm, were adopted, and these algorithms considered only 32 × 32 and 16 × 16 block sizes. When no on-chip SRAM is employed, i.e., the ME/DE processing unit communicates directly to the external memory, the demanded communication for real-time encoding reaches 92 GB/s, as demonstrated in Figure 6.6(a). Given that current memory technologies, as the embedded low-power memories, provide no more than 25.6 GB/s in ideal cases [53], this target performance is unfeasible. Considering the use of an on-chip SRAM to prefetch and store the SW used for the block-matching (360 kB), the communication is reduced to 9.3 GB/s. When the Level-C scheme is adopted in addition to the on-chip SRAM memory, the memory bandwidth drops to 3.1 GB/s, leading to a total communication reduction of 96%, as depicted in Figure 6.6(a). Even so, the energy consumption related to the external memory communication, and on-chip SRAM dynamic/static consumption is still high, and it hinders the implementation of 3D real-time systems, mainly focusing on handheld devices.

Regarding memory readings and writings, the work [36] estimated the number of DE/ME memory operations considering the 3D-HEVC encoding of HD 1,080p MVD videos with the 24 possible PU sizes. The number of encoded views estimated by [36] was defined in order to attempt a system capable of delivering nine views after the decoding process. 3D-HEVC requires 1.13×10^9 memory writings and 361×10^9 memory readings per second to encode three 1,080p views (texture plus depth maps).

Computational effort analyses considering 3D-HEVC ME/DE steps are incipient in the literature. Actually, some works as [54] evaluated ME computational effort considering a 2D approach with the HEVC Reference Software (HM—HEVC Test Model) [55] using the CTC for HEVC standard (2D videos) [56], but the presented results cannot be extrapolated to the ME/DE behavior in a 3D-HEVC

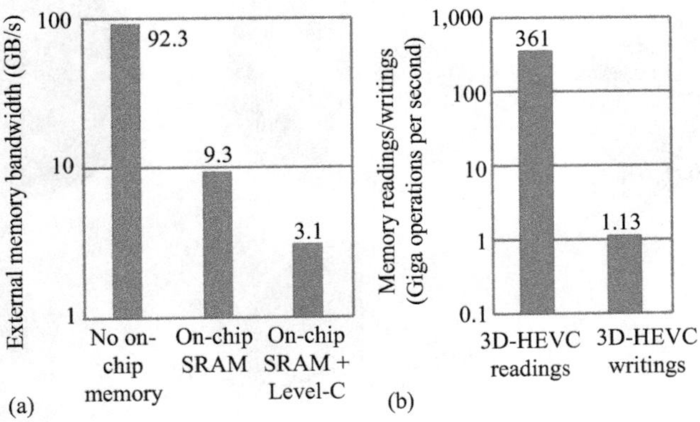

Figure 6.6 (a) ME/DE external memory bandwidth for three memory organizations (without on-chip memory, with on-chip SRAM, and with on-chip SRAM using the Level-C data reuse scheme) and (b) ME/DE memory reading and writing for 3D-HEVC

context along with MVD format. Therefore, this chapter presents an evaluation of the computational effort related to the ME and DE steps as well as the relation among ME, DE, and intra-frame prediction and the other tools used in the 3D-HTM. The experiments considered the random-access (RA) configuration [56], and two of the video sequences (*Balloons* and *Undo_Dancer*) defined in CTC [44] with two QP values (30 and 39). It is important to emphasize that, when considering the RA configuration, inter-frame prediction, inter-view prediction, and intra-frame prediction tools are available to encode texture and depth maps.

As one can notice in Figure 6.7, the time spent with each one of the channels (texture and depth maps) is similar considering an encoder configuration that uses the ME/DE steps, so that the texture used 51.1% of the encoding time, whereas depth maps used 48.9%.

Regarding the texture channel, a significant portion of the computational effort is due to the ME/DE steps, where ME is responsible for 40.95% of the spent time, DE uses 10.58% of encoding effort, and the other encoding tools use the remaining 48.47% of the time, as depicted in Figure 6.7. Considering the depth maps, the portion of computational effort due to the ME/DE is much smaller, where ME uses 6.33% of the time, DE uses 2.29% of the time, and the other encoding tools use 91.38% of the encoding time. This behavior is expected, since intra-prediction tools are available in the RA configuration and the depth maps tends to be more efficiently encoded when using the novel intra-frame prediction tools designed for this type of information, as discussed in the previous section.

Figure 6.7 also presents the total encoding time related to ME/DE and intra-frame prediction. Note that since ME/DE steps increase the complexity of the

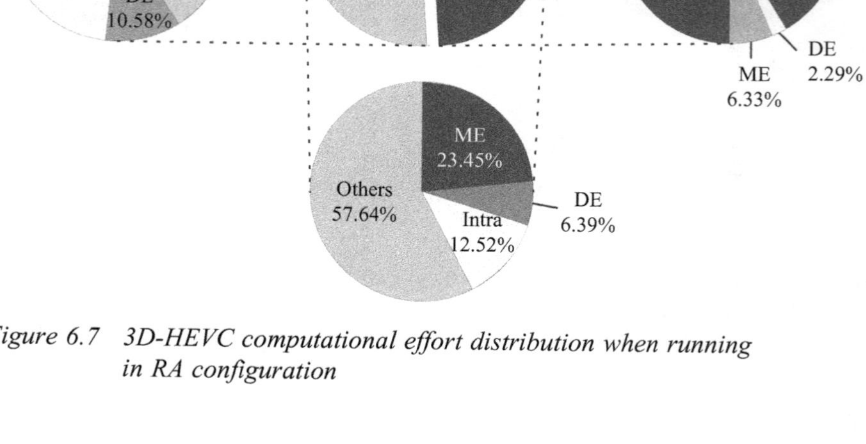

Figure 6.7 3D-HEVC computational effort distribution when running in RA configuration

texture encoding and the novel 3D-HEVC intra-encoding tools increases the complexity of the depth-map encoding, the time spent with predictions are significant considering all the encoding process. ME/DE predictions spent 29.84% of the time, where intra-frame prediction spent 12.52% of the time. The remained encoding tools spent 57.64% of the encoding time.

The memory and computational effort results presented and discussed in this section for both the novel depth 3D-HEVC intra-prediction tools and the 3D-HEVC ME/DE steps evidenced the need for the development of complexity-reduction strategies and VLSI designs focusing on both 3D-HEVC intra-prediction and inter-prediction. This way, the next sections present and discuss prominent works that cover this important research topic.

6.2 3D-HEVC real-time architecture for DMM-1 and DMM-4 intra-frame prediction modes

The work [37] proposed real-time scalable hardware architecture for the bipartition modes (DMM-1 and DMM-4) of the 3D-HEVC. For this purpose, a simplification in DMM-1 algorithm was performed to take better advantage of the hardware resources and to reduce the memory required for storing wedgelets patterns. Besides, the proposed architecture is capable of dealing with all block sizes supported in the 3D-HEVC DMMs (i.e., squared sizes from 4×4 to 32×32) and to reach different throughputs according to the application requirements.

6.2.1 DMM-1 algorithm simplification

The proposed architecture for the bipartition modes implements a simplified version of the DMM-1 algorithm by removing its refinement process while maintaining the DMM-4 algorithm unaltered. An analysis using the 3D-HTM 16.0 reference software and the CTC was performed to evaluate the impact of the DMM-1 algorithm without the refinement process. The obtained results for RA and AI encoder temporal configurations are presented in Table 6.2, where the Bjøntegaard difference (BD)-rate [57] indicates the impact on the synthesized views when removing the refinement process of the DMM-1 algorithm. These results demonstrate that the removal of the DMM-1 refinement process causes BD-rate increase of 0.09% and 0.25%, on average, for RA and AI encoder configurations, respectively.

Additionally, Table 6.3 presents the required memory bits when all wedgelets should be stored in a memory inside the architecture and when the wedgelets necessary for the DMM-1 refinement process are removed. The obtained results showed that when removing the refinement step, it is possible to achieve memory savings higher than 30% (on average). Thus, this work demonstrates that the removal of DMM-1 refinement process in a hardware design is capable of providing an interesting trade-off among hardware resources, encoding efficiency, and performance (since less memory access is performed) due to the large memory resources that should be spent to store all available refinement wedgelets and the impact on the encoding efficiency (demonstrated in Table 6.2).

Table 6.2 BD-rate results when removing the refinement process in the DMM-1 algorithm

Video	BD-rate (%)	
	Random access	All-intra
Balloons	0.16	0.21
Kendo	0.00	0.24
Newspaper_CC	0.19	0.40
GT_Fly	0.02	0.20
Poznan_Hall2	0.03	0.23
Poznan_Street	0.10	0.09
Undo_Dancer	0.08	0.25
Shark	0.15	0.41
Average	0.09	0.25

Table 6.3 Memory reduction when removing the refinement process in the DMM-1 algorithm

Block size	Memory (bit)		Memory reduction (%)
	All wedgelets	No refinement	
4 × 4	1,376	928	32.6
8 × 8	51,328	20,096	60.8
16 × 16	130,560	98,304	24.7
32 × 32	130,560	98,304	24.7
Total	313,824	217,632	30.6

6.2.2 Bipartition modes architecture

The Bipartition Core (BP-Core) was defined as the basic structure of the architecture developed by [37], which is replicated according to the encoding block size. The designed architecture was developed aiming to allow two levels of scalability: (i) different block sizes and (ii) different levels of throughput.

Each BP-Core receives a pixel value as input that carries the texture or depth information of the pixel. For the DMM-1 algorithm, the three steps are divided into prediction step, SAD step, and residue step. The Prediction step computes the average depth value of each region, SAD step computes the SAD of each wedgelet pattern, and Residue step computes the subtraction between the predicted samples and the original samples. The three stages of the DMM-4 algorithm are as follows: texture average stage, prediction stage, and residue stage. The texture average stage is responsible for computing the average value of texture block, the prediction stage subtracts the average value obtained in the previous stage from the stored texture pixel value, and the residue stage performs the subtraction between the predicted samples and the original samples. The designed architecture shares the modules of DMM-1 and DMM-4 to allow better resource usage.

Figure 6.8 illustrates a high-level block diagram of the proposed architecture for the bipartition modes of the 3D-HEVC. This scalable architecture is composed of (i) $K \times N \times N$ BP-Cores array; (ii) two memories for storing the wedgelets patterns of the DMM-1 algorithm (*patterns memory*) and the division values (*divider memory*); (iii) add trees for each $N \times N$ array; (iv) divider circuits for each $N \times N$ array, (v) K comparator circuits; and (vi) K register banks. This structure scales to all 3D-HEVC block sizes (i.e., 4×4, 8×8, 16×16, and 32×32) expanding the $N \times N$ BP-Core array. Additionally, this structure also enables to adapt the data throughput by varying the K factor.

6.2.3 Synthesis results

The designed architecture was described in VHDL and synthesized for application specific integrated circuit (ASIC) 65 nm standard cells technology. All synthesis results aim to achieve real-time processing with HD 1,080p@30fps considering one view.

Table 6.4 presents the synthesis results targeting different block sizes and using $K = 1$. As one can notice, the proposed architecture in this work is capable

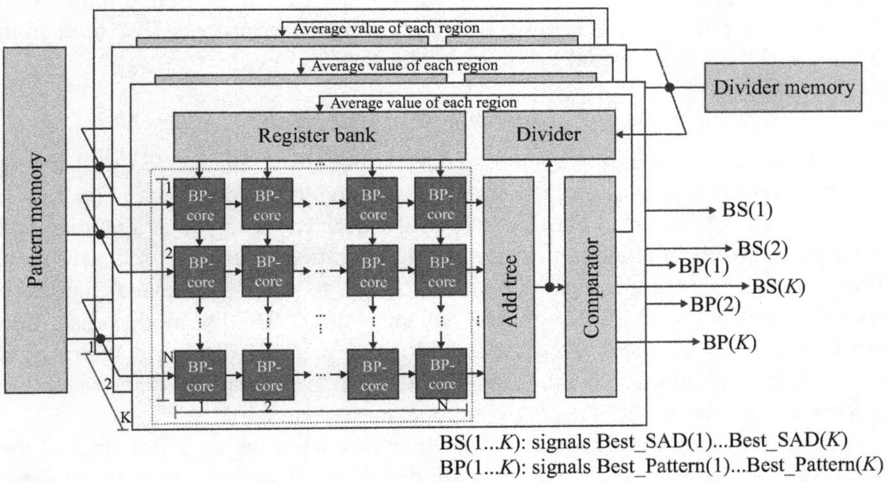

Figure 6.8 High-level block diagram of bipartition modes architecture

Table 6.4 Synthesis results targeting ST 65 nm standard cells

Block size	Area (gates)	Cycles per block	Frequency (MHz)	Latency (cycles)	Processing rate HD 1,080p (fps)	Power (mW)
4 × 4	4,070	134	515.5	134	30.1	22.2
8 × 8	15,749	650	632.9	650	30.0	90.9
16 × 16	49,628	818	198.8	818	30.0	133.1
32 × 32	211,950	878	53.2	878	30.0	166.5

of achieving real-time processing with HD 1,080p@30fps for all evaluated configurations. Moreover, it is possible to note that the gates and power do not grow linearly according to the block size increase. This occurs because the number of available wedgelets patterns and the required cycles for each block size also does not grow linearly, and consequently, the hardware resources keep the same behavior.

Besides, the results do not consider the other scalability axis that can be exploited in the designed architecture, whereas in 3D video coding, it is necessary to encode many views, the architecture allows increasing its throughput by a factor K, which multiplies by K the architecture performance.

6.3 3D-HEVC real-time architecture for DIS intra-frame prediction mode

The work [38] proposed a low-power and high-throughput hardware design for the DIS. A strategy was adopted to reduce the number of arithmetic operations and to avoid a costly process in the similarity criterion. This strategy can reduce the hardware resources and the power dissipation at the cost of negligible impacts on the encoding efficiency. The proposed architecture can process five ultra high definition (UHD) 2,160p@60fps views in real time.

6.3.1 Depth intra skip algorithm simplification

The DIS algorithm uses the synthesized view distortion change (SVDC) as the distortion metric to decide the best mode to encode a given depth map block [47]. However, the SVDC is a complex metric that uses rendering functionalities and requires a considerable amount of arithmetic operations, such as multiplications. Thus, the hardware implementation of the SVDC tends to use several hardware resources and the work [38] proposed replacing the SVDC by another distortion metric that allows a hardware-friendly design. The SAD distortion metric was chosen to be applied to the depth maps samples, which does not apply the rendering process and performs simpler arithmetic operations than SVDC.

Table 6.5 demonstrates the total arithmetic operations as a function of the block size for the SAD and SVDC distortion metrics, given an $N \times N$ block. From

Table 6.5 Arithmetic operations for an $N \times N$ block using SAD and SVDC

Distortion metric	**Arithmetic operations**			**Total**
	Subtractions	**Sums**	**Multiplications**	
SAD	N^2	$N^2 - N$	–	$2N^2 - N$
SVDC	$3N^2$	$2N^2 - N$	$2N^2$	$7N^2 - N$
Δ(SVDC-SAD)	$2N^2$	$N^2 - N$	$2N^2$	$5N^2 - N$

Table 6.6 BD-rate results using SAD as the similarity criterion

Video	Synthesized views BD-rate (%)
Balloons	0.23
Kendo	0.34
Newspaper_CC	0.26
GT_Fly	0.20
Poznan_Hall2	0.29
Poznan_Street	0.14
Undo_Dancer	0.02
Shark	0.21
Average	0.21

this analysis, one can conclude that the use of SAD metric is possible to reduce a significant amount of arithmetic operations and avoid the costly rendering process.

Besides, the trade-off between encoding efficiency and computational effort must be carefully considered, and the encoding efficiency impact of replacing the distortion metric was evaluated in 3D-HTM 16.0 using the CTC under AI encoder configuration. Table 6.6 presents the synthesized views BD-rate for each CTC video sequence when using the SAD distortion metric in DIS algorithm. According to the results, small losses in 3D-HEVC encoding efficiency are noticed, increasing the BD-rate of the synthesized views by 0.21%, on average.

In this way, the DIS algorithm simplification proposed in this work is capable of decreasing the computational effort while providing negligible encoding losses. Then, it was possible to conclude the SAD similarity criterion is friendlier than the SDVC for hardware design.

6.3.2 Depth intra skip architecture

The designed architecture was developed to support all block sizes of the 3D-HEVC DIS encoding tool (i.e., 8 × 8, 16 × 16, 32 × 32, and 64 × 64) in a parallel way. For that, DIS modules are responsible for processing and delivering the best SADs and the best DIS modes, one for each CU size, as demonstrated in Figure 6.9(a). In order to reach a high parallelism level, the proposed architecture works with a bottom-up approach, where the DIS modules processing of 16 × 16, 32 × 32, and 64 × 64 samples are composed by the processing of 8 × 8 blocks. This approach allows the reuse of input samples and partial SAD calculations, reducing the external-memory communication. Thus, the bottom-up processing strategy results in an efficient hardware design by allowing a better trade-off among hardware cost, power dissipation, and throughput.

Figure 6.9(b) illustrates a generic diagram of the DIS module. The main difference among the four DIS modules are in the control signals of each module, the I/O bit depth of partial SADs, the internal bit depth from the accumulators, and the

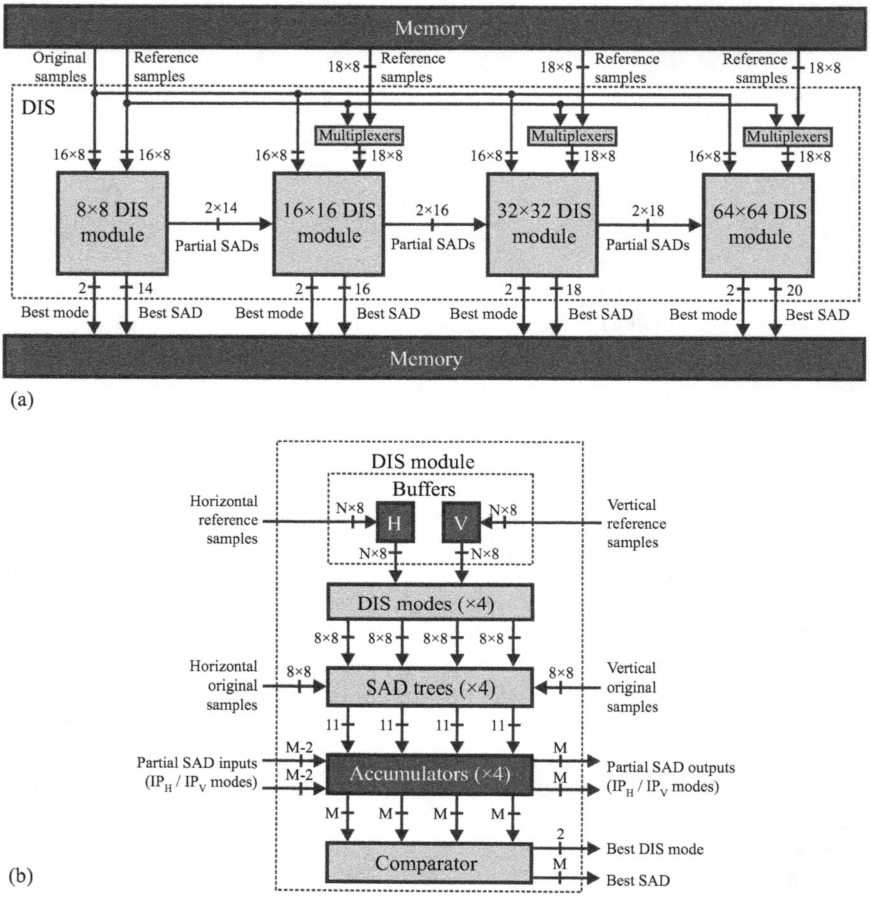

Figure 6.9 (a) High-level diagram of DIS hardware design and (b) generic diagram of the DIS modules

comparator. For instance, 8 × 8 blocks do not apply data reuse, and then, this module has no partial SAD inputs. However, it provides partial SADs and reference samples for the DIS modules that process larger blocks.

The data reuse scheme of larger block sizes by processing 8 × 8 blocks is demonstrated in Figure 6.10. The reuse of SAD values is possible for the IP_H and IP_V modes (see Section 6.1), while the reuse of input samples is possible for the IP_H mode. In Figure 6.10, light-gray-shaded squares represent the reuse of SAD values, while dark-gray-shaded squares represent the reuse of input samples, and white squares represent the blocks where no data reuse is possible. It is possible to note the reuse of SAD values reduces about 50% the arithmetic operations needed to process 16 × 16, 32 × 32, and 64 × 64 block sizes, when the IP_H and IP_V modes are considered.

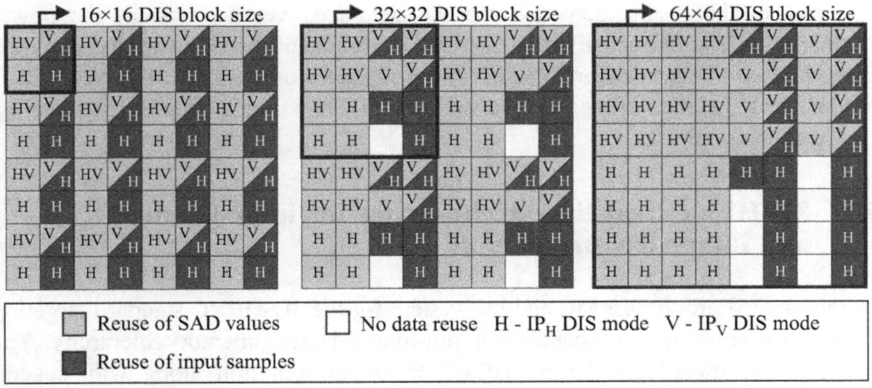

Figure 6.10 Data reuse scheme for blocks (a) 16 × 16, (b) 32 × 32, and (c) 64 × 64

Table 6.7 Synthesis results for ST 45 nm standard cells

Architecture		**Non-optimized**	**Optimized**	**Reduction (%)**
Total area (gates)		34.80k	35.37k	−1.64
1,080p@120fps	Frequency (MHz)	252.72	252.72	–
(five views)	Power (mW)	14.95	11.52	22.94
2,160@60fps	Frequency (MHz)	505.44	505.44	–
(five views)	Power (mW)	24.64	19.57	20.58

Although Figure 6.10 demonstrated the data reuse based on 8 × 8 blocks, the reuse of input samples in the hardware architecture is based on 8 × 8 blocks while the reuse of SAD values is based on the immediately smaller DIS module.

6.3.3 Synthesis results

The designed architecture was described in VHDL, and the synthesis results were generated targeting an ASIC 45 nm NanGate standard cells. Two versions of the proposed hardware architecture for the 3D-HEVC DIS encoding tool were developed. Table 6.7 presents the obtained results regarding the area, frequency, and power dissipation for these two versions, called *non-optimized* and *optimized*. The *non-optimized* version does not implement the data reuse scheme (SAD values and input samples) and low-power techniques, while the *optimized* version employs the data reuse scheme for SAD and input samples and implements the clock-gating technique.

The maximum frequency obtained in both versions of the proposed hardware design is superior to 550 MHz and it is enough to process five UHD 2,160p@60fps views. As the two versions present the same parallelism, both versions are able to reach real-time processing with the same operation frequency for HD and UHD resolutions. The results demonstrated the *optimized* version reduces the power dissipation in 20.58% considering five UHD 2,160p@60fps views when compared to the *non-optimized* version. The hardware resource usage of the two versions is similar, and the data reuse scheme and the clock-gating technique increase by 1.64% of the total area.

6.4 3D-HEVC real-time architecture for inter-frame and inter-view predictions

The work [36] proposes a ME and DE system for the 3D-HEVC standard targeting low-energy consumption based on a run-time adaptive memory hierarchy. The designed system adopts the 3D-HEVC FCO configuration along with several optimizations to deal with inter-channel and inter-view redundancies aiming at reducing the computational effort required by the ME/DE system. The memory hierarchy scheme employed window-based prefetching, data reuse, subsampling, and dynamic voltage scaling (DVS) in order to reduce the on-chip memory energy. The next subsections detail the strategies presented by [36] focusing on the ME/DE 3D-HEVC processes.

Due to the large number of memory accesses and the high-energy consumption related to the ME/DE processes in 3D-HEVC, the work [36] defined a set of hardware-oriented constraints and proposes new hardware-friendly algorithms focusing on complexity and memory communication reductions before proposing its energy-aware system itself. These solutions include (i) a simplification in the used block sizes (only 16×16 and 32×32 blocks are supported); (ii) a simplification in the ME BMA using TZS without predictors and raster; (iii) the definition of HOTZS static scheduling to provide a more regular memory access pattern for the processing units; (iv) the definition of HDS algorithm as an alternative to the TZS algorithm for the DE process to efficiently explore the DE behavior; (v) the use of SAD instead of sum of absolute transformed differences [47] in fractional motion estimation (FME) step to encode texture frames; (vi) the support to bidirectional prediction [47] was disabled; (vii) the search was limited to one reference frame per direction (both in ME and DE); (viii) depth-map subsampling (from 8 bit down to 4 bit); (ix) the use of texture-based motion vector inheritance (TMVI) to explore the correlation between the texture picture and the depth map for the BV; (x) the definition of depth-based dynamic SW resizing (DSWR) that dynamically reduces the SW used in the texture ME and DE processes using the depth information; and (xi) a sector-level DVS in the memories.

All these hardware constraints and new algorithms caused a BD-rate increase of 25.63% for synthesized views while the encoding computational effort was reduced by 59% when the simplifications were implemented in the 3D-HTM. Although a high impact on BD-rate is noticed, the solution is aligned with video coding for mobile systems, where lower resolutions are considered. However, the high impact in BD-rate indicates that other strategies must be considered in future works to allow a low-power and high-throughput ME/DE system.

6.4.1 The energy-aware motion and disparity estimation system

An overview of the energy-aware ME/DE system developed by [36] can be observed in Figure 6.11. As one can notice, the system is composed of three main hardware units, one responsible for processing the BV and two for the DVs. The BV is the first view to be encoded to maintain compatibility with 2D-HEVC-based systems (it does not depend on other views). The other views are DVs, i.e., the encoding process of those views depends on the information provided from other views. The BV unit has one ME processing unit per channel (texture and depth), whereas each dependent-view unit has one ME and one DE processing unit per channel. Also, each processing unit is fed from a dedicated on-chip SRAM memory.

Figure 6.11 summarizes the techniques applied to each processing unit and memory in the white balloons. Given the BV should process the texture channel before the depth channel in order to maintain the HEVC compatibility, the Texture unit was used to provide information related to the ME result to the depth-map unit and completely skip the search process in the depth maps under specific situations. In the DVs, the inter-channel correlation is explored, adapting the SW of the texture searches for both disparity and motion displacements from the depth-channel information (enhanced by the FCO configuration). The arrows in Figure 6.11 represent auxiliary information flow used by the proposed algorithms.

The developed system was implemented in a four-stage macro-pipeline at the frame level, and also in a four-stage macro-pipeline at the CTU level, as depicted in Figure 6.12. At the frame level, the first stage processes texture for the BV and the second processes depth for the BV. After, the third stage processes ME and DE for depth maps of all DVs (in parallel). Finally, the fourth stage processes ME and DE for texture pictures of all DVs. At the CTU level, each CTU contains four 32×32 PUs and sixteen 16×16 PUs to be processed (according to the used hardware constraints, as explained before). This way, each 64×64 CTU is divided into four 32×32 PUs, which are sequentially processed. Along with each 32×32 CU, the four 16×16 CUs located inside it are processed simultaneously in order to allow data reuse from the memories. At the beginning of the process, each 32×32 block and both its respective upper 16×16 blocks begin to be processed in parallel. Immediately after, the bottom 16×16 blocks start to be processed in parallel, as detailed in Figure 6.12.

As the number of cycles spent in each macro-pipeline stage at frame level or CTU level varies according to data characteristics, clock gating was used to reduce the energy consumption. The architecture includes dedicated processing units capable of processing 16 lines in parallel, where each line is composed of 16 or 32 samples, depending on the block size. In other words, 16 candidate blocks are evaluated simultaneously. This parallelism was defined based on a statistical analysis of the number of candidate blocks evaluated during the BMA execution from the 3D-HTM implementation considering the proposed hardware constraints. Then, 240 candidates were defined as the target performance (average-case design

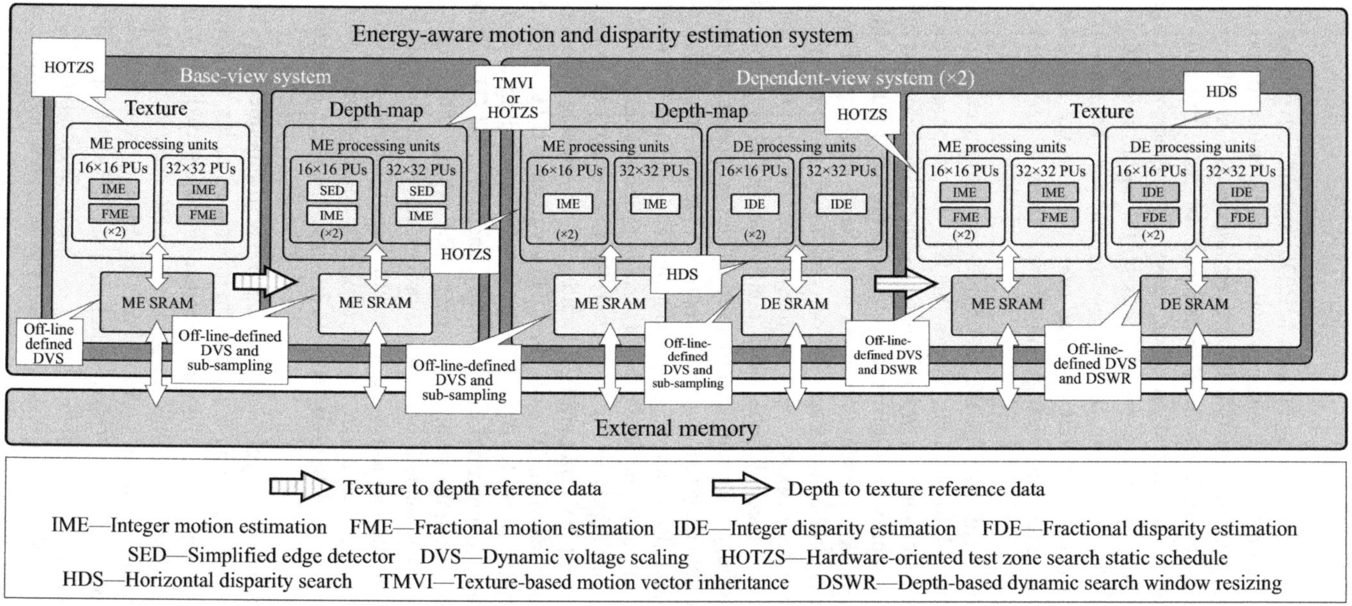

Figure 6.11 Energy-aware ME/DE system

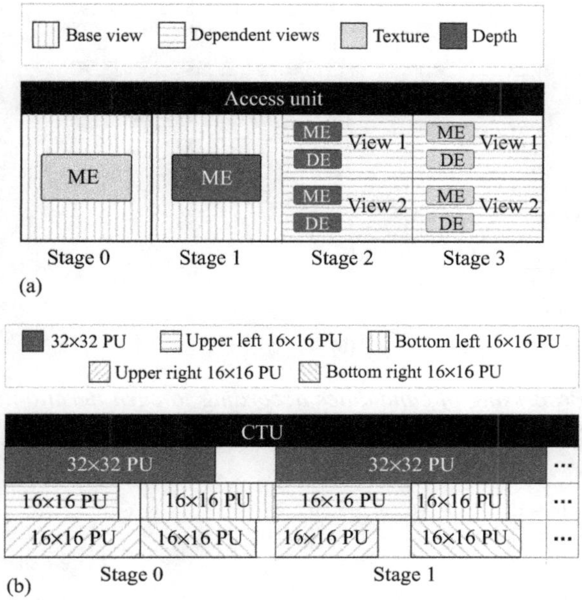

Figure 6.12 Macro-pipelines (a) at the frame level and (b) at the CTU level

strategy). Fractional ME/DE steps are applied to the best candidate found by the integer ME/DE in order to improve the coding efficiency of texture pictures. The next subsection presents the hardware-oriented BMA and schedule proposed in [36].

6.4.2 Proposed hardware-oriented algorithms

The hardware-oriented algorithms proposed in [36] and previously cited are presented in this section. The first algorithm is the HOTZS static scheduling that was developed in order to provide a more regular memory access pattern for the processing units. As observed in Figure 6.13, the TZS is composed of multiple iterations that comprise a crescent number of candidate blocks per iteration, while the HOTZS rearranges the processing schedule to process 16 candidates per iteration.

The second hardware-oriented algorithm proposed in [36] is the HDS, presented in Figure 6.14. HDS explore the typical horizontal camera disposition used to capture different views in a 3D video. Then the horizontal displacements tend to be much more relevant than vertical displacements. HDS is composed of three steps, as an alternative to the TZS algorithm for the DE process. As one can notice in Figure 6.14, the first two steps consist of horizontal searches, and the third step is a refinement step. The start point of the first step is centered on the collocated block position. Finally, the third step performs a 2D simple 1-sample search around the best matching obtained from the second step. As a benefit of the HDS algorithm, the DE SW is reduced (from 192 $\times$ 192 to 192 $\times$ 96, considering an search range (SR) = 64).

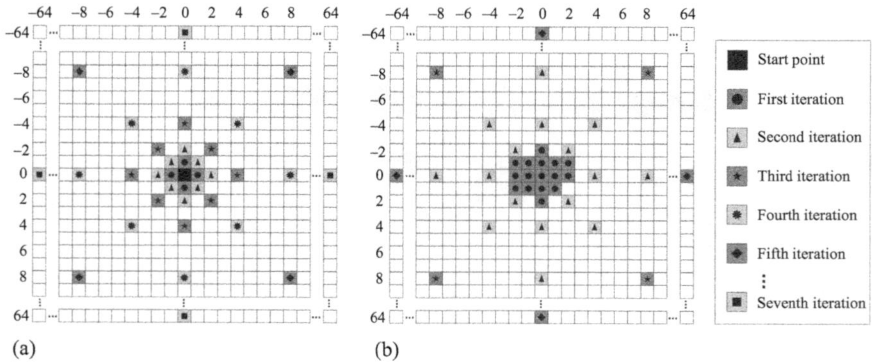

Figure 6.13 Processing of candidates according to each iteration: (a) TZS and (b) proposed HOTZS

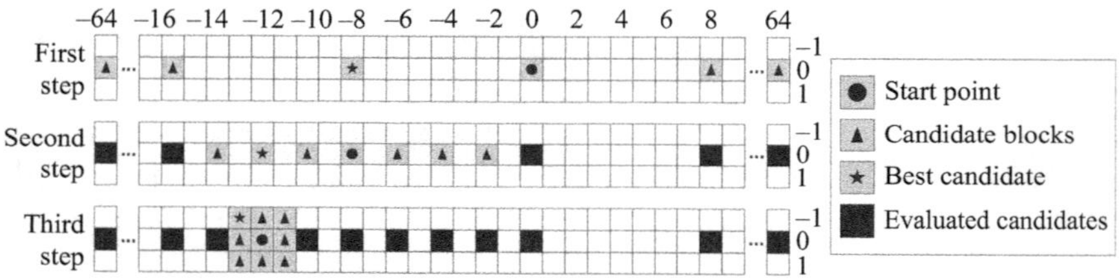

Figure 6.14 Proposed horizontal disparity search algorithm

The third hardware-oriented algorithm proposed in [36] is the TMVI which intends to explore the correlation between the texture picture and the depth maps for the BV. Once different motion vectors can lead to good matching in depth-map homogeneous regions, the motion vector selected during the texture encoding is inherited of the collocated block for the depth block in the case of a homogeneous region. Basically, the proposed TMVI calculates the SED metric [58] to define if the depth block under processing is a sharp-edge or homogeneous region. If the depth block was considered a homogeneous region, the motion vector from the texture channel is used. If a sharp-edge was detected, the HOTZS is normally applied to the depth map ME. When the same motion vector is reused, a complete search is skipped, and the memory associated with the depth ME at the BV may be scaled to idle mode.

Finally, the fourth hardware-oriented proposed algorithm proposed in [36] is the DSWR that dynamically reduces the SW used in the texture ME and DE processes using the depth information. Basically, the DSWR takes into account the block/object depth within the scene to reduce the SW, in contrary to typical implementations that consider a fixed SW along the whole ME/DE process. As shown in Figure 6.15, the proposed DSWR scales the active SW (shaded area in

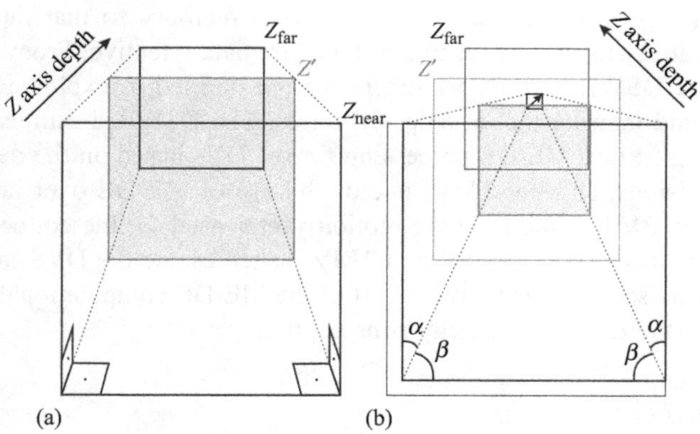

Figure 6.15 Active search window: (a) typical ME/DE implementations and (b) Depth-based dynamic search window resizing algorithm

Figure 6.15) according to a perspective projection. Basically, the algorithm adopts a large SW as larger are the motion vectors, the closer the object is. In conclusion, DSWR considers the disparity/motion vectors selected during the depth-maps ME/DE to displace the active SW by applying a projection distortion. In Figure 6.15, the Z_{near} and Z_{far} correspond to the depth-sample values of the closest and farthest objects in the scene, respectively. Z' corresponds to the current block depth value to be considered for the DSWR.

6.4.3 The on-chip memory design and management based on channel aspects

The energy-aware 3D-HEVC ME/DE system developed by [36] employs four different types of on-chip memories using a similar memory organization for both ME and DE memories and both texture and depth channels. Basically, as the ME on-chip memories cover an SW of 192 × 192 samples, i.e., ME SW is larger than the DE SW (192 × 66 samples), the first are composed of more sectors, columns, and rows. The other difference among memories is that the sectors for depth-map memories are smaller than texture memories due to the depth map subsampling technique also proposed in [36]. By using depth-map subsampling, it was possible to reduce in 50% the depth memories sizes.

The bit-depth subsampling for depth maps (subsampling from 8 bit down to 4 bit) was evaluated using four subsampling patterns in order to reduce the size of on-chip memories and exploit the no complex-textured regions of the depth-map content. Based on the reached results, the four less significant bits were removed at the cost of a drop in the BD-rate, as previously discussed.

The work [36] also implemented a sector-level DVS in the memories with the goal to reduce leakage energy on idle sectors. A power-state matrix was used to perform the power-state management through sleep circuitries (SCs) distributed in each specific position related to a sector. Each SC is responsible for actuating on the voltage (V_{dd}) of a specific sector of a given memory so that three different power states are switched, where two of them are data retentive. From these power states, the use DSWR and TMVI schemes were defined to exploit inter-channel correlations and manage the on-chip memories. The DSWR dynamically reduces the SW during texture ME/DE processing for the DVs based on the depth ME/DE encoding decisions, and the TMVI avoids the use of ME/DE over homogeneous regions for the BV by inheriting the motion vector used for the collocated texture block. In conclusion, both DSWR and TMVI schemes use the DVS to disable the memory sectors so necessary, saving part of the ME/DE computational effort, and, consequently, the respective energy consumption.

6.4.4 Hardware results

The work [36] presented the reached results separately according to the adaptive memories scenarios and processing units. To estimate the energy consumption associated with memory implementation for leakage, reading, writing, and power-state transition, the NVSim [59] simulator was used along with the 45 nm CMOS technology power models [60] focusing on SRAM memories. Also, a memory simulator capable of emulating the behavior of the on-chip memories was developed in C# [61], targeting to collect the numbers for each memory sector: active-mode cycles, sleep-mode cycles, and transitions; and the total number of readings and writings for each memory line. All memory requests to be used in the C# simulator were extracted using the 3D-HTM reference software during the ME/DE processing.

Thus, from the trace files and the C# simulator, different memory scenarios were evaluated and a scenario using a memory hierarchy with run-time adaptive management was selected, since it reached the lowest power dissipation (1.10 W), which represents a reduction of 56.5% in power dissipation when compared with other solutions.

The power dissipation related to the processing units was obtained from synthesis results using the Cadence RTL Compiler 11.10 tool [62] and NanGate 45 nm standard cells [63]. The synthesis results were presented for each design considering ME/DE tools and texture/depth channels, as shown in Table 6.8.

Table 6.8 Hardware design synthesis results

Architecture	**Area (μm²)**	**Instances**	**No. of gates (× 10⁶)**	**Power (W)**		
				Leakage	**Dynamic**	**Total**
ME—texture	8,542,735	3	6.817	0.147	0.984	1.131
ME—depth map	2,797,576	3	2.232	0.047	0.422	0.469
DE—texture	6,388,479	2	5.098	0.123	0.509	0.631
DE—depth map	2,220,003	2	1.772	0.041	0.152	0.193
Total	51,237,897	–	40.888	0.908	5.539	6.447

The texture-dedicated architecture used more on-chip area compared to depth maps ones since the last has a smaller internal bit depth. The full hardware design related to the processing units has 51.2 mm^2 on-chip area (40.9M gates, considering normalization w.r.t. NAND-2X1 gate) while presenting a total dissipation of 6.45 W@100 MHz.

The complete energy consumption per AU at the system level indicates that the implementation presents an average consumption of only 0.107 J per AU. Therefore, at the system level, the proposed energy-aware motion and DE system reduced the total energy consumption by 71.1% when all the proposed techniques are considered.

6.5 Conclusions

The previous generation of 3D videos was not appealing enough to justify the support for them in the handheld devices and to maintain this support on other electronic devices, like TVs. The huge amount of data to be transmitted, and limited view synthesis performance resulted in visual discomfort and poor-user experience [64], which culminated on a momentary stop in the manufacture of those devices [65]. In last years, contrasting with this scenario, emerging 3D-video-related technologies covering 3D-movies, virtual reality, augmented reality, mixed reality, and other technologies, have gained place again in the consumer market, especially in handheld devices. These technologies are driven by devices like Microsoft RealSense 3D [4], Structure Sensor [5], and Stereolabs ZED [6]. Most of these devices had focused on a texture plus depth approach, which is the most efficient way to represent/encode 3D-videos nowadays [7]. In this context, 3D-HEVC was published in 2015 adopting the MVD format and novel encoding tools to efficiently deal with MV videos captured by multiple cameras and improved view-synthesis performance.

Some works focusing on VLSI designs and the real-time processing of 3D videos can be found in the literature, as discussed in this chapter. Despite the 3D-HEVC being the most efficient way available to encode 3D-videos, video coding in a scenario with multiple views remains to be challenging even considering the 3D-HEVC and dedicated VLSI designs. Then, a few works are available in the literature relating hardware designs targeting specifically the 3D-HEVC tools. Three works were selected to be detailed in this chapter, two of them focusing on the new 3D-HEVC intra-frame prediction tools (DMMs and DIS) and one focusing on the 3D-HEVC inter-frame and inter-view predictions. These three architecture are capable of processing at least HD 1,080p@30fps videos in real time. For that, these works developed solutions with different approaches to achieve high throughput with the minimum drawback in area usage and power dissipation. Table 6.9 presents a summary of the results reached by these three works.

The architecture for the bipartition modes [37] uses a strategy to reduce the memory requirements and shares the resources in the execution of DMM-1 and

Table 6.9 Summary of the presented real-time 3D-HEVC architecture

Work	Sanchez [37]	Afonso [38]	Afonso [36]
Encoding tools	DMM-1 and DMM-4	DIS	ME and DE
Technology	VLSI/65 nm	VLSI/NanGate 45 nm	VLSI/NanGate 45 nm
Maximum views for HD 1,080p@30fps	1*	40	3
Gates	211,950**	35,370	40.888M
Frequency (MHz)	53.2**	505.44	100
Total power dissipation	166.5 mW**	19.57 mW	7.55 W
BD-rate increase (%)	0.25	0.21	25.63
Main contributions	Parallelism exploiting from scalable architecture	Change of similarity metric, reuse of samples, and SAD calculations	Hardware-constraints, hardware-friendly algorithms, inter-channel exploration

*The number of views is scalable by adding hardware modules.
**Results consider only the hardware module to process the 32 × 32 block size.

DMM-4 to achieve smaller power dissipation and higher throughput. The results demonstrated an area usage of 211,950 gates with a power dissipation of 166.5 mW at 53.2 MHz considering the 32 × 32 block size, as presented in Table 6.9.

The designed architecture for DIS encoding tool also uses techniques to reduce the power dissipation, such as reuse of information and clock-gating. The architecture results demonstrated an area usage of 35,370 gates with a power consumption of 19.57 mW at an operating frequency of 505.44 MHz, as presented in Table 6.9.

The real-time architecture targeting the ME and DE of 3D-HEVC presented in [36] is the only VLSI design in the literature focusing on the 3D-HEVC modules. This architecture explores the memory/processing/energy issues in a context with texture plus depth format and multiple views. The energy-aware ME/DE system proposed by [36] dissipates 7.55 W while processing three HD 1,080p views (three texture views and their associated depth maps) at 30 fps in real time. The impact in the compression was evaluated using the 3D-HEVC CTC and pointed to a total increase of 25.63% in BD-rate.

The works presented in Table 6.9 have a very distinct level of complexity, where the architecture presented in [37,38] target the new intra-frame prediction encoding modes for depth maps, the work presented in [36] targets a complete ME/DE system to handle with texture and depth-maps and including the memory management. These differences naturally impact in the reached results, where work [36] used a much higher area and consumes much more energy than the works [37,38], even with the low power strategies used in [36].

Although the works presented in [36–43] and other works in the literature contribute to the development of 3D-HEVC real-time systems, a vast space for new researches and future works considering both algorithm simplification and hardware architecture design to achieve high-throughput, low-power dissipation, and low impact on encoding efficiency still can be exploited. As previously presented, 3D-HEVC inherited many modules from HEVC (transform, quantization, filters, entropy coding, etc.) and the existing solutions in the literature can be used in 3D-HEVC implementations. However, 3D video-coding applications process more data leading to higher performance need that require new dedicated hardware designs.

Besides, we can highlight MVD is an interesting and promising approach to deal with 3D videos, and the next generation of 3D video coding tends to adopt it. With new 2D video-coding standards being developed, such as versatile video coding (VVC) [66], a proposal of a 3D extension with MVD format is expected. Since VVC and its 3D extension tend to support new encoding tools and larger block sizes (i.e., new intra-prediction modes, new interpolation filters, larger CTU block sizes, larger transform block sizes, etc.), it is expected that a series of new challenges will emerge during the upcoming years.

Acknowledgements

The authors thank their institutions in Brazil for supporting the development of this work, including the Federal University of Pelotas (UFPel), the Federal University of Rio Grande do Sul (UFRGS) and the Sul-rio-grandense Federal Institute (IFSul). The authors also thank the Brazilian research support agencies that financed this research: the Foundation for Research Support of Rio Grande do Sul (FAPERGS), the National Council for Scientific and Technological Development (CNPq) and the Coordination for the Improvement of Higher Education Personnel (CAPES).

References

[1] Cisco, Cisco Visual Networking Index: Forecast and Methodology, 2016–2021. Retrieved Nov. 9, 2018: https://www.cisco.com/c/en/us/solutions/collateral/service-provider/visual-networking-index-vni/complete-white-paper-c11-481360.html.

[2] Google Glass Technology. Retrieved Nov. 9, 2018: https://googleglassesbus237.weebly.com/.

[3] Microsoft HoloLens Technology. Retrieved Nov. 9, 2018: https://www.microsoft.com/en-us/hololens.

[4] Intel RealSense Technology. Retrieved Nov. 9, 2018: https://www.intel.com/content/www/us/en/architecture-and-technology/realsense-overview.html.

[5] Structure Inc. Sensor. Retrieved Nov. 9, 2018: https://structure.io/.

[6] Stereolabs ZED. Retrieved Nov. 9, 2018: https://www.stereolabs.com/zed/.

[7] G. Tech, Y. Chen, K. Muller, J. Ohm, A. Vetro, and Y. Wang, "Overview of the Multiview and 3D Extensions of High Efficiency Video Coding," *Transactions on Circuits and Systems for Video Technology (TCSVT)*, v. 26, n. 1, pp. 35–49, 2016.
[8] Joint Video Team (JVT) of ISO/IEC MPEG & ITU-T VCEG, "Joint Draft 8.0 on multiview video coding," JVT-AB204, 2008.
[9] International Telecommunication Union, ITU-T Recommendation H.264: Advanced Video Coding for Generic Audiovisual Services, Aug. 2017.
[10] G. Sullivan, J. Boyce, Y. Chen, J. Ohm, C. Segall, and A. Vetro, "Standardized Extensions of High Efficiency Video Coding (HEVC)," *Journal of Selected Topics in Signal Processing*, v. 7, n. 6, pp. 1001–1016, 2013.
[11] B. Zatt, M. Shafique, S. Bampi, and J. Henkel, 3D Video Coding for Embedded Devices: Energy Efficient Algorithms and Architectures. Springer Science, New York, NY, 2013.
[12] B. Zatt, M. Shafique, F. Sampaio, L. Agostini, S. Bampi, and J. Henkel, "Run-time adaptive energy-aware motion and disparity estimation in multi-view video coding," *in 2011 48th ACM/EDAC/IEEE Design Automation Conf. (DAC)*, New York, NY, USA, Jun. 5–9, 2011, pp. 1026–1031.
[13] L. Ding, W. Chen, P. Tsung, *et al.*, "A 212 MPixels/s 4096 2160p Multiview Video Encoder Chip for 3D/Quad Full HDTV Applications," *IEEE Journal of Solid-State Circuits*, v. 45, n. 1, pp. 46–58, 2010.
[14] M. Choi, I. Chang, and J. Kim, "High Performance and Hardware Efficient Multiview Video Coding Frame Scheduling Algorithms and Architectures," *Transactions on Circuits and Systems for Video Technology (TCSVT)*, v. 23, n. 8, pp. 1312–1321, 2013.
[15] B. Zatt, M. Shafique, S. Bampi, and J. Henkel, "Multi-level pipelined parallel hardware architecture for high throughput motion and disparity estimation in multiview video coding," in Design, Automation & Test in Europe Conference, Grenoble, France, 2011. *Design, Automation & Test*, Grenoble, France, 2011.
[16] W. Liu, J. Li, and Y. B. Cho, "A Novel Architecture for Parallel Multi-view HEVC Decoder on Mobile Device," *EURASIP Journal on Image and Video Processing*, 2017.
[17] P. Kauff, N. Atzpadin, C. Fehn, *et al.*, "Depth Map Creation and Image-based Rendering for Advanced 3DTV Services Providing Interoperability and Scalability," *Signal Processing: Image Communication (SPIC)*, v. 22, n. 2, pp. 217–234, 2007.
[18] B. Min, Z. Xu, and R. Cheung, "A Fully Pipelined Hardware Architecture for Intra Prediction of HEVC," *Transactions on Circuits and Systems for Video Technology (TCSVT)*, v. 27, n. 12, pp. 2702–2713, 2017.
[19] M. Correa, B. Zatt, M. Porto, and L. Agostini, "High-throughput HEVC intrapicture prediction hardware design targeting UHD 8K videos," *in International Symposium on Circuits and Systems (ISCAS)*, Baltimore, MD, USA, May 2017.

[20] D. Palomino, F. Sampaio, L. Agostini, S. Bampi, and S. Susin, "A memory aware multiplierless VLSI architecture for the complete intra prediction of the HEVC emerging standard," *in International Conference on Image Processing (ICIP)*, Orlando, FL, USA, Oct. 2012.

[21] G. Pastuszak, and A. Abramowski, "Algorithm and Architecture Design of the H.265/HEVC Intra Encoder," *Transactions on Circuits and Systems for Video Technology (TCSVT)*, v. 26, n. 1, pp. 210–222, 2016.

[22] M. Perleberg, V. Afonso, R. Conceição, *et al.*, "Energy and Rate-Aware Design for HEVC Motion Estimation Based on Pareto Efficiency," *Journal of Integrated Circuits and Systems (JICS)*, v. 13, pp. 1–12, 2018.

[23] G. Sanchez, B. Zatt, M. Porto, and L. Agostini, "Hardware-friendly HEVC Motion Estimation: New Algorithms and Efficient VLSI Designs Targeting High Definition Videos," *Analog Integrated Circuits and Signal Processing*, v. 82, n. 1, pp. 135–146, 2015.

[24] E. Kalali, and I. Hamzaoglu, "Approximate HEVC Fractional Interpolation Filters and Their Hardware Implementations," *Transactions on Consumers Electronics (TCE)*, v. 64, n. 3, pp. 285–291, 2018.

[25] Y. Fan, L. Huang, B. Hao, and X. Zeng, "A Hardware-Oriented IME Algorithm for HEVC and Its Hardware Implementation," *Transactions on Circuits and Systems for Video Technology (TCSVT)*, v. 28, n. 8, pp. 2048–2057, 2018.

[26] E. Kalali, E. Ozcan, O. Yalcinkaya, and I. Hamzaoglu, "A Low Energy HEVC Inverse Transform Hardware," *Transactions on Consumer Electronics (TCE)*, v. 60, n. 4, pp. 754–761, 2014.

[27] B. Lee, and M. Kim, "A CU-Level Rate and Distortion Estimation Scheme for RDO of Hardware-Friendly HEVC Encoders Using Low-Complexity Integer DCTs," *Transactions on Image Processing (TIP)*, v. 25, n. 8, pp. 3787–3800, 2016.

[28] L. Braatz, B. Zatt, D. Palomino, L. Agostini, and M. Porto, "High-throughput and low-power integrated direct/inverse HEVC quantization hardware design," *in International Symposium on Circuits and Systems (ISCAS)*, Florence, Italy, May 2018.

[29] J. Goebel, G. Paim, L. Agostini, B. Zatt, and M. Porto, "An HEVC multi-size DCT hardware with constant throughput and supporting heterogenous CUs," *in International Symposium on Circuits and Systems (ISCAS)*, Montreal, Canada, 2016.

[30] F. Ramos, J. Goebel, B. Zatt, M. Porto, and S. Bampi, "Low-power hardware design for the HEVC binary arithmetic encoder targeting 8K videos," *in Symposium on Integrated Circuits and Systems Design (SBCCI)*, Belo Horizonte, Brazil, Sep. 2016.

[31] D. Kim, J. Moon, and S. Lee, "Hardware implementation of HEVC CABAC encoder," *in International SoC Design Conference (ISOCC)*, Gyeongju, South Korea, 2015.

[32] Y. Zhang, and C. Lu, "A Highly-Parallel Hardware Architecture of Table-Based CABAC Bit Rate Estimator in HEVC intra encoder," *Transactions*

on Circuits and Systems for Video Technology (TCSVT), v. 29, n. 5, pp. 1544–1558, 2019.
[33] W. Shen, Y. Fan, Y. Bai, *et al.*, "A Combined Deblocking Filter and SAO Hardware Architecture for HEVC," *Transactions on Multimedia (TMM)*, v. 18, n. 6, pp. 1022–1033, 2016.
[34] S. Cho, H. Kim, H. Y. Kim, and M. Kim, "Efficient In-Loop Filtering Across Tile Boundaries for Multi-Core HEVC Hardware Decoders With 4 K/8 K-UHD Video Applications," *Transactions on Multimedia (TMM)*, v. 17, n. 6, pp. 778–791, 2015.
[35] W. Zhou, J. Zhang, X. Zhou, Z. Liu, and X. Liu, "A High-Throughput and Multi-Parallel VLSI Architecture for HEVC Deblocking Filter," *Transactions on Multimedia (TMM)*, v. 18, n. 6, pp. 1034–1047, 2016.
[36] V. Afonso, R. Conceição, M. Saldanha, *et al.*, "Energy-Aware Motion and Disparity Estimation System for 3D-HEVC with Run-Time Adaptive Memory Hierarchy," *Transactions on Circuits and Systems for Video Technology (TCSVT)*, v. 29, n. 6, pp. 1878–1892, 2019.
[37] G. Sanchez, C. Marcon, and L. Agostini, "Real-Time Scalable Hardware Architecture for 3D-HEVC Bipartition Modes," *Journal of Real-Time Image Processing (JRTIP)*, v. 13, n. 1, pp. 71–83, 2017.
[38] V. Afonso, A. Susin, L. Audibert, *et al.*, "Low-power and high-throughput hardware design for the 3D-HEVC depth intra skip," in *IEEE International Symposium on Circuits and Systems (ISCAS)*, Baltimore, MD, USA, 2017.
[39] G. Sanchez, C. Marcon, and L. Agostini, "High efficient architecture for 3D-HEVC DMM-1 decoder targeting 1080p videos," *in International Symposium on Circuits and Systems (ISCAS)*, Florence, Italy, May 2018.
[40] G. Sanchez, M. Saldanha, M. Porto, B. Zatt, and L. Agostini, "Real-time simplified edge detector architecture for 3D-HEVC depth maps coding," *in International Conference on Electronics, Circuits and Systems (ICECS)*, Monte Carlo, Monaco, Feb. 2017.
[41] M. Perleberg, V. Afonso, R. Conceição, *et al.*, "A power-efficient and high-throughput hardware design for 3D-HEVC disparity estimation," *in Symposium on Integrated Circuits and Systems Design (SBCCI)*, Bento Gonçalves, Brazil, Nov. 2018.
[42] W. Ahmad, M. Martina, and G. Masera, "Complexity and Implementation Analysis of synthesized view distortion estimation architecture in 3D high efficiency video coding," *in International Conference on 3D Imaging (IC3D)*, Liege, Belgium, 2015.
[43] F. Amish, and E. Bourennane, "An Efficient Hardware Solution for 3D-HEVC Intra-prediction," *Journal of Real-Time Image Processing (JRTIP)*, pp. 1–13, 2017.
[44] K. Muller, and A. Vetro, "Common test conditions of 3DV core experiments," ISO/IEC JTC1/SC29/WG11 MPEG2011/N12745, Jan. 2014.
[45] S. Gopalakrishna, M. M. Hannuksela, and M. Gabbouj, "Flexible coding order for 3D video extension of H.265/HEVC," *in Picture Coding Symposium (PCS)*, 2013.

[46] G. Sullivan, J. Ohm, W. Han, and T. Wiegand, "Overview of the High Efficiency Video Coding (HEVC) Standard," *Transactions on Circuits and Systems for Video Technology (TCSVT)*, v. 22, n. 12, pp. 1649–1668, 2012.

[47] 3D-HEVC Test Model. Retrieved Nov. 2018: https://hevc.hhi.fraunhofer.de/svn/svn_3DVCSoftware/tags/HTM-16.0/.

[48] H. Liu, and Y. Chen, "Generic segment-wise DC for 3D-HEVC depth intra coding," *in Proc. IEEE International Conference on Image Processing (ICIP)*, 2014, pp. 3219–3222.

[49] G. Sanchez, J. Silveira, L. Agostini, and C. Marcon, "Performance Analysis of Depth Intra Coding in 3D-HEVC," *Transactions on Circuits and Systems for Video Technology (TCSVT)*, Early Access, Date of Publication: 2018.

[50] G. Sanchez, M. Saldanha, G. Balota, B. Zatt, M. Porto, and L. Agostini, "DMMFast: A Complexity Reduction Scheme for Three-Dimensional High-Efficiency Video Coding Intraframe Depth Map Coding," *Journal of Electronic Imaging*, v. 24, n. 2, pp. 023011, 2015.

[51] G. Sanchez, L. Agostini, and C. Marcon, "Complexity reduction by modes reduction in RD-list for intra-frame prediction in 3D-HEVC depth maps," *in: International Symposium on Circuits and Systems (ISCAS)*, 2017.

[52] C. Chen, C. Huang, Y, Chen, and L. Chen, "Level C+ Data Reuse Scheme for Motion Estimation With Corresponding Coding Orders," *IEEE Transactions on Circuits and Systems for Video Technology*, v. 16, n. 4, pp. 553–558, 2006.

[53] Micron Technology Inc., "272b: x64 Mobile LPDDR4 SDRAM features," MT53B384M64D4, 2014.

[54] V. Afonso, H. Maich, L. Audibert, *et al.*, "Hardware Implementation for the HEVC Fractional Motion Estimation Targeting Real-Time and Low-Energy," *Journal of Integrated Circuits and Systems*, v. 11, n. 2, pp. 106–120, 2016.

[55] HEVC Test Model. Retrieved Nov. 2018: https://hevc.hhi.fraunhofer.de/svn/svn_3DVCSoftware/tags/.

[56] F. Bossen, "Common test conditions and software reference configurations," ITU-T SG16 WP3 and ISO/IEC JTC1/SC29/WG11, Document: JCTVC-L1100, Jan. 2013.

[57] G. Bjøntegaard, "Calculation of average PSNR differences between RD curves," VCEG-M33, Austin, TX, USA, Apr. 2–4, 2001.

[58] G. Sanchez, M. Saldanha, G. Balota, B. Zatt, M. Porto, and L. Agostini. "Complexity reduction for 3D-HEVC depth maps intra-frame prediction using simplified edge detector algorithm," *in IEEE International Conference on Image Processing (ICIP)*, 2014.

[59] NVSim. [Online]. Available: http://nvsim.org/.

[60] B. Wang, J. Zhou, and T. Kim. "SRAM Devices and Circuits Optimization Toward Energy Efficiency in Multi-Vth CMOS," *Microelectronics Journal*, v. 46, n. 3, pp. 265–272. 2015. [Online]. Available: http://www.sciencedirect.com/science/article/pii/S0026269214003528.

[61] ViTech. 3D-HEVC Memory Simulator. [Online]. Available: https://wp.ufpel.edu.br/vitech/en/downloads/.

[62] Cadence. Encounter RTL Compiler. [Online]. Available: https://www.cadence.com/content/cadencewww/global/en_US/home/training/all-courses/84441.html.
[63] NanGate. FreePDK45 Generic Open Cell Library, Retrieved Nov. 2018. http://projects. si2.org/openeda.si2.org/projects/nangatelib.
[64] F. Pereira, and E. Silva, "Efficient plenoptic imaging representation: Why do we need it?," *in 2016 IEEE Int'l Conf. Multimedia and Expo (ICME)*, Seattle, WA, USA, Jul. 11–15, 2016.
[65] Lifewire. 3D TV Dies – Is it Really the End?, Retrieved Nov. 2018. https://www.lifewire.com/ why-3d-tv-died-4126776.
[66] B. Bross, J. Chen, and S. Liu, "Versatile video coding (Draft 2)," ISO/IEC doc. JVET-K1001, Ljubljana, SI, 10–18, Jul, 2018.

Chapter 7

Frame memory compression for high-resolution video coding

Yibo Fan[1], Chenhao Gu[1], Tianwen Yang[1] and Ruixue Lei[1]

This chapter presents a frame memory compression method used in video coding. Frame memory compression compresses the data to be stored in the frame memory in order to reduce the external bandwidth and the related power consumption, as shown in Figure 7.1.

When the pixels of a motion-compensated frame have to be written into external DRAM, the frame memory compression engine compresses those pixels. During motion estimation (ME), the frame memory compression engine decompresses the compressed pixels of previous frames and passes them to the video codec.

As shown in Figure 7.2, most of the frame memory compression algorithms are composed of three stages: prediction, entropy coding and memory organization [1], which are respectively introduced in Sections 7.1, 7.2 and 7.3.

7.1 Prediction

For the prediction stage, previous research can be classified into two categories, spatial domain and frequency domain.

In the spatial domain prediction, the current pixel is predicted using adjacent pixels to reduce spatial redundancy with various methods, such as differential pulse-code modulation (DPCM), intra-mode referenced in-block prediction, hierarchical average and copy (HAC) and hierarchical minimum and difference (HMD). Spatial domain prediction is based on the fact that successive pixels usually show significant correlations.

In the frequency domain prediction, many transform methods, such as modified Hadamard transform (MHT) and discrete cosine transform (DCT), are utilized for de-correlation, followed by quantization to eliminate high-frequency components.

[1]State Key Laboratory of ASIC and System, Fudan University, Shanghai, China

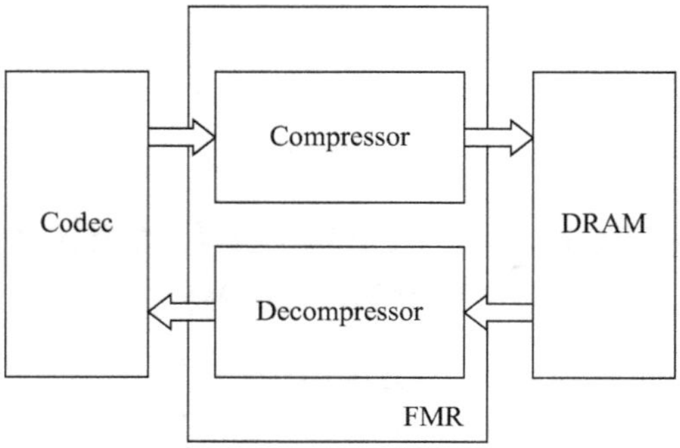

Figure 7.1 The high-level block diagram for the frame memory compression system. FMR, Frame memory recompression

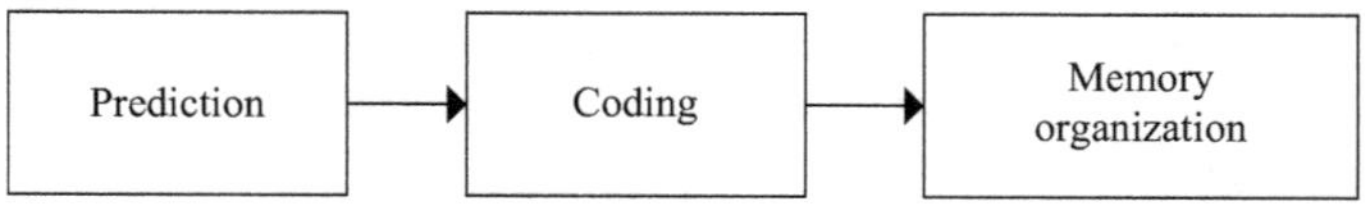

Figure 7.2 The process flow of frame memory recompression

7.1.1 Differential pulse-code modulation

DPCM is a procedure of taking a subtraction between the current pixel and the adjacent pixel and using the difference to represent the data as

$$P_{pred_n} = P_{n-1}$$
$$Res_n = P_n - P_{n-1}$$

where P_n is the nth pixel, P_{pred_n} is the prediction value of P_n and Res_n is the difference of P_n and P_{n-1}. Due to the correlation between adjacent pixels, the subtraction results concentrate on the zero with a high possibility [2].

In Figure 7.3, a simple DPCM prediction example of an 8×8 block is shown. The pixel value of the top-left pixel is stored without subtraction since this sample is used to start the process of decoding this block. For the next seven samples of first column, their prediction direction is vertical and the residuals are the difference between the current sample and the sample in the previous row. For the remaining 56 samples of the second–seventh columns, their prediction direction is horizontal and the residuals are the difference between the current sample and the sample in previous columns [3]. After the subtraction operation, these residuals are passed to the entropy coding stage for further process. On the other hand, the decoding phase applies the reverse process of the encoding phase.

Instead of using only one scan direction, the adaptive DPCM (ADPCM) proposed in [4] selects the best mode among eight scan directions. Figure 7.4 shows

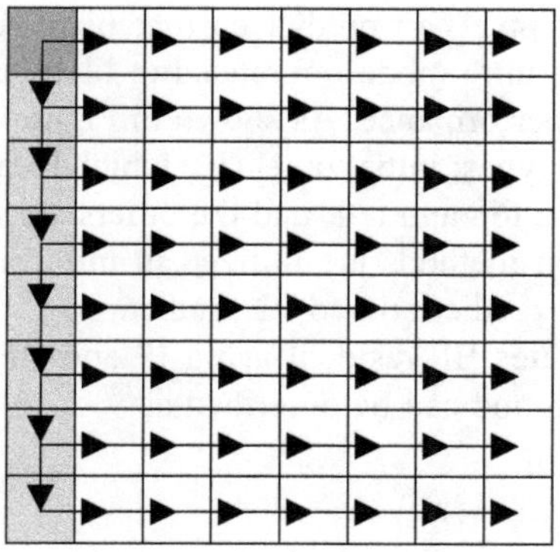

Figure 7.3 DPCM example of an 8 × 8 block

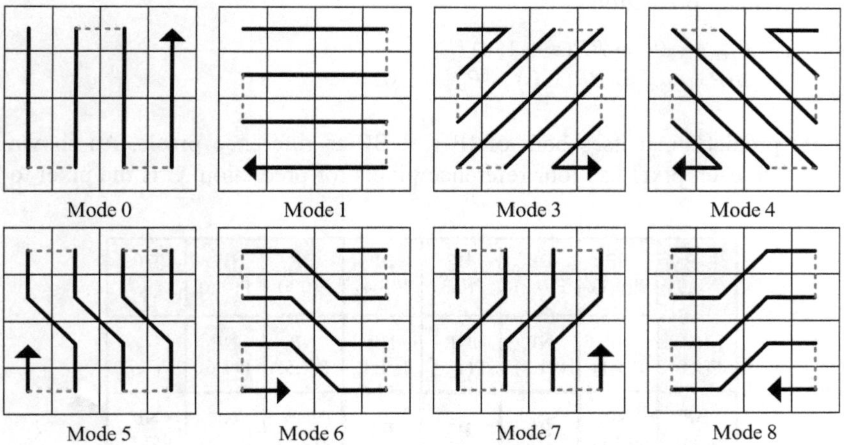

Figure 7.4 Eight scan modes for ADPCM

the proposed eight different scan orders and the arrowed lines show the scan order. The best mode is chosen according to H.264 intra-prediction results. Note that H.264 4 × 4 intra-prediction has nine different modes and Mode 2 (the DC mode) is in excluded in Figure 7.4 because the DC mode has no information for order selection. By choosing the best mode, the difference between adjacent pixels is smaller compared to algorithms using only one scan direction so that the data can be represented by a smaller number of bits. For example, Mode 0 is suitable for an image with vertical stripes, while Mode 1 is suitable for horizontal stripes.

7.1.2 Intra-mode referenced in-block prediction

In [5], an intra-mode referenced in-block prediction algorithm is provided. Instead of calculating the prediction value using only one neighboring pixel, this method utilizes several adjacent pixels to predict current pixel. By using more adjacent pixels for prediction, the intra-mode referenced in-block prediction algorithm can gain a better prediction performance. As shown in Figure 7.5, the input pixels can be categorized into three types: initial pixel (IP) which located in (0,0), basic pixels (BP) which located in the top and left, and the others are normal pixels (NP).

Two main prediction methods are utilized in intra-mode referenced in-block prediction algorithm: P1 prediction and P2 prediction.

P1 prediction generates BP residual using IP and BP to do vertical and horizontal DPCM prediction and can be described as

P1 prediction:

1. Vertical prediction

$$\begin{aligned} P_{pred}(0,x) &= BP(0,x-1) \\ BP_{Residual} &= BP(0,x) - P_{pred}(0,x) \end{aligned}$$

2. Horizontal prediction

$$\begin{aligned} P_{pred}(x,0) &= BP(x-1,0) \\ BP_{Residual} &= BP(x,0) - P_{pred}(x,0) \end{aligned}$$

For P2 prediction, it uses both of IP and BP as reference pixels. As shown in Figure 7.5, each pixel has four reference pixels for prediction. C is the pixel to be

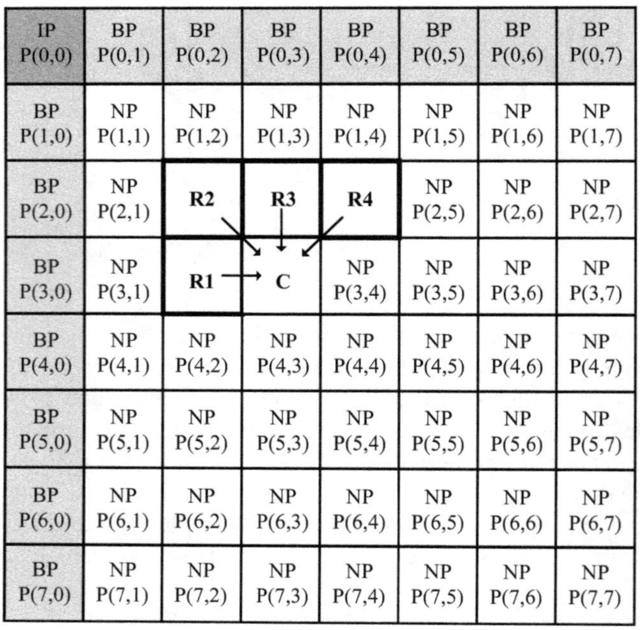

Figure 7.5 Three pixel types of intra-mode referenced in-block prediction algorithm

Table 7.1 Prediction mode

Mode	Prediction (C=)
0	$R1$
1	$R3$
2	$(R1 + R2)/2$
3	$(R3 + R4)/2$
4	$(R1 + R4)/2$
5	$(R1 + R3)/2$
6	$((R1 + R2)/2 + R3)/2$
7	$((R1 + R2)/2 + (R3 + R4)/2)/2$

Table 7.2 Mode mapping for H.265 intra-mode and prediction mode

Intra-mode	Prediction mode	Intra-mode	Prediction mode
Planner	4,5,7	11–18	0,2,6
DC	4,5,7	19–26	1,5,6
2–10	0,4,5	27–34	1,3,4

predicted, $R1$–$R4$ are neighboring reference pixels. The detailed function of P2 is shown as below.

P2 prediction:

$$C(x,y) = f\left(R1_{(x-1,y)} R2_{(x-1,y-1)}, R3_{(x,y-1)} R4_{(x+1,y-1)}\right)$$

$$NP_{residual}(x,y) = NP(x,y) - C(x,y)$$

The detailed prediction function $f()$ is listed in Table 7.1. Three candidate modes are selected according to the intra-mode from video encoder, and one best NP prediction is chosen after mode decision. Table 7.2 shows a mode-mapping example.

7.1.3 Block truncation coding

Block truncation coding (BTC) is a fairly simple and effective image compression technique at moderate bitrates, which compress every partitioned image block independently. The BTC is based on the use of a moment preserving quantizer to preserve statistical moments of the input and output, especially the block mean and the standard deviation [6].

A basic BTC algorithm is introduced in [7]. The image is firstly divided into $n \times n$ pixel blocks (in [7], $n = 4$) and then the blocks are coded individually, each into a two-level signal. $X_1, X_2, \ldots, X_m$ are the values of the pixels in a block of the original picture ($m = n^2$).

The first and second moments and the sample variance are

$$\overline{X} = \frac{1}{M}\sum_{i=1}^{m} X_i;$$

$$\overline{X^2} = \frac{1}{M}\sum_{i=1}^{m} X_i^2;$$

$$\overline{\sigma^2} = \overline{X^2} - \overline{X}^2.$$

And a threshold, X_{th}, and two output levels, a and b are defined.

$$if\ X_i < X_{th}\ output = a;$$

$$if\ X_i \geq X_{th}\ output = b;$$

$$i = 1, 2, \ldots, m.$$

In [7], $X_{th} = \overline{X}$. The output levels a and b are found by solving the following equations:

$$q = number\ of\ X_i's\ greater\ than\ X_{th}$$

$$m\overline{X} = (m - q)a + qb$$

$$m\overline{X^2} = (m - q)a^2 + qb^2$$

Solving for a and b:

$$a = \overline{X} - \overline{\sigma}\sqrt{\left[\frac{q}{m - q}\right]} *$$

$$b = \overline{X} + \overline{\sigma}\sqrt{\left[\frac{m - q}{q}\right]} *$$

Each block is then described by the values of $\overline{X}$, $\overline{\sigma}$ and an $n \times n$ bit plane consisting of 1s and 0s. The bit plane indicates whether pixels are above or below X_{th}. When the length of $\overline{X}$ and $\overline{\sigma}$ are equal to 2, the data rate is 2 bit/pixel. The receiver calculates a and b from equation * to reconstruct the image block and assigns these values to pixels according to the bit plane.

7.1.4 Hierarchical minimum and difference

HMD algorithm features the advantage that only adds operations are needed in the decompression process [8]. Figure 7.6 shows an HMD for 8 × 8 block.

In HMD, the minimum pixel value is first searched within each 2 × 2 block, which is called "min2 × 2." Then the difference between each pixel value and the corresponding min2 × 2 is calculated, which is represented as "diffpixel." Next, for each 4 × 4 block, there are four min2 × 2, and the minimum value among four min2 × 2s is found to be called "min4 × 4." Then the difference between each min2 × 2 and the relevant min4 × 4 is calculated to be called "diff2 × 2."

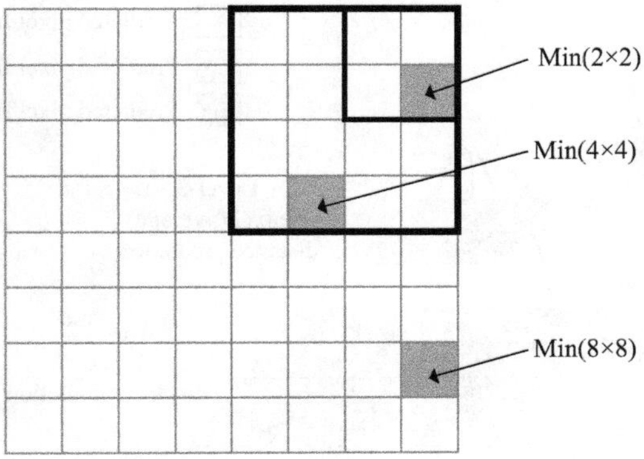

Figure 7.6 HMD example for an 8 × 8 block

This process is repeated for bigger blocks. Finally, variable length coding (VLC) and packing are performed to compress the min and diff.

In the decompression process, the value of each pixel is performed by adding up the differences, as described in *.

$$\begin{aligned} pixel[a][b][c] &= min8 \times 8 + difference \\ &= min8 \times 8 + diff4 \times 4[a] + diff4 \times 4[a][b] + diffpixel[a][b][c] \; * \end{aligned}$$

where a, b and c are the index for 4 × 4 block, 2 × 2 block and pixel, respectively.

7.1.5 Hierarchical average and copy

HAC is a hierarchical prediction method based on pixel averaging and copying [9]. The HAC prediction algorithm is represented in Figure 7.7 for an 8 × 8 block.

There are four different prediction methods in HAC: (1) horizontal average prediction (HAP), predicted by the average of left and right pixel; (2) vertical average prediction (VAP), predicted by the average of upper and bottom pixel; (3) horizontal copy prediction (HCP), predicted by the horizontally adjacent pixel; and (4) vertical copy prediction (VCP), predicted by the vertically adjacent pixel. The average prediction and copy prediction are defined as follows:

$$P_{pred,avg_n} = \frac{P_{n-1} + P_{n+1} + 1}{2}$$

$$P_{pred,copy_n} = P_{n-1}$$

The arrow tail represents the source pixel and the arrowhead predicts the destination pixel. The pixel which is predicted by the average of the values of 2 pixel is pointed to by two arrows. If a pixel is pointed to by one arrow, the pixel is predicted by copying the neighboring pixel. Most of the pixels are predicted by HAP or VAP except for the boundary pixels of the block, which are predicted by HCP or VCP. The pixels in level-2 and level-3 are those subsampled from their respective lower level.

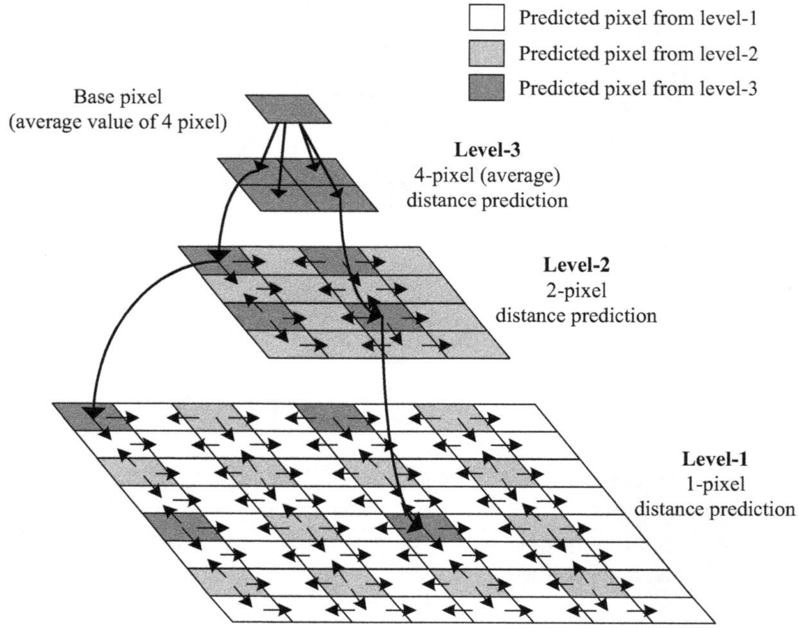

Figure 7.7 HAC example of an 8 × 8 block

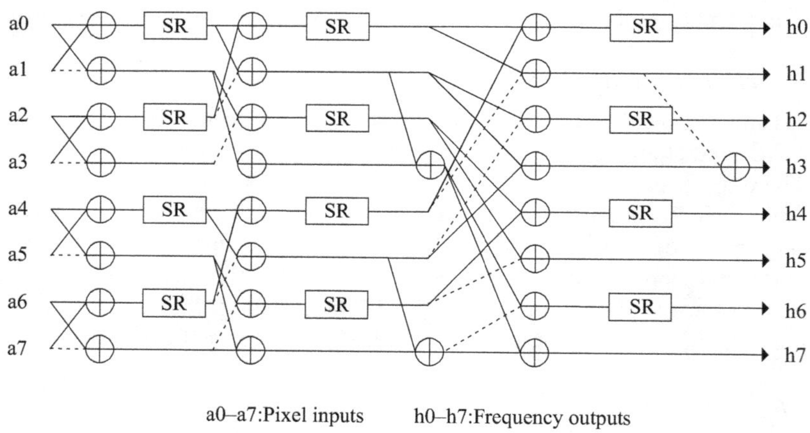

Figure 7.8 MHT algorithm

7.1.6 Modified Hadamard transform

MHT is a relatively simple transform method compared to DCT and discrete wavelet transform (DWT). Figure 7.8 illustrates the MHT. Addition and subtraction are represented by solid and dashed lines, respectively.

In [10], recompression is performed on 1 × 8 pixel arrays obtained from every row of 8 × 8 blocks. 1 × 8 pixel arrays are input one by one and one-dimensionally transformed into frequency components. MHT uses only addition/subtraction and 1-bit shift right operation but produces frequency components efficiently. Therefore, the hardware design is quite straightforward. That is, only adders and wiring are required because the shift operation is hardwired.

7.1.7 *Discrete cosine transform*

A DCT is a Fourier-related transform similar to the discrete Fourier transform, using only real numbers. A DCT is a finite sequence of data points in terms of a sum of cosine functions oscillating at different frequencies [11]. DCTs are applied in many fields, such as lossy compression of audio and images, and spectral methods for the numerical solution of partial differential equations.

The DCT used in video compression is two-dimensional and typically $N \times N$. The result is an $N \times N$ transform coefficient array in which the (0,0) element (top-left) is the DC (zero-frequency) component, and entries with increasing vertical and horizontal index values represent higher vertical and horizontal spatial frequencies.

A DCT is used in [12]. Figure 7.9 shows the proposed EC algorithm.

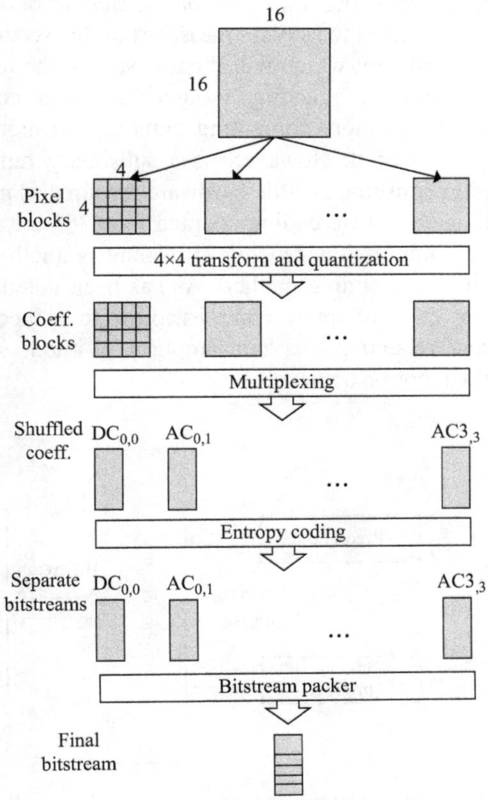

Figure 7.9 The encoding process of the proposed EC algorithm

The input 16 × 16 block is split into 4 × 4 subblocks and each of them is transformed using a DCT and gets quantized individually. After that, the transformed coefficients are packed to a final bitstream. Once the bitstream is generated, it can be written to the external memory.

7.2 Entropy coding

Most frame compression algorithms introduce a coding module to further compress pixel data rather than directly store it after prediction. Basically, a coding module contains two parts—a certain coding method and a strategy for data packaging. Since the former is decisive on the performance of a coding module, this section focuses on different coding methods and a simple comparison between them.

7.2.1 Coding methods in frame compression

The typical structure of a frame-compression system with a coding module is illustrated in the block diagram in Figure 7.10. The input of a coding module comes from the prediction module, and in some cases, straight from the original reference frames. All these inputs, including raw pixels or coefficients or differences between pixels, are collectively referred to as pixel messages in this section. The output of a coding module is the bitstream which will then be sent to the memory.

With the introduction of a coding module, a higher compression rate is obtained, but at the cost of more computing units and memory resource for the hardware implementation of it. However, as a subsidiary function, frame compression is supposed to consume as little hardware resource as possible. Therefore, those methods such as arithmetic coding adopted in H.265 are too complicated to be adopted in frame compression. Coding efficiency is another issue to be considered when deciding on a coding method. As has been noted at the start of this chapter, the main objective of frame compression is to reduce bandwidth to the external memory and related power consumption. Without satisfactory coding efficiency, this goal cannot be achieved.

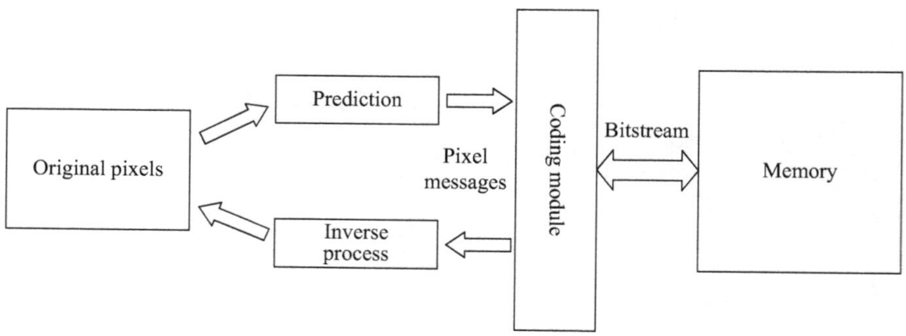

Figure 7.10 A frame-compression system with a coding module

Facing the trade-off between hardware consumption and coding efficiency, even though there are various kinds of coding methods, only a few of them can be utilized in frame compression. Currently, existing coding methods adopted in frame compression can be generally divided into three groups—Huffman coding, Golomb coding and self-defined coding methods, such as SFL (semifixed length) coding and SBT (significant bit truncation) coding. All these methods can be implemented easily while maintaining a good coding efficiency.

7.2.2 Huffman coding

The first group is Huffman coding. Huffman coding is a famous and widely used method for data compression. It was proposed by David A. Huffman in 1952 when he was studying the minimum-redundancy codes. The basic idea of Huffman coding is to assign shorter codewords to the symbols with larger probabilities of occurrence. Thus, before the encoding process, a scan of the entire set of symbols needs to be performed to acquire the exact probability distribution of every symbol. After that, each symbol will be encoded in Huffman's way following the steps below:

1. Create a queue of the symbols sorted by their probabilities in descending or ascending order. Every symbol will be represented by its probability, so there is no specific symbol in the queue. This queue is called "message ensemble" in [13].
2. Find the two least probable symbols in the ensemble and create a composite symbol whose probability is the sum of the probabilities of them. This newly created symbol is also represented by its probability and it will take the place of the two symbols from which it is created. The new ensemble containing the composite symbol is called "auxiliary ensemble." Its member should be rearranged if necessary so that the symbols are again ordered according to their possibilities.
3. The final digits of the two least probable symbols will be assigned as "0" or "1," one for each, because from Huffman's perspective, the codes of the two least probable symbols in every ensemble should have equal length and be identical except for their last digit; otherwise, it would not be an optimum code.
4. The new ensemble will be considered and processed exactly as the original ensemble is; therefore, steps 2 and 3 will be repeated over and over again. When the number of members in the most recently formed auxiliary ensemble is reduced to one, the coding procedure is complete.

Here is an illustration of the process above. Assume there is a sequence of characters distributed as Figure 7.11(a) shows. Figure 7.11(b) illustrates how to combine two symbols to form a single composite symbol and create auxiliary ensembles step by step. Since bit "0" and bit "1" are used to represent the two least probable symbols of every auxiliary ensemble, a symbol can be uniquely represented by the bit string from the final ensemble to the original ensemble.

Symbol	E	T	O	A	N	I	R
Probability	0.25	0.2	0.2	0.15	0.1	0.05	0.05

(a)

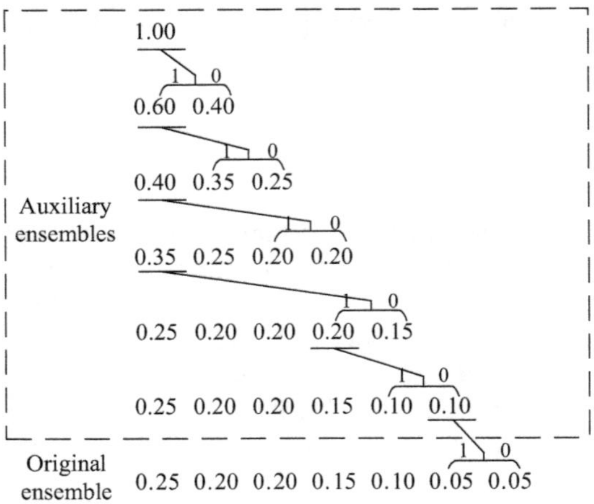

(b)

Symbol	Code
E	10
T	01
O	00
A	110
N	1111
I	11101
R	11100

(c)

Figure 7.11 (a) Probability distribution, (b) Huffman coding procedure and (c) Huffman table

The mapping relationship between symbols and bit strings is stored in a special data structure named Huffman table as shown in Figure 7.11(c). Due to the Huffman table, the decoding of Huffman codes will become quite easy—just select the right table containing the string to be decoded (if more than one table is used) and then output the corresponding symbol.

The major advantage of Huffman coding is that it brings a higher coding efficiency than most of the other coding methods in most cases. Besides, Huffman codes are prefix-free, which means there is no whole codeword that is the initial segment of any other codeword in Huffman codes set. As illustrated in Figure 7.12, a string of prefix-free codewords is uniquely decodable while a string of non-prefix-free codewords can be decoded into different combinations of symbols. With this property, Huffman codes can be easily packed because no additional indication is necessary to specify where a message code begins and ends once the starting point of a sequence of messages is known [13].

Despite many advantages, the procedures of Huffman coding consume so much hardware resource that original Huffman coding is nearly impractical to be adopted in frame compression. Another reason that it is not used is the difficulty in getting the definite probability distribution of pixel messages to be encoded, especially in real time video compression. To overcome these problems, hardware implementations of Huffman coding in frame compression usually use one or more static Huffman tables to bypass statistical analysis of pixel messages. These static

Prefix-free codes	
Symbol	Code
E	10
T	01
O	00
A	110
N	1111
I	11101
R	11100

Input: NEAR
Bitstream: 11111011011100
Output: NEAR

Non-prefix-free codes	
Symbol	Code
E	10
T	01
O	00
A	101
N	1111
I	11101
R	11100

Input: NEAR
Bitstream: 11111010111100
Output: NEAR or NATR or NEENO

Figure 7.12 Prefix-free codes and non-prefix-free codes

Table 7.3 Static Huffman table in [2]

Diff.	***Y***	**Diff.**	***Y***	**Diff.**	***Y***	**Diff.**	***Y***
0	00	−6	101110	+11	1101101	−17	11011100
−1	010	+6	101111	−12	1111100	+17	11011101
+1	100	−7	110100	+12	1111101	−18	11011110
−2	1010	+7	110101	−13	01101100	+18	11011111
+2	1100	−8	111100	+13	01101101	−19	11111100
−3	01110	+8	111101	−14	01101110	+19	11111101
+3	10110	−9	0110100	+14	01101111	−20	11111110
−4	11100	+9	0110101	−15	01111010	+20	11111111
+4	11101	−10	0111100	+15	01111011	Others	–
−5	011000	+10	0111110	−16	01111110	–	–
+5	011001	−11	1101100	+16	01111111	–	–

tables are based on the pre-analysis of different video sequences with typical features rather than the sequences to be processed. As shown in Table 7.3, every symbol in the static Huffman table corresponds to a static codeword. This compromise makes it possible to apply Huffman coding to frame compression, but surely the coding efficiency is not that high compared with the original Huffman coding.

The number of static Huffman tables is an issue in this implementation. Only using one table cannot achieve a similarly good performance on different sequences, while using multiple tables involves a larger requirement for memory elements, a decision strategy to select a table for each sequence or each frame and probably more bits to indicate which table is used. Hence, before practicing this adapted Huffman coding method, this issue must be carefully thought about.

Here are some works using static Huffman tables in frame compression, as shown in Table 7.4. Work [2] uses one Huffman table to compress the differential

Table 7.4 Summary of works using static Huffman tables

Work	Process before coding	Data to be encoded	Number of tables	Strategy to select a table	Method to indicate the used table
[2]	DAP	Differences of adjacent pixels	2	One for luminance block, one for chrominance block	N/A
[12]	DCT and quantization	Quantized coefficients	15	Each table for coefficients of each frequency (except DC components)	Different frequency components packed together in zigzag scan order; need bits to denote the bit length of each frequency components
[14]	Average pixel value calculation	Original pixels	8	According to the average value of the encoded block	Three bits to indicate the used table

data of luminance blocks and another table to compress the differential data of chrominance blocks. The algorithm in Work [12] uses different Huffman tables to compress the coefficients of a 16 × 16 block after it is transformed using DCT, one table for components of one frequency. Work [14] proposes a context adaptive solution named reference frame context adaptive variable-length coder, using eight static Huffman tables to compress the reference frames. The average pixel value of a 4 × 4 luminance block is used to determine which Huffman table is used to compress the block.

7.2.3 Golomb coding

The second group is Golomb coding. Golomb coding is also a popular data-compression method which has been adopted by some famous algorithms, such as lossless JPEG (JPEG-LS) and H.264. It was developed by Solomon W. Golomb in the 1960s and primarily designed for encoding symbols which are geometrically distributed. Compared with Huffman coding, Golomb coding is not based on the probability distribution of symbols. Even though the distribution pattern is totally unknown or the symbol list is infinite, Golomb coding can still be performed because the codewords of it are independent of the probability distribution.

Actually, the Golomb codes are a family of codes parameterized by an integer $m > 0$. With the parameter m, any integer n can be represented using two numbers—q and r. As defined in the following formulas, when n is divided by m, q is the quotient and r is the remainder ($\lfloor x \rfloor$ is the largest integer smaller than or equal to x):

$$q = \left\lfloor \frac{n}{m} \right\rfloor$$

$$r = n - qm$$

In Golomb coding, q is represented with q 1s followed by a 0, which is known as the unary code. The remainder r can take on the values in 0, 1, 2, …, $m-1$. If m is a power of 2, r will be represented in its $\log_2 m$-bit binary form. If not, r can still be represented using $\log_2 m$-bit, as shown in Table 7.5. To further reduce the number of bits required, the first $2^{\lceil \log_2 m \rceil} - m$ values of r can be represented in its $(\lceil \log_2 m \rceil - 1)$-bit binary form and the rest of values can be represented using the $\lceil \log_2 m \rceil$-bit binary form of $r + 2^{\lceil \log_2 m \rceil} - m$ ($\lceil x \rceil$ is the smallest integer greater than or equal to x).

Golomb–Rice coding refers to an adaptive scheme of Golomb coding. In Golomb–Rice coding, the parameter m is a power of two, which makes Golomb–Rice codes convenient to be used since multiplication and division by 2 can be implemented efficiently by shifts. Table 7.6 shows Golomb–Rice codes for $m = 2$, 4, 8, 16.

Table 7.5 Golomb codes for m = 1, 2, 3, 4

$m = 1$		$m = 2$		$m = 3$		$m = 4$	
n	**Codeword**	n	**Codeword**	n	**Codeword**	n	**Codeword**
0	0	0	00	0	00	0	000
1	10	1	01	1	010	1	001
2	110	2	100	2	011	2	010
3	1110	3	101	3	100	3	011
4	11110	4	1100	4	1010	4	1000
5	111110	5	1101	5	1011	5	1001
6	1111110	6	11100	6	1100	6	1010
7	11111110	7	11101	7	11010	7	1011
8	111111110	8	111100	8	11011	8	11000
9	1111111110	9	111101	9	11100	9	11001
10	11111111110	10	1111100	10	111010	10	11010

Table 7.6 Golomb–Rice codes for m = 2, 4, 8, 16

$m = 2$		$m = 4$		$m = 8$		$m = 16$	
n	**Codeword**	n	**Codeword**	n	**Codeword**	n	**Codeword**
0	00	0	000	0	0000	0	00000
1	01	1	001	1	0001	1	00001
2	100	2	010	2	0010	2	00010
3	101	3	011	3	0011	3	00011
4	1100	4	1000	4	0100	4	00100
5	1101	5	1001	5	0101	5	00101
6	11100	6	1010	6	0110	6	00110
7	11101	7	1011	7	0111	7	00111
8	111100	8	11000	8	10000	8	01000
9	111101	9	11001	9	10001	9	01001
10	1111100	10	11010	10	10010	10	01010

Before applying Golomb–Rice coding to frame compression, there are two conditions that should be satisfied. First, since Golomb–Rice coding assigns shorter codes to smaller integers, not all kinds of pixel messages could be encoded efficiently with Golomb–Rice coding. It is suitable only for those messages which have a distribution that smaller integers have a larger probability of occurrence, such as differences between pixels and coefficients after transformation. Second, Golomb–Rice coding is used to encode symbols with a nonnegative integral value. However, the input pixel messages are usually negative. To convert these negative symbols to positive symbols, a method named Rice remapping is usually adopted. The remapping rule is defined as follows:

$$\begin{aligned} q' &= 2|q| \quad \textit{if } q \geq 0 \\ &= 2|q| - 1 \quad \textit{if } q < 0 \end{aligned}$$

Obviously, the LSB of the remapped symbol indicates whether the original symbol is positive or not. According to the LSB, the inverse conversion is easy to be performed because what needs to be done is only to shift the remapped symbol to the right by one bit and then if it is negative, obtain the two's complement of it. Therefore, this remapping method takes a low hardware cost. Another method is to encode symbols by encoding their absolute values and then adding a sign bit at the end of the Golomb codewords to indicate whether the encoded symbols are negative or not. This method is simple too and in general, the codewords using this method are shorter than the codewords using the remapping method.

A lot of works use Golomb–Rice Coding in frame compression, as shown in Table 7.7. Work [4] uses Golomb–Rice coding to encode differences after DPCM operation. The required number of bits for each block is pre-calculated to determine the best k ($k = \log_2 m$) value and quantization parameter Q. Work [15] introduces an algorithm using Golomb–Rice coding to compress AC coefficients after line-by-line MHT. Similarly, the length of Golomb–Rice codes for different k will be calculated before encoding to decide the best k value for each line. In work [16], Golomb–Rice coding is used to compress prediction errors after gradient-adjusted prediction (GAP) and a more complicated measure to estimate the best k is applied.

Exp-Golomb coding is another coding method based on Golomb coding. Also parameterized by a positive integer k, namely, order k, the codewords of exp-Golomb coding are quite different from the original Golomb codewords. As shown in Table 7.8, with the increasing of the order k, small values have longer codewords on average while big values have shorter codewords. Therefore, a big k is usually used to encode values from a large range.

Here are some works using exp-Golomb coding, as shown in Table 7.7. In work [5], exp-Golomb coding is used to encode prediction errors with an absolute value bigger than 32. Work [18] applies order k exp-Golomb coding to encode the quotient part of prediction errors when the error is divided by 2^k and the quotient is bigger than 3. A prediction scheme is used in [18] to deduce the k of the current pixel according to neighboring encoded pixels' information.

Table 7.7 Summary of works using Golomb–Rice coding or exp-Golomb coding

Work	Coding method	Process before coding	Data to be encoded	Rice-remapping/ additional sign bit	Strategy for selecting a parameter *k*
[4]	Golomb–Rice	DPCM	Differences of adjacent pixels	Rice-remapping	The same as work [17] except that the *k* is chosen between 1 and 2
[15]	Golomb–Rice	MHT	AC coefficients	Rice-remapping	Calculate the required bits to select the best *k* from 1 to 5
[16]	Golomb–Rice	GAP	Prediction errors	Rice-remapping	Track the *k* values (from 1 to 3) of past prediction and use the average as the *k* value for coding
[5]	Exp-Golomb	In-block prediction	Prediction errors	Additional sign bit	N/A
[18]	Exp-Golomb	Directional prediction	The quotient part of prediction errors	Additional sign bit	Deduce *k* using the correlations of *k* among neighboring pixels

Table 7.8 Codewords of exp-Golomb coding for m = 1, 2, 3, 4

$k = 1$		$k = 2$		$k = 3$		$k = 4$	
n	**Codeword**	*n*	**Codeword**	*n*	**Codeword**	*n*	**Codeword**
0	0	0	00	0	000	0	0000
1	100	1	01	1	001	1	0001
2	101	2	1000	2	010	2	0010
3	11000	3	1001	3	011	3	0011
4	11001	4	1010	4	10000	4	0100
5	11010	5	1011	5	10001	5	0101
6	11011	6	110000	6	10010	6	0110
7	1110000	7	110001	7	10011	7	0111
8	1110001	8	110010	8	10100	8	100000
9	1110010	9	110011	9	10101	9	100001
10	1110011	10	110100	10	10110	10	100010

7.2.4 SFL coding and SBT coding

The last group is several specific coding methods. Compared with Huffman coding and Golomb coding, the most significant feature of these methods is that the lengths of their codewords are not variable but semifixed, i.e., fixed in a partial range or a small group of values.

In [1,3,5,19], similar SFL coding methods are proposed. Basically, SFL coding consists of two steps: first separating the pixel messages into small groups, and then performing fixed length coding according to the local feature of each group. For instance, in [19], before applying SFL coding, a differential process is performed between column adjacent pixels which will concentrate the differences near zero. Just like adapted Huffman coding, SFL codes are generated and decoded simply according to a coding table, as shown in Table 7.9. Since the difference "0" accounts for the highest possibility of occurrence and the differences in the range [−4, 3] account for a large possibility of occurrence, only 1 bit is used to represent "0" and 4 bit to the rest. When the difference is out of the range [−4, 3], 4-bit characteristic codes "1111" followed by the 9-bit binary representation of the difference is used.

SBT coding is another simple coding method with SFL codes. It is used to compress the prediction errors in [9]. The basic concept of SBT coding is to express all of the prediction errors in a group with as many significant bits truncated as the information of each prediction error is preserved [9]. As shown in Figure 7.13, eight prediction errors are represented in the two's complement notation.

Table 7.9 SFL codes in [19]

Diff.	Code	Diff.	Code	Diff.	Code
−4	1011	−1	1000	2	1101
−3	1010	0	0	3	1110
−2	1001	1	1100	Exc.	1111+difference

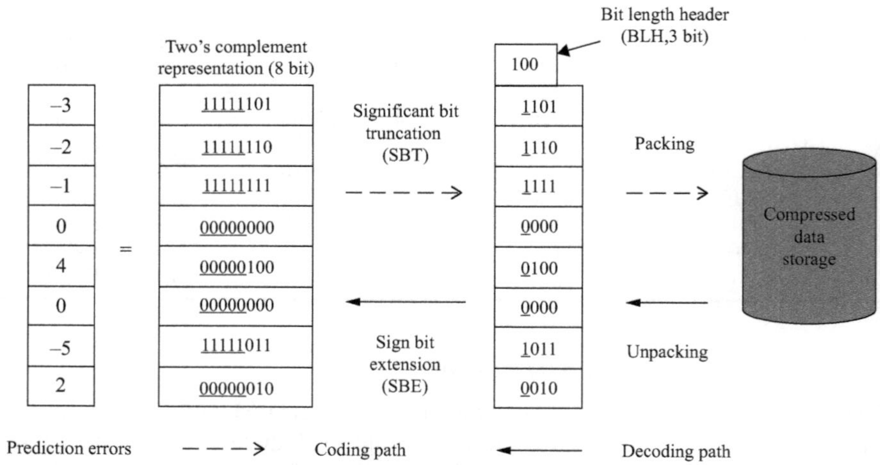

Figure 7.13 SBT coding and decoding path

The underlined part of each prediction error can be regarded as an extension of its sign bit, which can be truncated without causing any loss of information. Then, the truncated prediction errors are packed with the bit length header (BLH) representing the bit length of each prediction error. As a result, 64 bits for eight prediction errors are compressed into 35 bits in this example. The decoding process is exactly the inverse of the encoding process. According to the value of BLH, each truncated prediction error is recovered to an 8-bit value by extending its most significant bit (MSB).

The biggest advantage of SBT coding is that it can be implemented using a parallel architecture. In VLC, codewords of different lengths are usually packed together into a bitstream without explicit boundary information. Therefore, the starting point of an arbitrary codeword in the middle of the bitstream remains unknown until all the preceding codewords are decoded. As a result, VLC codewords must be decoded one by one, which makes it difficult to implement a parallel architecture of the decoder and results in a low throughput and a high-decoding latency. While in fixed-length coding, this problem does not exist because all codewords have the same length and it is easy to clarify where a codeword begins and ends. But fixed-length coding is not adopted in frame compression because of its low-coding efficiency. If the lengths of codewords are semifixed as SBT coding, a parallel decoder can be implemented without losing much coding efficiency.

7.3 Memory organization

When the prediction and coding schemes mentioned above are finished, the external DRAM is applied to store the huge amounts of the compressed data of the reference frames. Bandwidth and power limitation then become the two main obstacles while accessing those reference frames. Efforts are made to increase the efficiency of bandwidth [5,8,20,21], save the DRAM traffic to reduce the power consumption [18,19,22–25], minimize the DRAM access latency or save the memory space [26].

7.3.1 Works on bandwidth efficiency

7.3.1.1 Address table compression and reference memory access style of IP module and ME module

In [8] (for H.264), they compressed the start address of the compressed blocks and packed it in a certain packing scheme shown in Figure 7.14(A). First, the length of the compressed data is measured and quantized as shown in Figure 7.14(B). Because the compressed length (CL) changes smoothly, the quantized length would be compressed based on this locality property.

The start address consists of the row address and column address, and the row address is divided into the base and offset address. The base address means the address of the first block in every row of the uncompressed frame, which would be stored in a base address table. For other blocks in the same row, the offset address relative to the base address is stored.

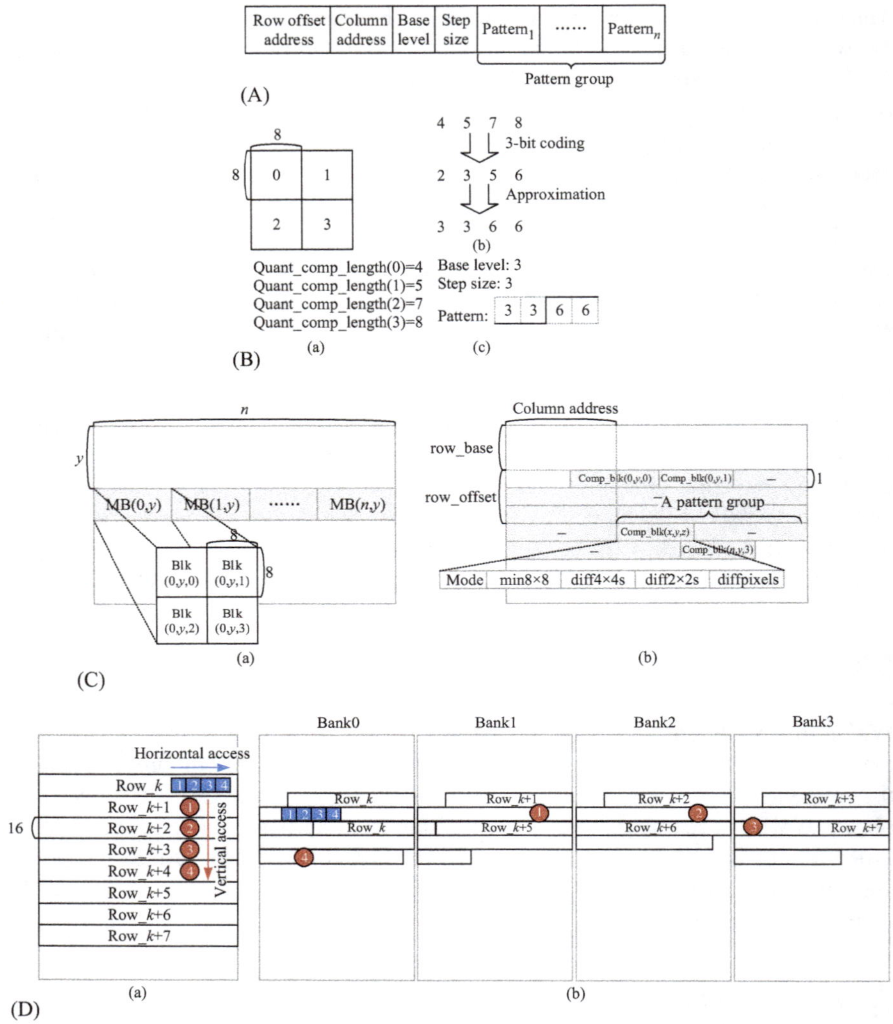

Figure 7.14 Address table compression and reference memory access style of IP module and ME module: (A) address packing scheme; (B) conceptual illustration of pattern-based compression: (a) four compression unit blocks, (b) coding and approximation and (c) code components; (C) start address: (a) original frame and (b) compressed block placement in SDRAM; (D) frame memory placement scheme for four-bank SDRAM: (a) original frame and (b) placement in SDRAM

As shown in Figure 7.14(D), every fourth MB row of the original frame is mapped to each bank. Thus, the row change of adjacent vertical direction access can be reduced.

7.3.1.2 Pixels duplication and luma–chroma correlated mapping

In [20] (for H.264), a pixel duplication and luma–chroma correlated mapping scheme is proposed to increase bandwidth efficiency and reduce bank conflict (Figure 7.15). Reference pixels are duplicated so that one pixel can be stored in two different words with different starting pixel positions. In this way, word crossings can be avoided. The luma pixel and its correlated chroma pixels are mapped correspondingly into different banks; thus, bank conflict between luma block and chroma block can be avoided.

7.3.1.3 TLB addressing and fixed addressing

In [5] (for H.265), they use three banks and one small on-chip SRAM to store reference pixels: PR bank, TR bank, TLB bank and a TLB SRAM. TLB is used to

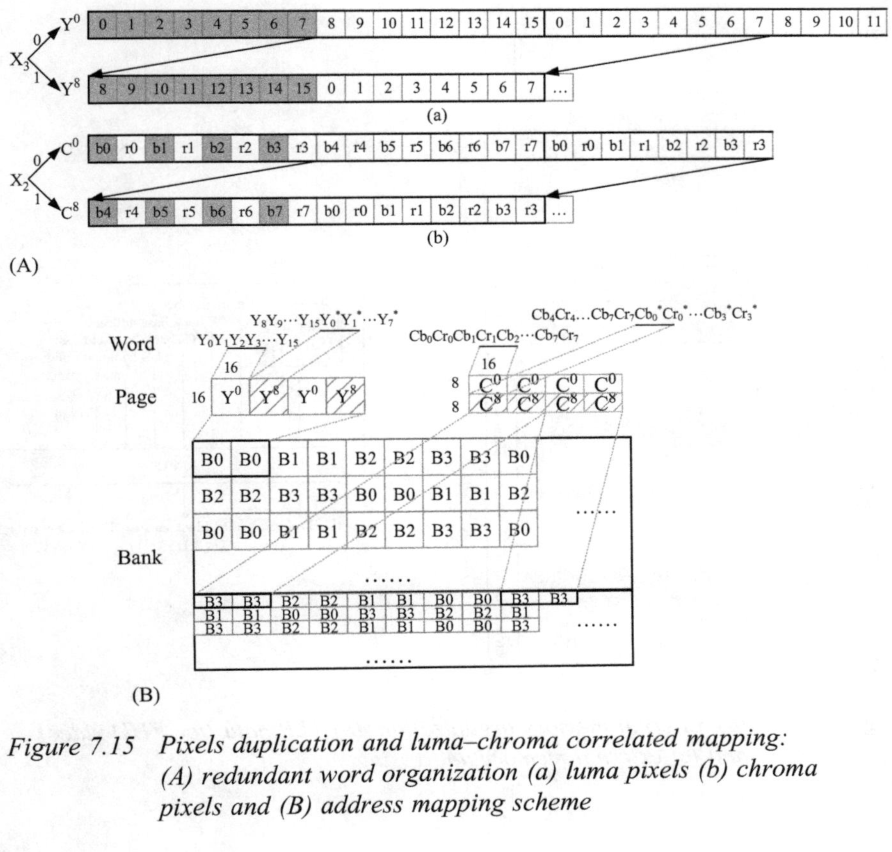

Figure 7.15 Pixels duplication and luma–chroma correlated mapping: (A) redundant word organization (a) luma pixels (b) chroma pixels and (B) address mapping scheme

record the start address and length of each 8 × 8 block in PR bank (Figure 7.16). Fixed length data, such as TR and TLB data, are addressed by their base address and block index. Variable data such as PR data is byte aligned and use the TLB addressing method.

7.3.1.4 Dual-mode memory addressing (DMMA): direct block memory access (DBMA) and relative block memory access (RBMA)

In [21] (for H.264), the dual-mode memory addressing method is proposed to deal with the concatenated allocation of storage entry. Each frame has a specified word length (WL) of a pixel after recompression. In direct block memory access (DBMA) mode, WL_DBMA is calculated as below:

$$WL\_DBMA_i = \lceil \log_2(CL\_Max_i + 1) \rceil$$
$$CL\_Max_i = \max_{j=1,\ldots,N} \{CL(B_{ij})\}$$

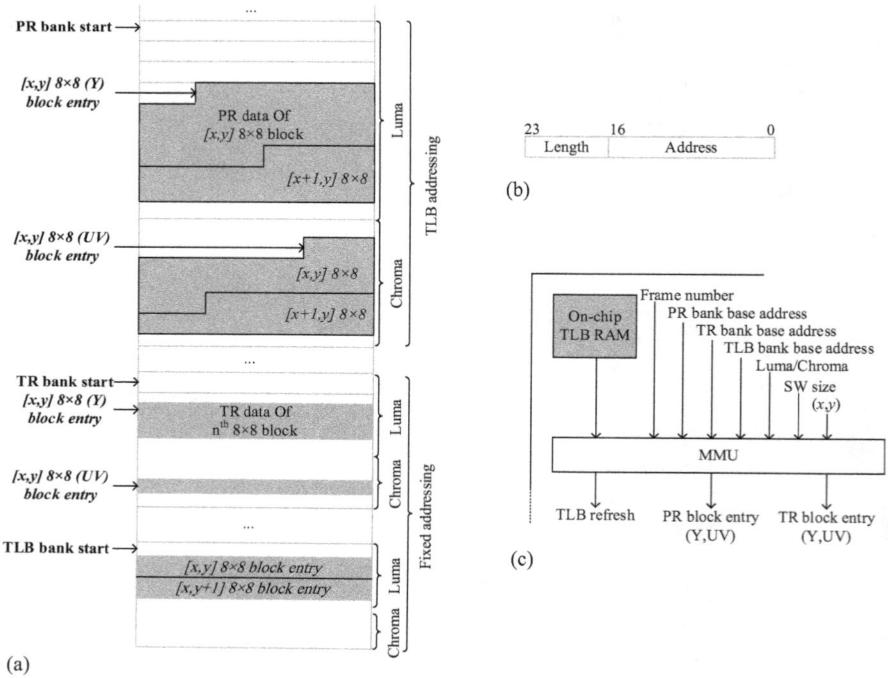

Figure 7.16 (a) External memory organization, (b) TLB data (for FHD video) and (c) external memory addressing

When the difference among CLs of every block in a frame is small enough, a more compact WL (WL_RBMA) would be used to get the start address of each MB of the frames.

$$WL\_RBMA_i = \begin{cases} \lceil \log_2 |N_{MIN_i}| \rceil + 1, if |N_{MIN_i}| > P_{MAX_i} \\ \lceil \log_2 (P_{MAX_i} + 1) \rceil + 1, if |N_{MIN_i}| < P_{MAX_i} \end{cases}$$

$$N_{MIN_i} = \begin{cases} \min S_{N_i}, \text{if} S_{N_i} \neq \emptyset \\ 0, \text{if} S_{N_i} = \emptyset \end{cases}$$

where $S_{N_i} = \{RL(B_{ij}) | RL(B_{ij}) < 0, j = 1, \ldots, N\}$

$$P_{MAX_i} = \begin{cases} \max S_{P_i}, if S_{P_i} \neq \emptyset \\ 0, if S_{P_i} = \emptyset \end{cases}$$

where $S_{P_i} = \{RL(B_{ij}) | RL(B_{ij}) \geq 0, j = 1, \ldots, N\}$

At last, WL is equal to the smaller one between WL_DBMA and WL_RBMA.

7.3.2 Works on power consumption

7.3.2.1 Different-level memory mapping for VCR-LFRC (variable-compression-ratio lossless frame recompression)

In [22] (for H.265), a two-level memory mapping is proposed to save memory traffic for variable CR (compression ratio) encoding. As shown in Figure 7.17, the lower level is each partition and the higher level is a p-group, composed of several horizontally or vertically consecutive partitions. After recompression, p-groups-level memory mapping is unchanged, while all the partitions are stored compactly inside each p-group. Furthermore, the burst access method could be used to alleviate the overhead in memory traffic on average.

Earlier in [23–25] (for H.264), similar group level memory mapping methods are proposed. For example, in [23], as shown in Figure 7.18, in each reference frame, every 128 × 64 pixel is stored in the same row address of DRAM. They are split into four banks (64 × 32 pixel), each of which consists of eight groups (16 × 16 pixel). Moreover, each group contains eight vertically adjacent partitions (16 × 2). After compression, the mapping of row, bank and group addresses are preserved, while the compressed partitions are stored compactly inside a group. To reduce the overhead of memory access, the compressed data of the upper half (group 0–3) are stored stick to the lower boundary, while that of the lower half (group 4–7) are stored stick to the upper boundary. Thus a vertical burst access method can be used even if it crosses the group boundary.

7.3.2.2 Main memory and auxiliary memory

In [19] (for H.265), two regions of memory are used to work in parallel with SFL codes. The main memory is divided into 2,025 partitions, and each partition uses

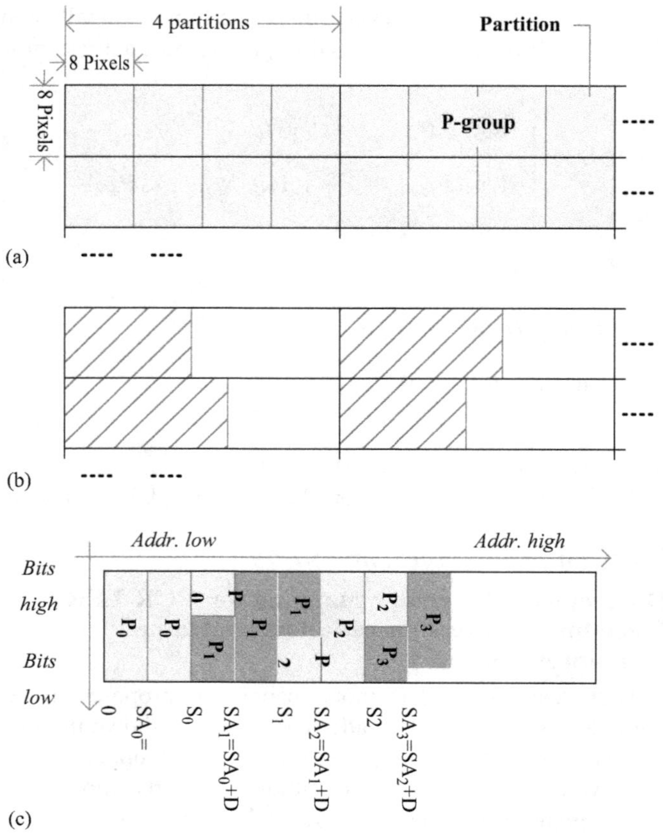

Figure 7.17 Two-level memory mapping to support random access of frame data compressed in variable CR: (a) example memory mapping of the original frame data. A partition consists of 8 × 8 samples. A p-group consists of 4 × 1 partition. (b) P-group-level memory mapping is preserved after recompression. Inside each p-group, compressed partitions are stored compactly to one end of the p-group. (c) Partition-level memory mapping inside a p-group. PN stands for the compressed data of the Nth partition. The SAs of the partitions can be derived by accumulating the DSs

128 words (fixed-length) to store an encoded 32 × 32 block. The first 127 words (32 bits per word) of the encoded blocks are stored in the main memory, and the remaining words (variable-length) are all stored in the auxiliary memory to reduce memory fragmentation. The 128th word in the main memory is used to store a pointer to the auxiliary memory.

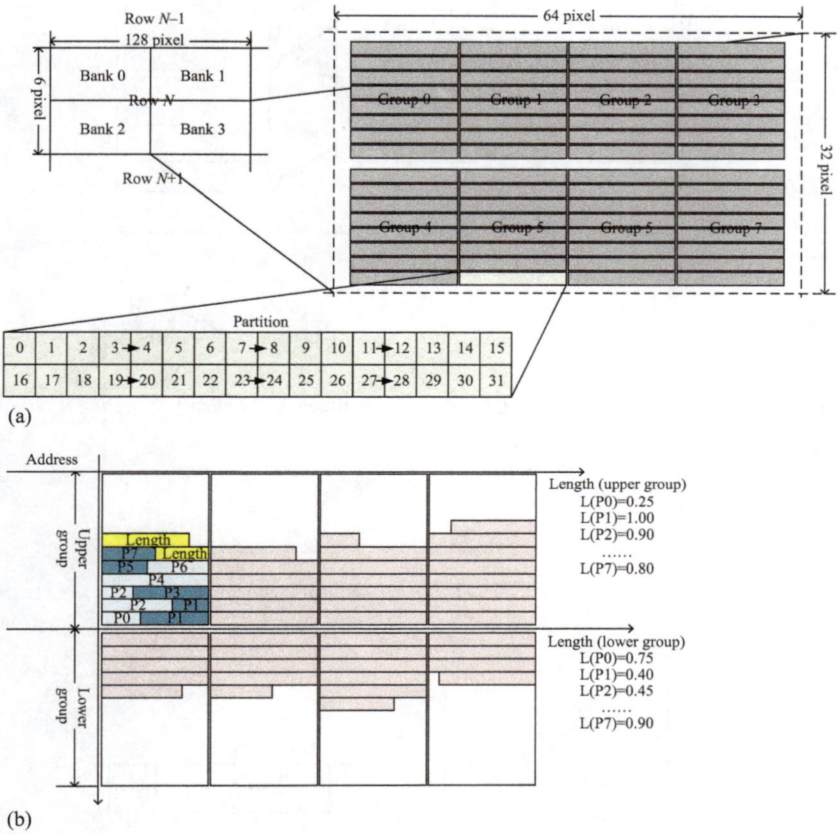

Figure 7.18 (a) Memory organizations for reference frames and (b) compressed mapping of partitions

7.3.2.3 Partition group table-based compression storage

In [18], a partition group table (PGT)-based compression storage scheme is proposed. As shown in Figure 7.19, every two horizontally adjacent 16×16 partitions constitute a PG, and every PG corresponds to a PGT item to record address mapping information. Within each PGT item, partition group offset (PGO) records the starting address of the PG; CSF means compression skip flag of the chroma components in a PG; length records the optimal burst length; MF indicates whether to use the compression storage scheme.

7.3.3 Other works

7.3.3.1 MB-column-based memory mapping

In [26] (H.264), an MB-column-based memory mapping scheme is proposed to fully eliminate bank-access conflicts. As shown in Figure 7.20, reference frames

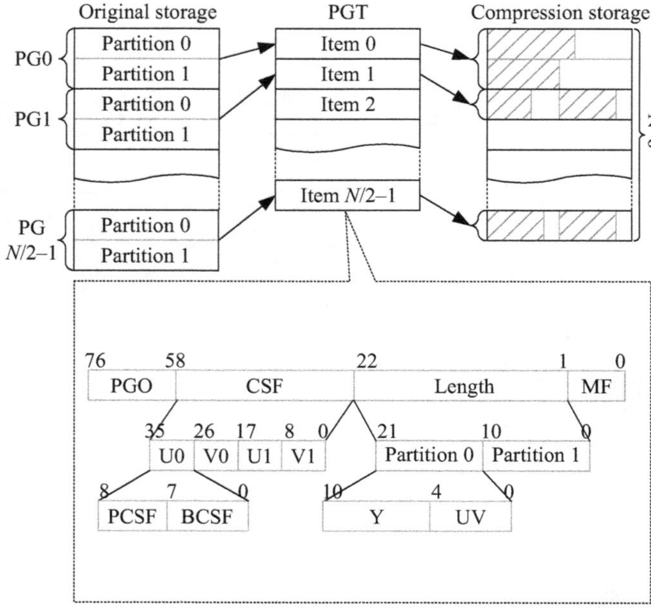

Figure 7.19 PGT-based memory mapping

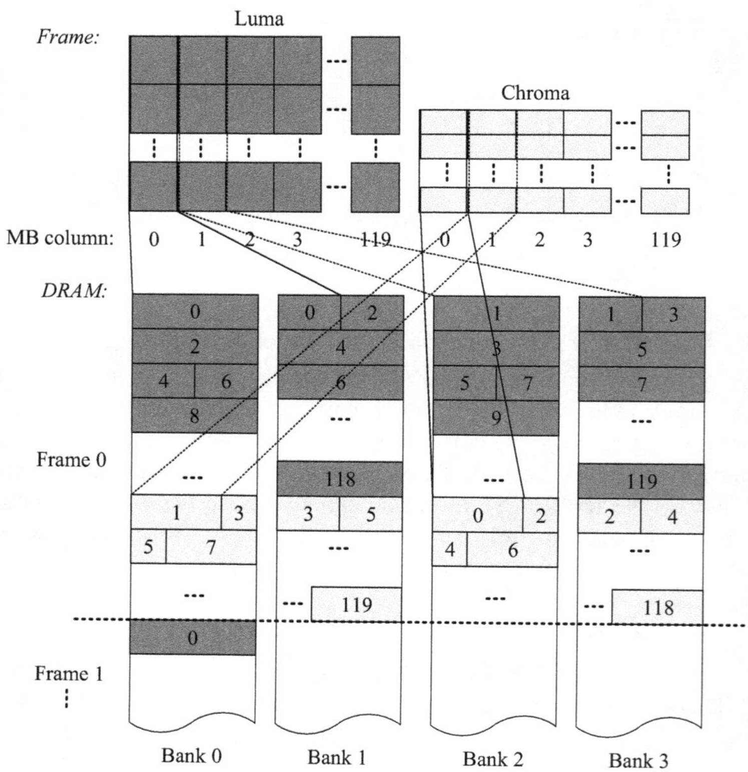

Figure 7.20 MB-column-based mapping (in case of full HD)

are split into MB-column-aligned regions, so accesses to DRAM are also aligned to MB columns. Luma data of even-numbered MB-columns is mapped to bank 0 and 1, while that of odd-numbered MB-columns are mapped to bank 3 and 4. For chroma data, it is mapped in the opposite way. Both MC and DF would access bank 0 and 1 first to collect luma or chroma pixels, followed by bank 2 and 3.

References

[1] L. Guo, D. Zhou and S. Goto, "A New Reference Frame Recompression Algorithm and Its VLSI Architecture for UHDTV Video Codec," in *IEEE Transactions on Multimedia*, vol. 16, no. 8, pp. 2323–2332, 2014.

[2] T. Song and T. Shimamoto, "Reference Frame Data Compression Method for H.264/AVC," in *IEICE Electronics Express*, vol. 4, no. 3, pp. 121–126, 2007.

[3] D. Silveira, G. Povala, L. Amaral, B. Zatt, L. Agostini and M. Porto, "A low-complexity and lossless reference frame encoder algorithm for video coding," *2014 IEEE International Conference on Acoustics, Speech and Signal Processing (ICASSP)*, Florence, 2014, pp. 7358–7362.

[4] Y. Lee, C. Rhee and H. Lee, "A new frame recompression algorithm integrated with H.264 video compression," *2007 IEEE International Symposium on Circuits and Systems*, New Orleans, LA, 2007, pp. 1621–1624.

[5] Y. Fan, Q. Shang and X. Zeng, "In-Block Prediction-Based Mixed Lossy and Lossless Reference Frame Recompression for Next-Generation Video Encoding," in *IEEE Transactions on Circuits and Systems for Video Technology*, vol. 25, no. 1, pp. 112–124, 2015.

[6] K. Yoo, C. Han, U. Kang, and K. Sohn, "Embedded compression algorithm using error-aware quantization and hybrid DPCM/BTC coding," *2011 IEEE International Conference on Multimedia and Expo*, Barcelona, 2011, pp. 1–6.

[7] E. Delp and O. Mitchell, "Image Compression Using Block Truncation Coding," in *IEEE Transactions on Communications*, vol. 27, no. 9, pp. 1335–1342, 1979.

[8] S. Lee, M. Chung, S. Park and C. Kyung, "Lossless Frame Memory Recompression for Video Codec Preserving Random Accessibility of Coding Unit," in *IEEE Transactions on Consumer Electronics*, vol. 55, no. 4, pp. 2105–2113, 2009.

[9] J. Kim and C. Kyung, "A Lossless Embedded Compression Using Significant Bit Truncation for HD Video Coding," in *IEEE Transactions on Circuits and Systems for Video Technology*, vol. 20, no. 6, pp. 848–860, 2010.

[10] T. Y. Lee, "A New Frame-Recompression Algorithm and Its Hardware Design for MPEG-2 Video Decoders," in *IEEE Transactions on Circuits and Systems for Video Technology*, vol. 13, no. 6, pp. 529–534, 2003.

[11] M. Narasimha and A. Peterson, "On the Computation of the Discrete Cosine Transform," in *IEEE Transactions on Communications*, vol. 26, no. 6, pp. 934–936, 1978.

[12] W.-Y. Chen, L.-F. Ding, P.-K. Tsung and L.-G. Chen, "Architecture design of high performance embedded compression for high definition video coding," *2008 IEEE International Conference on Multimedia and Expo*, Hannover, 2008, pp. 825–828.

[13] D. A. Huffman, "A Method for the Construction of Minimum-Redundancy Codes," in *Proceedings of the IRE*, vol. 40, no. 9, pp. 1098–1101, 1952.

[14] D. Silveira, M. Porto and L. Agostini, "A lossless approach for external memory bandwidth reduction in video coding systems and its VLSI architecture," *2013 IEEE International Conference on Multimedia and Expo (ICME)*, San Jose, CA, 2013, pp. 1–6.

[15] T. L. Bao Yng, B. Lee and H. Yoo, "A Low Complexity and Lossless Frame Memory Compression for Display Devices," in *IEEE Transactions on Consumer Electronics*, vol. 54, no. 3, pp. 1453–1458, August 2008.

[16] Y. Hwang, M. Lyu and C. Lin, "A Low-Complexity Embedded Compression Codec Design With Rate Control for High-Definition Video," in *IEEE Transactions on Circuits and Systems for Video Technology*, vol. 25, no. 4, pp. 674–687, 2015.

[17] T. L. Bao Yng, B. Lee and H. Yoo, "A Low Complexity and Lossless Frame Memory Compression for Display Devices," in *IEEE Transactions on Consumer Electronics*, vol. 54, no. 3, pp. 1453–1458, 2008.

[18] X. Lian, Z. Liu, W. Zhou and Z. Duan, "Lossless Frame Memory Compression Using Pixel-Grain Prediction and Dynamic Order Entropy Coding," in *IEEE Transactions on Circuits and Systems for Video Technology*, vol. 26, no. 1, pp. 223–235, 2016.

[19] D. Silveira, G. Povala, L. Amaral, B. Zatt, L. Agostini and M. Porto, "A real-time architecture for reference frame compression for high definition video coders," *2015 IEEE International Symposium on Circuits and Systems (ISCAS)*, Lisbon, 2015, pp. 842–845.

[20] P. Zhang, W. Gao, D. Wu and D. Xie, "An efficient reference frame storage scheme for H.264 HDTV decoder," *2006 IEEE International Conference on Multimedia and Expo*, Toronto, ON., 2006, pp. 361–364.

[21] Y. H. Lee, C. C. Chen, and Y.-L. You, "Design of VLSI Architecture of Autocorrelation-Based Lossless Recompression Engine for Memory-Efficient Video Coding Systems," in *Springer Circuits, Systems, and Signal Processing*, vol. 33, no. 2, pp. 459–482, 2014.

[22] D. Zhou, L. Guo, J. Zhou and S. Goto, "Reducing power consumption of HEVC codec with lossless reference frame recompression," *2014 IEEE International Conference on Image Processing (ICIP)*, Paris, 2014, pp. 2120–2124.

[23] D. Zhou, J. Zhou, X. He, *et al.*, "A 530 Mpixels/s 4096 × 2160@60fps H.264/AVC High Profile Video Decoder Chip," in *IEEE Journal of Solid-State Circuits*, vol. 46, no. 4, pp. 777–788, April 2011.

[24] D. Zhou, L. Guo, J. Zhou and S. Goto, "Constant Rate Bandwidth of HEVC codec with lossless reference frame recompression," *2014 IEEE International Conference on Image Processing (ICIP)*, Paris, 2014, pp. 2120–2124.
[25] X. Bao, D. Zhou and S. Goto, "A lossless frame recompression scheme for reducing DRAM power in video encoding," *Proceedings of 2010 IEEE International Symposium on Circuits and Systems*, Paris, 2010, pp. 677–680.
[26] P. Chao and Y.-L. Lin, "Reference frame access optimization for ultra high resolution H.264/AVC decoding," *2008 IEEE International Conference on Multimedia and Expo*, Hannover, 2008, pp. 1441–1444.

Chapter 8
Discrete transform approximations for video coding

Diego Felipe Gomes Coelho[1], Vitor de Andrade Coutinho[3], Thiago Lopes Trugillo da Silveira[2], Renato J. Cintra[1,3], Fábio Mariano Bayer[4], Arjuna Madanayake[5], and Vassil Simeonov Dimitrov[1]

8.1 Introduction

Discrete transforms are central elements in image and video-compression codecs. Several discrete transforms have been employed in the context of image and video compression, such as the Hadamard transform, the Haar transform, the Walsh–Hadamard transform, the discrete Fourier transform (DFT), and the discrete Tchebichef transform, to name a few [1–3].

Among the many existing discrete transforms, the discrete cosine transform (DCT) of type II—hereafter termed only DCT—is widely regarded a major discrete transform in the context of image and video compression [2,4–6]. This is due to its capability for approximating the Karhunen–Loève transform (KLT) when input signals follow a highly correlated stationary Markov-I process [2].

The KLT is a data-dependent discrete transform, and its definition requires the eigendecomposition of the autocorrelation matrix of the input signal [2]. Consequently, the KLT possesses a high computational complexity. The usefulness of the DCT comes from the fact that it is not data-dependent as the KLT, and its associated arithmetic complexity is remarkably smaller. The adoption of the DCT for image and video processing is particularly suitable because natural images and

[1]Department of Electrical and Computer Engineering, University of Calgary, Calgary, Canada
[2]Programa de Pós-Graduação em Computação, Universidade Federal do Rio Grande do Sul, Porto Alegre, Brazil
[3]Signal Processing Group, Departamento de Estatística, Universidade Federal de Pernambuco, Recife, Brazil
[4]Departamento de Estatística, Universidade Federal de Santa Maria, Santa Maria, Brazil
[5]Department of Electrical and Computer Engineering, The University of Akron, Akron, OH, United States

video frames are, in general, well modeled as a highly correlated stationary Markov-I process. In such a case, the DCT behaves as an asymptotic approximation for the KLT and performs near optimality.

Because of its relevance in image and video compression, the DCT has received significant attention from the scientific community, and several fast algorithms have been proposed for the DCT computation [2,3]. Among such algorithms, we highlight the following: the Feig–Winograd fast algorithm for 1D and 2D DCT [7], the Loeffler decomposition [8], the Chen factorization [9], the Arai algorithm [10], and the split-radix algorithm [11]. The intense development of fast algorithms for the DCT computation generated methods capable of computing the DCT at the minimum or near minimum arithmetic complexity. In particular, the Loeffler algorithm is especially relevant because it achieves the mathematically minimum lower bound for the DCT multiplicative complexity, which was demonstrated by Heideman [12].

The increasing demand for storage and transmission of image and video revealed a critical need for efficient multimedia-compression methods. Such fact has motivated researchers to investigate novel transforms offering a compromise between performance and arithmetic complexity. A specifically relevant approach is the design of approximate transforms based on the DCT or other usual transforms. Although those transforms cannot provide exact computation, they are able to furnish *approximate* results that have practical significance at a considerably smaller arithmetic and computational cost.

Studies in neuroscience have shown that the human brain is robust to sensorial perception distortion [13,14]. This is a characteristic of neural networks, i.e., to perform interpolation of missing data [15,16]. This fact suggests that image and video contents do not need to be perfectly computed to be meaningfully interpreted by the human brain. Therefore, approximate methods can compress and recover video content that is not an exact match to the original but can still be correctly perceived by the human brain without noticeable quality degradation. In this scenario, low-complexity DCT approximations are good candidates to provide video-compressing methods that can still result in quality content at a significantly smaller hardware, power, and time usage.

The signed DCT (SDCT) [17], whose coefficients are simply the sign of the DCT coefficients, is regarded as the pioneering DCT approximation. Because the SDCT is such a crude approximation, some mathematical properties of the exact DCT are lost in the approximation process. For example, although not an extremely serious issue, the SDCT lacks orthogonality. However, it fully demonstrates that extreme quantization or crude approximation of the DCT can still result in transformation matrices that can be good enough for image and video compression. One of the reasons for such robustness is the fact that the DCT computation is not the only source of errors in a complete compression/decompression system. In fact, other subsystems, such as the quantization step, introduce a number of error components [2,18]. Thus, the precision of the transformation step can—and should—be reduced to the error floor affected by the other parts of the system.

In general, approximate transforms substitute irrational coefficients in the definition of the exact transform by carefully selected low-complexity rational quantities. This drastically reduces the arithmetic complexity by eliminating multiplications and possibly minimizing the number of additions. In terms of hardware implementation, such elimination of multiplications often affects a reduction of resource consumption, such as energy, gate count, registers, and chip area. Not only that, it often also leads to circuit designs capable of higher throughput and operating frequency.

In this chapter, we review several relevant low-complexity transforms for image and video compression. We present metrics for evaluating DCT approximate transforms and search for optimal transforms in terms of the evaluation metrics. The optimal transforms are then submitted to hardware implementation and compared to the standard approximations for high-efficiency video coding (HEVC).

8.2 Literature review

After the introduction of the SDCT in 2001, several other transforms have been proposed for approximating the DCT. In [19], Lengwehasatit and Ortega (LO) proposed a scalable structure to approximate the DCT which resulted in a class of DCT approximations. The method drawback is that most of the resulting transforms require a high additive complexity. Nevertheless, one of the LO approximations, often referred to as the level-one approximation, possesses a remarkably low arithmetic complexity compared to its peer transforms. The LO level-one approximation [19] is frequently considered as a benchmark transform for performance comparisons [20–25].

In 2008, Bouguezel, Ahmad, and Swamy (BAS) started a series of papers proposing new transforms in the context of image compression [20,21]. The series of approximations by BAS continued until 2013, totaling eight transforms, denoted in this work as $\mathbf{T}_{\mathrm{BAS}_1}$ [20], $\mathbf{T}_{\mathrm{BAS}_2}$ [21], $\mathbf{T}_{\mathrm{BAS}_3}$ [26], $\mathbf{T}_{\mathrm{BAS}_4}$ [22], $\mathbf{T}_{\mathrm{BAS}_5}$ [24] (for $a = 1$), $\mathbf{T}_{\mathrm{BAS}_6}$ [24] (for $a = 0$), $\mathbf{T}_{\mathrm{BAS}_7}$ [24] (for $a = 1/2$), and $\mathbf{T}_{\mathrm{BAS}_8}$ [23]. The BAS transforms were designed based on inspection [20,21,26], single variable matrix parametrization of existing approximations [24], and derivations based on other low-complexity matrices [23] (Tables 8.1 and 8.2).

Several DCT approximations were proposed based on the judicious substitution of the required irrational quantities in the DCT. The work in [27] proposes a new DCT approximation based on scaling the irrational quantities required by the DCT followed by a rounding operation. The work in [28] derives approximations for DCT, DFT, and discrete Hartley transform. Such transforms are based on the application of the rounding function to the original transform matrix. The transform proposed in [29] is obtained through inspection and possesses a low complexity when compared to the state-of-the-art DCT approximations.

The work in [30] proposes several DCT approximations based on integer functions, where several transforms are catalogued and compared. The transforms

Table 8.1 BAS series of transforms

Transform matrix	Orthogonal?	Description
$\mathbf{BAS}_1 = \begin{bmatrix} 1 & 1 & 1 & 1 & 1 & 1 & 1 & 1 \\ 1 & 1 & 1 & 0 & 0 & -1 & -1 & -1 \\ 1 & 1 & -1 & -1 & -1 & -1 & 1 & 1 \\ 1 & 0 & -1 & 0 & 0 & 1 & 0 & -1 \\ 1 & -1 & -1 & 1 & 1 & -1 & -1 & 1 \\ 1 & -1 & 1 & 0 & 0 & -1 & 1 & -1 \\ 1 & -1 & 1 & -1 & -1 & 1 & -1 & 1 \\ 1 & -1 & 1 & -1 & 1 & -1 & 1 & -1 \end{bmatrix}$	No	BAS [20]
$\mathbf{BAS}_2 = \begin{bmatrix} 1 & 1 & 1 & 1 & 1 & 1 & 1 & 1 \\ 1 & 1 & 0 & 0 & 0 & 0 & -1 & -1 \\ 1 & \frac{1}{2} & -\frac{1}{2} & -1 & -1 & -\frac{1}{2} & \frac{1}{2} & 1 \\ 0 & 0 & -1 & 0 & 0 & 1 & 0 & 0 \\ 1 & -1 & -1 & 1 & 1 & -1 & -1 & 1 \\ 1 & -1 & 0 & 0 & 0 & 0 & 1 & -1 \\ \frac{1}{2} & -1 & 1 & -\frac{1}{2} & -\frac{1}{2} & 1 & -1 & \frac{1}{2} \\ 0 & 0 & 0 & -1 & 1 & 0 & 0 & 0 \end{bmatrix}$	Yes	BAS [21]
$\mathbf{BAS}_3 = \begin{bmatrix} 1 & 1 & 1 & 1 & 1 & 1 & 1 & 1 \\ 1 & 1 & 0 & 0 & 0 & 0 & -1 & -1 \\ 1 & 1 & -1 & -1 & -1 & -1 & 1 & 1 \\ 0 & 0 & -1 & 0 & 0 & 1 & 0 & 0 \\ 1 & -1 & -1 & 1 & 1 & -1 & -1 & 1 \\ 1 & -1 & 0 & 0 & 0 & 0 & 1 & -1 \\ 1 & -1 & 1 & -1 & -1 & 1 & -1 & 1 \\ 0 & 0 & 0 & -1 & 1 & 0 & 0 & 0 \end{bmatrix}$	Yes	BAS [26]
$\mathbf{BAS}_4 = \begin{bmatrix} 1 & 1 & 1 & 1 & 1 & 1 & 1 & 1 \\ 1 & 1 & 1 & 1 & -1 & -1 & -1 & -1 \\ 2 & 1 & -1 & -2 & -2 & -1 & 1 & 2 \\ 2 & 1 & -1 & -2 & 2 & 1 & -1 & -2 \\ 1 & -1 & -1 & 1 & 1 & -1 & -1 & 1 \\ 1 & -1 & -1 & 1 & -1 & 1 & 1 & -1 \\ 1 & -2 & 2 & -1 & -1 & 2 & -2 & 1 \\ 1 & -2 & 2 & -1 & 1 & -2 & 2 & -1 \end{bmatrix}$	Yes	BAS [22]

in [30] were obtained through the application of several integer functions such as truncation and rounding in its different modes of operation. Tables 8.3–8.6 show all the different transforms listed in [30], which were obtained via truncation, round-half-to-even, and round-half-toward-zero functions, respectively [30]. In [31], the authors propose a new class of DCT approximations based on the Feig–Winograd fast

Table 8.2 BAS series of transforms

Transform matrix	Orthogonal?	Description
$\mathbf{BAS}_5 = \begin{bmatrix} 1 & 1 & 1 & 1 & 1 & 1 & 1 & 1 \\ 1 & 1 & 0 & 0 & 0 & 0 & -1 & -1 \\ 1 & 1 & -1 & -1 & -1 & -1 & 1 & 1 \\ 0 & 0 & 1 & 0 & 0 & -1 & 0 & 0 \\ 1 & -1 & -1 & 1 & 1 & -1 & -1 & 1 \\ 0 & 0 & 0 & 1 & -1 & 0 & 0 & 0 \\ 1 & -1 & 0 & 0 & 0 & 0 & 1 & -1 \\ 1 & -1 & 1 & -1 & -1 & 1 & -1 & 1 \end{bmatrix}$	Yes	BAS [24]
$\mathbf{BAS}_6 = \begin{bmatrix} 1 & 1 & 1 & 1 & 1 & 1 & 1 & 1 \\ 1 & 1 & 0 & 0 & 0 & 0 & -1 & -1 \\ 1 & 0 & 0 & -1 & -1 & 0 & 0 & 1 \\ 0 & 0 & 1 & 0 & 0 & -1 & 0 & 0 \\ 1 & -1 & -1 & 1 & 1 & -1 & -1 & 1 \\ 0 & 0 & 0 & 1 & -1 & 0 & 0 & 0 \\ 1 & -1 & 0 & 0 & 0 & 0 & 1 & -1 \\ 0 & -1 & 1 & 0 & 0 & 1 & -1 & 0 \end{bmatrix}$	Yes	BAS [24]
$\mathbf{BAS}_7 = \begin{bmatrix} 1 & 1 & 1 & 1 & 1 & 1 & 1 & 1 \\ 1 & 1 & 0 & 0 & 0 & 0 & -1 & -1 \\ 1 & \frac{1}{2} & -\frac{1}{2} & -1 & -1 & -\frac{1}{2} & \frac{1}{2} & 1 \\ 0 & 0 & 1 & 0 & 0 & -1 & 0 & 0 \\ 1 & -1 & -1 & 1 & 1 & -1 & -1 & 1 \\ 0 & 0 & 0 & 1 & -1 & 0 & 0 & 0 \\ 1 & -1 & 0 & 0 & 0 & 0 & 1 & -1 \\ \frac{1}{2} & -1 & 1 & -\frac{1}{2} & -\frac{1}{2} & 1 & -1 & \frac{1}{2} \end{bmatrix}$	Yes	BAS [24]
$\mathbf{BAS}_8 = \begin{bmatrix} 1 & 1 & 1 & 1 & 1 & 1 & 1 & 1 \\ 1 & 1 & 1 & 1 & -1 & -1 & -1 & -1 \\ 1 & 1 & -1 & -1 & -1 & -1 & 1 & 1 \\ 1 & 1 & -1 & -1 & 1 & 1 & -1 & -1 \\ 1 & -1 & -1 & 1 & 1 & -1 & -1 & 1 \\ 1 & -1 & -1 & 1 & -1 & 1 & 1 & -1 \\ 1 & -1 & 1 & -1 & -1 & 1 & -1 & 1 \\ 1 & -1 & 1 & -1 & 1 & -1 & 1 & -1 \end{bmatrix}$	Yes	BAS [23]

algorithm. The new DCT approximations are obtained through a nonlinear discrete optimization problem.

Although all the previous video codecs have concentrated in using only 8-point blocks, the new HEVC standard requires transforms for various blocksizes: 4×4, 8×8, 16×16, and 32×32. This demand for transforms of larger blocklengths requires

Table 8.3 Selected transforms: SDCT, RDCT, and Lengwehasatit–Ortega

Transform matrix	Orthogonal?	Reference
$\mathbf{T}_{\mathrm{SDCT}} = \begin{bmatrix} 1 & 1 & 1 & 1 & 1 & 1 & 1 & 1 \\ 1 & 1 & 1 & 1 & -1 & -1 & -1 & -1 \\ 1 & 1 & -1 & -1 & -1 & -1 & 1 & 1 \\ 1 & -1 & -1 & -1 & 1 & 1 & 1 & -1 \\ 1 & -1 & -1 & 1 & 1 & -1 & -1 & 1 \\ 1 & -1 & 1 & 1 & -1 & -1 & 1 & -1 \\ 1 & -1 & 1 & -1 & -1 & 1 & -1 & 1 \\ 1 & -1 & 1 & -1 & 1 & -1 & 1 & -1 \end{bmatrix}$	No	SDCT [17]
$\mathbf{T}_{\mathrm{RDCT}} = \begin{bmatrix} 1 & 1 & 1 & 1 & 1 & 1 & 1 & 1 \\ 1 & 1 & 1 & 0 & 0 & -1 & -1 & -1 \\ 1 & 0 & 0 & -1 & -1 & 0 & 0 & 1 \\ 1 & 0 & -1 & -1 & 1 & 1 & 0 & -1 \\ 1 & -1 & -1 & 1 & 1 & -1 & -1 & 1 \\ 1 & -1 & 0 & 1 & -1 & 0 & 1 & -1 \\ 0 & -1 & 1 & 0 & 0 & 1 & -1 & 0 \\ 0 & -1 & 1 & -1 & 1 & -1 & 1 & 0 \end{bmatrix}$	Yes	RDCT [27]
$\mathbf{T}_{\mathrm{LO}} = \begin{bmatrix} 1 & 1 & 1 & 1 & 1 & 1 & 1 & 1 \\ 1 & 1 & 1 & 0 & 0 & -1 & -1 & -1 \\ 1 & \frac{1}{2} & -\frac{1}{2} & -1 & -1 & -\frac{1}{2} & \frac{1}{2} & 1 \\ 1 & 0 & -1 & -1 & 1 & 1 & 0 & -1 \\ 1 & -1 & -1 & 1 & 1 & -1 & -1 & 1 \\ 1 & -1 & 0 & 1 & -1 & 0 & 1 & -1 \\ \frac{1}{2} & -1 & 1 & -\frac{1}{2} & -\frac{1}{2} & 1 & -1 & \frac{1}{2} \\ 0 & -1 & 1 & -1 & 1 & -1 & 1 & 0 \end{bmatrix}$	Yes	Lengwehasatit–Ortega [19]

efficient DCT approximations and algorithms for hardware and software implementations. A number of 16-point DCT approximations obtained directly from the exact 16-point DCT have been proposed in the literature due to its importance given in the HEVC standard [32–35]. The works in [36,37] used a scalable procedure to obtain larger DCT approximations from smaller transforms. In this way, the work in [36] proposed a structure that can furnish all the DCT approximations required by the HEVC standard. This suggests that if we have at hand a good 8-point DCT approximation, we can possibly obtain good 16- and 32-point DCT approximations. Therefore, by considering the availability of a scalable methodology, we focus our attention to 8-point DCT approximations, and the scalable method proposed in [36] can be used to produce larger size DCT approximations.

Table 8.4 Selected transforms based on the truncation function

Transform matrix	Orthogonal?	Reference
$\tilde{\mathbf{T}}_{\mathrm{T}_1} = \begin{bmatrix} 1 & 1 & 1 & 1 & 1 & 1 & 1 & 1 \\ 1 & 1 & 0 & 0 & 0 & 0 & -1 & -1 \\ 1 & 0 & 0 & -1 & -1 & 0 & 0 & 1 \\ 1 & 0 & -1 & 0 & 0 & 1 & 0 & -1 \\ 1 & -1 & -1 & 1 & 1 & -1 & -1 & 1 \\ 0 & -1 & 0 & 1 & -1 & 0 & 1 & 0 \\ 0 & -1 & 1 & 0 & 0 & 1 & -1 & 0 \\ 0 & 0 & 1 & -1 & 1 & -1 & 0 & 0 \end{bmatrix}$	No	Introduced in [38]
$\mathbf{T}_{\mathrm{T}_1} = \begin{bmatrix} 1 & 1 & 1 & 1 & 1 & 1 & 1 & 1 \\ 1 & 1 & 1 & 0 & 0 & -1 & -1 & -1 \\ 1 & 0 & 0 & -1 & -1 & 0 & 0 & 1 \\ 1 & 0 & -1 & -1 & 1 & 1 & 0 & -1 \\ 1 & -1 & -1 & 1 & 1 & -1 & -1 & 1 \\ 1 & -1 & 0 & 1 & -1 & 0 & 1 & -1 \\ 0 & -1 & 1 & 0 & 0 & 1 & -1 & 0 \\ 0 & -1 & 1 & -1 & 1 & -1 & 1 & 0 \end{bmatrix}$	Yes	Introduced in [38]
$\mathbf{T}_{\mathrm{T}_2} = \begin{bmatrix} 1 & 1 & 1 & 1 & 1 & 1 & 1 & 1 \\ 2 & 1 & 1 & 0 & 0 & -1 & -1 & -2 \\ 1 & 0 & 0 & -1 & -1 & 0 & 0 & 1 \\ 1 & 0 & -2 & -1 & 1 & 2 & 0 & -1 \\ 1 & -1 & -1 & 1 & 1 & -1 & -1 & 1 \\ 1 & -2 & 0 & 1 & -1 & 0 & 2 & -1 \\ 0 & -1 & 1 & 0 & 0 & 1 & -1 & 0 \\ 0 & -1 & 1 & -2 & 2 & -1 & 1 & 0 \end{bmatrix}$	Yes	Introduced in [38]
$\mathbf{T}_{\mathrm{T}_3} = \begin{bmatrix} 1 & 1 & 1 & 1 & 1 & 1 & 1 & 1 \\ 2 & 1 & 1 & 0 & 0 & -1 & -1 & -2 \\ 2 & 0 & 0 & -2 & -2 & 0 & 0 & 2 \\ 1 & 0 & -2 & -1 & 1 & 2 & 0 & -1 \\ 1 & -1 & -1 & 1 & 1 & -1 & -1 & 1 \\ 1 & -2 & 0 & 1 & -1 & 0 & 2 & -1 \\ 0 & -2 & 2 & 0 & 0 & 2 & -2 & 0 \\ 0 & -1 & 1 & -2 & 2 & -1 & 1 & 0 \end{bmatrix}$	Yes	Introduced in [38]
$\mathbf{T}_{\mathrm{T}_4} = \begin{bmatrix} 2 & 2 & 2 & 2 & 2 & 2 & 2 & 2 \\ 3 & 2 & 2 & 0 & 0 & -2 & -2 & -3 \\ 3 & 1 & -1 & -3 & -3 & -1 & 1 & 3 \\ 2 & 0 & -3 & -2 & 2 & 3 & 0 & -2 \\ 2 & -2 & -2 & 2 & 2 & -2 & -2 & 2 \\ 2 & -3 & 0 & 2 & -2 & 0 & 3 & -2 \\ 1 & -3 & 3 & -1 & -1 & 3 & -3 & 1 \\ 0 & -2 & 2 & -3 & 3 & -2 & 2 & 0 \end{bmatrix}$	Yes	Introduced in [38]

Table 8.5 Selected transforms based on the round-half-to-even function

Transform matrix	Orthogonal?	Reference
$\tilde{\mathbf{T}}_{\mathrm{E}_1} = \begin{bmatrix} 1 & 1 & 1 & 1 & 1 & 1 & 1 & 1 \\ 1 & 1 & 0 & 0 & 0 & 0 & -1 & -1 \\ 1 & 0 & 0 & -1 & -1 & 0 & 0 & 1 \\ 1 & 0 & -1 & 0 & 0 & 1 & 0 & -1 \\ 1 & -1 & -1 & 1 & 1 & -1 & -1 & 1 \\ 0 & -1 & 0 & 1 & -1 & 0 & 1 & 0 \\ 0 & -1 & 1 & 0 & 0 & 1 & -1 & 0 \\ 0 & 0 & 1 & -1 & 1 & -1 & 0 & 0 \end{bmatrix}$	No	Introduced in [38]
$\mathbf{T}_{\mathrm{E}_1} = \begin{bmatrix} 1 & 1 & 1 & 1 & 1 & 1 & 1 & 1 \\ 1 & 1 & 1 & 0 & 0 & -1 & -1 & -1 \\ 1 & 0 & 0 & -1 & -1 & 0 & 0 & 1 \\ 1 & 0 & -1 & -1 & 1 & 1 & 0 & -1 \\ 1 & -1 & -1 & 1 & 1 & -1 & -1 & 1 \\ 1 & -1 & 0 & 1 & -1 & 0 & 1 & -1 \\ 0 & -1 & 1 & 0 & 0 & 1 & -1 & 0 \\ 0 & -1 & 1 & -1 & 1 & -1 & 1 & 0 \end{bmatrix}$	Yes	Introduced in [38]
$\mathbf{T}_{\mathrm{E}_2} = \begin{bmatrix} 1 & 1 & 1 & 1 & 1 & 1 & 1 & 1 \\ 1 & 1 & 1 & 0 & 0 & -1 & -1 & -1 \\ 1 & 1 & -1 & -1 & -1 & -1 & 1 & 1 \\ 1 & 0 & -1 & -1 & 1 & 1 & 0 & -1 \\ 1 & -1 & -1 & 1 & 1 & -1 & -1 & 1 \\ 1 & -1 & 0 & 1 & -1 & 0 & 1 & -1 \\ 1 & -1 & 1 & -1 & -1 & 1 & -1 & 1 \\ 0 & -1 & 1 & -1 & 1 & -1 & 1 & 0 \end{bmatrix}$	Yes	Introduced in [38]
$\mathbf{T}_{\mathrm{E}_3} = \begin{bmatrix} 1 & 1 & 1 & 1 & 1 & 1 & 1 & 1 \\ 2 & 1 & 1 & 0 & 0 & -1 & -1 & -2 \\ 1 & 1 & -1 & -1 & -1 & -1 & 1 & 1 \\ 1 & 0 & -2 & -1 & 1 & 2 & 0 & -1 \\ 1 & -1 & -1 & 1 & 1 & -1 & -1 & 1 \\ 1 & -2 & 0 & 1 & -1 & 0 & 2 & -1 \\ 1 & -1 & 1 & -1 & -1 & 1 & -1 & 1 \\ 0 & -1 & 1 & -2 & 2 & -1 & 1 & 0 \end{bmatrix}$	Yes	Introduced in [38]
$\mathbf{T}_{\mathrm{E}_4} = \begin{bmatrix} 1 & 1 & 1 & 1 & 1 & 1 & 1 & 1 \\ 2 & 1 & 1 & 0 & 0 & -1 & -1 & -2 \\ 2 & 1 & -1 & -2 & -2 & -1 & 1 & 2 \\ 1 & 0 & -2 & -1 & 1 & 2 & 0 & -1 \\ 1 & -1 & -1 & 1 & 1 & -1 & -1 & 1 \\ 1 & -2 & 0 & 1 & -1 & 0 & 2 & -1 \\ 1 & -2 & 2 & -1 & -1 & 2 & -2 & 1 \\ 0 & -1 & 1 & -2 & 2 & -1 & 1 & 0 \end{bmatrix}$	Yes	Introduced in [38]
$\mathbf{T}_{\mathrm{E}_5} = \begin{bmatrix} 2 & 2 & 2 & 2 & 2 & 2 & 2 & 2 \\ 3 & 2 & 1 & 1 & -1 & -1 & -2 & -3 \\ 2 & 1 & -1 & -2 & -2 & -1 & 1 & 2 \\ 2 & -1 & -3 & -1 & 1 & 3 & 1 & -2 \\ 2 & -2 & -2 & 2 & 2 & -2 & -2 & 2 \\ 1 & -3 & 1 & 2 & -2 & -1 & 3 & -1 \\ 1 & -2 & 2 & -1 & -1 & 2 & -2 & 1 \\ 1 & -1 & 2 & -3 & 3 & -2 & 1 & -1 \end{bmatrix}$	Yes	Introduced in [38]

Table 8.6 Selected transforms based on the round-half-toward-zero function

Transform matrix	Orthogonal?	Reference
$\mathbf{T}_{\mathrm{H}_1} = \begin{bmatrix} 1 & 1 & 1 & 1 & 1 & 1 & 1 & 1 \\ 2 & 1 & 1 & 0 & 0 & -1 & -1 & -2 \\ 1 & 0 & 0 & -1 & -1 & 0 & 0 & 1 \\ 1 & 0 & -2 & -1 & 1 & 2 & 0 & -1 \\ 1 & -1 & -1 & 1 & 1 & -1 & -1 & 1 \\ 1 & -2 & 0 & 1 & -1 & 0 & 2 & -1 \\ 0 & -1 & 1 & 0 & 0 & 1 & -1 & 0 \\ 0 & -1 & 1 & -2 & 2 & -1 & 1 & 0 \end{bmatrix}$	Yes	Introduced in [38]
$\mathbf{T}_{\mathrm{H}_2} = \begin{bmatrix} 1 & 1 & 1 & 1 & 1 & 1 & 1 & 1 \\ 2 & 1 & 1 & 0 & 0 & -1 & -1 & -2 \\ 2 & 0 & 0 & -2 & -2 & 0 & 0 & 2 \\ 1 & 0 & -2 & -1 & 1 & 2 & 0 & -1 \\ 1 & -1 & -1 & 1 & 1 & -1 & -1 & 1 \\ 1 & -2 & 0 & 1 & -1 & 0 & 2 & -1 \\ 0 & -2 & 2 & 0 & 0 & 2 & -2 & 0 \\ 0 & -1 & 1 & -2 & 2 & -1 & 1 & 0 \end{bmatrix}$	Yes	Introduced in [38]

8.3 Design of approximate transforms

In general, a low-complexity DCT approximation $\hat{\mathbf{C}}$ for image and video compression can be written as the product of two matrix elements: (i) a low-complexity matrix $\mathbf{T}$ and (ii) a orthogonalizing matrix $\mathbf{S}$, such that

$$\hat{\mathbf{C}} = \mathbf{S} \cdot \mathbf{T}. \tag{8.1}$$

Only the low-complexity transform $\mathbf{T}$ is actually implemented in hardware. Such matrix usually possesses coefficients that are suitable for efficient hardware implementation, such as small fixed integer quantities.

The orthogonalizing term $\mathbf{S}$ is directly obtained from $\mathbf{T}$ according to

$$\mathbf{S} = \sqrt{\mathrm{diag}\ (\mathbf{T} \cdot \mathbf{T}^{\mathrm{T}})^{-1}}, \tag{8.2}$$

where $\sqrt{\cdot}$ denotes the matrix square root and *diag*() returns the diagonal matrix formed with the elements of the main diagonal in its matrix argument. Note that $\hat{\mathbf{C}}$ is uniquely determined by $\mathbf{T}$, i.e., given $\mathbf{T}$, we can obtain $\mathbf{S}$ and therefore the approximation $\hat{\mathbf{C}}$. In such conditions, we say that $\hat{\mathbf{C}}$ is the DCT approximation induced by $\mathbf{T}$.

8.3.1 Design parameters

The transformation matrices employed for image and video coding share two main properties: they are real-valued and invertible. Such properties come from the fact that, in video and image coding, (i) the computation is usually performed over real arithmetic and (ii) the compressed data often is required to be reconstructed. The exact DCT clearly satisfies such requirements. Because the DCT matrix is real-valued, it is possible to process image/video without the need of complex-valued operations, as it is the case with data compression based on the DFT.

The invertibility of the DCT also plays a very important role in multimedia compression. Such property guarantees the recovery of compressed media. Furthermore, not only the DCT is invertible, but it is orthogonal, i.e., its inverse is its transpose matrix. The orthogonality property [2] of the DCT allows for the *reuse* of the same hardware both in the compression and recovering stages. Orthogonality not only facilitates savings in chip area and resources but also promotes the improvement of performance measurements in metrics such as maximum operating frequency and throughput. Note, however, that orthogonality is not an indispensable property. In fact, near orthogonal transforms can be also suitable for hardware reuse, when the inverse matrix can be meaningfully approximated by the transpose of the approximation. Because of that, transforms that are good for hardware implementation of video compression schemes are usually orthogonal or *near* orthogonal.

A near orthogonal matrix $\hat{\mathbf{C}}$ is one that, although its rows and columns are not orthonormal to each other, the matrix $\hat{\mathbf{C}}\ \hat{\mathbf{C}}^{\mathrm{T}}$ is "almost" a diagonal matrix. This concept was discussed in [17] and further explored in [30,31]. An associate measure is the deviation from orthogonality, which is based on the deviation from diagonality [30]. The deviation from orthogonality is given by

$$\delta(\hat{\mathbf{C}} \cdot \hat{\mathbf{C}}^{\mathrm{T}}) = 1 - \frac{\|\mathrm{diag}\left(\hat{\mathbf{C}} \cdot \hat{\mathbf{C}}^{\mathrm{T}}\right)\|_{\mathrm{F}}^{2}}{\|\hat{\mathbf{C}} \cdot \hat{\mathbf{C}}^{\mathrm{T}}\|_{\mathrm{F}}^{2}}, \tag{8.3}$$

where $\mathrm{diag}(\cdot)$ returns a diagonal matrix with the diagonal elements of its argument and $\|\cdot\|_{\mathrm{F}}$ is the Frobenius norm [2].

Since the DCT is an orthogonal transform, many authors aimed at deriving DCT approximations that are also orthogonal. But this is too severe a requirement for a transform to be useful in the context of image/video coding. For the sake of reference, the very first DCT approximation, the SDCT is not orthogonal with a deviation from orthogonality of 0.20. Many other transforms that are not orthogonal have proven to be useful in the context of image and video coding [30,31]. We adopt the following criterion for establishing orthogonality and near orthogonality for a given DCT approximation $\hat{\mathbf{C}}$: if $\delta(\hat{\mathbf{C}} \cdot \hat{\mathbf{C}}^{\mathrm{T}}) = 0$, then it is said to be orthogonal; if $0 < \delta(\hat{\mathbf{C}} \cdot \hat{\mathbf{C}}^{\mathrm{T}}) \leq 0.20$, then it is said to be near-orthogonal; otherwise it is non-orthogonal.

To sum it up, a good transform is (i) invertible and (ii) orthogonal or near-orthogonal. In this work, these are the main properties that will guide our design and analysis. Only matrices meeting the constraints (i) and (ii) will be considered.

8.3.2 Scalable transforms

In [36], Jridi, Alfalou, and Meher (JAM) proposed a method for obtaining a large DCT approximation for image and video coding from smaller transforms. The JAM method is similar to the Chen's DCT [9] definition. However, it exploits redundancies which allow concise and highly parallelizable hardware implementations [36].

The JAM method generates an $N \times N$ matrix transformation $\mathbf{T}_N$ by combining two instances of a given low-complexity $N/2$-point transform $\mathbf{T}_{N/2}$. The N-point transform matrix is defined by

$$\mathbf{T}_N = \frac{1}{\sqrt{2}}\mathbf{M}_N^{\text{per}} \begin{bmatrix} \mathbf{T}_{\frac{N}{2}} & \mathbf{0}_{\frac{N}{2}} \\ \mathbf{0}_{\frac{N}{2}} & \mathbf{T}_{\frac{N}{2}} \end{bmatrix} \mathbf{M}_N^{\text{add}}, \tag{8.4}$$

where $\mathbf{0}_{N/2}$ is the null matrix of order $N/2$. Matrices $\mathbf{M}_N^{\text{add}}$ and $\mathbf{M}_N^{\text{per}}$ are defined as

$$\mathbf{M}_N^{\text{add}} = \begin{bmatrix} \mathbf{I}_{\frac{N}{2}} & \bar{\mathbf{I}}_{\frac{N}{2}} \\ \bar{\mathbf{I}}_{\frac{N}{2}} & -\mathbf{I}_{\frac{N}{2}} \end{bmatrix}$$

and

$$\mathbf{M}_N^{\text{per}} = \begin{bmatrix} \mathbf{P}_{N-1,\frac{N}{2}} & \mathbf{0}_{1,\frac{N}{2}} \\ \mathbf{0}_{1,\frac{N}{2}} & \mathbf{P}_{N-1,\frac{N}{2}} \end{bmatrix},$$

where $\mathbf{I}_{N/2}$ and $\bar{\mathbf{I}}_{N/2}$ are, respectively, the identity and counter-identity matrices of order $N/2$, and $\mathbf{P}_{N-1,N/2}$ is an $(N-1) \times (N/2)$ matrix whose row vectors are given by

$$\mathbf{P}_{N-1,N/2}^{(i)} = \begin{cases} \mathbf{0}_{1,\frac{N}{2}}, & \text{if } i = 1, 3, 5, \ldots, \quad N-1, \\ \mathbf{I}_{\frac{N}{2}}^{(i/2)}, & \text{if } i = 0, 2, 4, \ldots, \quad N-2. \end{cases}$$

The permutation matrix $\mathbf{M}_N^{\text{per}}$ does not introduce any arithmetic cost, whereas $\mathbf{M}_N^{\text{add}}$ contributes with only N additions. Furthermore, the irrational scaling factor $1/\sqrt{2}$ of (8.4) can be merged into the image/video compression quantization step [36]. Let $A(\cdot)$ and $S(\cdot)$ return the additive cost and the bit-shifting count of its matrix argument. The arithmetic cost of the underlying N-point transform is given by the following recursions:

$$A(\mathbf{T}_N) = 2 \cdot A(\mathbf{T}_{N/2}) + N, \tag{8.5}$$

$$S(\mathbf{T}_N) = 2 \cdot S(\mathbf{T}_{N/2}). \tag{8.6}$$

8.4 Evaluation metrics

A relevant figure of merit for evaluating a DCT approximation $\hat{\mathbf{C}}$ is its arithmetic complexity. Because there is no irrational quantity required in the computation of DCT approximations, the arithmetic complexity of an approximation $\hat{\mathbf{C}}$ can be fully described through the counting of additions $A(\hat{\mathbf{C}})$ and bit-shifting operations $S(\hat{\mathbf{C}})$. Note that the complexity associated with the diagonal term in an approximation $\hat{\mathbf{C}} = \mathbf{S} \cdot \mathbf{T}$ can be neglected in the context of image and video compression.

This is because the diagonal $\mathbf{S}$ can be incorporated in the quantization matrix [31,38], representing no further computational complexity for the whole compression/decompression system.

Besides the arithmetic complexity inherent to a DCT approximation, different metrics have been proposed in order to evaluate the image quality resulting from reconstruction algorithms and methods [2,4]. Measures such as mean square error (MSE) and coding gain can be used to assess DCT approximations. However, such metrics do not capture an important DCT characteristic relevant for image and video compression: the vector basis functions of the transformation matrix (its rows) do not contribute equally to the final compressed media. In video compression scenario, the higher frequency DCT components are generally discarded by the quantization step, and only the low-frequency components are retained [2,19,23]. This is because the signal energy of natural images is usually concentrated on the first DCT coefficients. As a consequence, the matrix rows associated to higher frequencies do not have a large impact on the approximation performance. In this work, we propose a generalization of the error-related metrics to better capture this phenomenon according to a weight function to balance the importance of each matrix row.

Let us introduce a weighting diagonal matrix given by

$$\mathbf{W} = diag\left([w_0, w_1, \ldots, w_{N-1}]^{\mathrm{T}}\right), \tag{8.7}$$

where $w_0, w_1, \ldots, w_{N-1}$ are real constants. At this point, we do not specify the values of the constants $w_0, w_1, \ldots, w_{N-1}$. However, the importance of the first rows is represented by the constraint $w_0 > w_1 > \cdots > w_{N-1}$. The specific values of $w_0, w_1, \ldots, w_{N-1}$ will be defined and justified later.

8.4.1 The weighted total error energy

The total error energy evaluates the error between two matrices through the Euclidean distance between its rows [27]. This measure consists the Frobenius norm [39] of the difference [31] of two matrices. However, it does not capture the different importance between the rows of each DCT approximation. We proposed the following modified formulation considering the weighting matrix:

$$\epsilon_{\mathbf{W}}(\hat{\mathbf{C}}) = \pi \cdot N \cdot \|\mathbf{W} \cdot (\mathbf{C} - \hat{\mathbf{C}})\|_{\mathrm{F}}^2, \tag{8.8}$$

where $\mathbf{W}$ is given by (8.7) and $\mathbf{C}$ is the DCT matrix as given by [2, p. 74]. Note that the weighted total error energy is backward compatible with the usual total error energy. In fact, if $\mathbf{W} = (1/N) \cdot \mathrm{diag}\left(\begin{bmatrix} 1 & 1 & \cdots & 1 \end{bmatrix}^{\mathrm{T}}\right)$, the weighted total error energy collapses to the usual total error energy.

8.4.2 The weighted mean square error

Let $\mathbf{R_{xx}}$ be the autocorrelation matrix of a Markov-I stationary process with correlation coefficient ρ and $tr(\cdot)$ be the trace function which returns the sum of main diagonal elements of its matrix argument. The autocorrelation matrix is given by

$$[\mathbf{R_{xx}}]_{(i,j)} = \rho^{|i-j|}, \tag{8.9}$$

where $i,j = 0, 1, \ldots, N-1$ [1,2].

The mean square error (MSE) evaluates the difference in the energy spread between a DCT approximation and the exact DCT after a series following a Markov-I process is submitted to compression. Because the low-frequency components are the most relevant for compression scenarios, we want to give more importance to the energy contained in the first coefficients. This is achieved by quantifying the weighted metric:

$$\text{MSE}_{\mathbf{W}}(\hat{\mathbf{C}}) = \text{tr}(\sqrt{\mathbf{W}} \cdot (\mathbf{C} - \hat{\mathbf{C}}) \cdot \mathbf{R}_{\mathbf{xx}} \cdot (\mathbf{C} - \hat{\mathbf{C}})^{\mathrm{T}} \cdot \sqrt{\mathbf{W}}). \tag{8.10}$$

In order to specify $\mathbf{R}_{\mathbf{xx}}$, we follow the assumption that the input data is a Markov-I process and the correlation coefficient ρ is equal to 0.95 [2]. Notice that with the introduced weight, the first coefficients returned by the trace function are weighted according to the square root of the elements of $\mathbf{W}$. Additionally, the usual MSE [2] is a particular case of the proposed $\text{MSE}_{\mathbf{w}}$ when the weights are constant $w_k = 1/N$, for $k = 0, 1, \ldots, N-1$.

8.4.3 *The unified transform coding gain*

The unified transform coding gain provides a measure for quantifying how good a transform is to compress signals that follows an autoregressive model of order one [40–42]. It is derived from the usual transform coding gain as in [2]. Let $\mathbf{g}_k$ and $\mathbf{h}_k$ be the kth row of $\hat{\mathbf{C}}^{\mathrm{T}}$ and $\hat{\mathbf{C}}$, respectively. The unified transform coding gain, originally defined by Katto [40], is equivalent to

$$C_g(\hat{\mathbf{C}}) = 10 \cdot \log_{10}\left[\prod_{k=0}^{N-1} \frac{1}{(A_k \cdot B_k)^{1/N}}\right] \quad \text{(dB)}, \tag{8.11}$$

where $A_k = su\left[(\mathbf{h}_k \cdot \mathbf{h}_k^{\mathrm{T}}) \circ \mathbf{R}_{\mathbf{xx}}\right]$, $\text{su}(\cdot)$ returns the sum of elements of its matrix argument, the operator $\circ$ denotes the element-wise matrix product, $B_k = \|\mathbf{g}_k\|_2^2$, and $\|\cdot\|_2$ is the usual $L2$ vector norm [2]. In order to quantify the importance of each row, we scale the rows of the transform when used in the unified coding gain. This is translated into modifying the terms A_k and B_k, which result in $\tilde{A}_k = \text{su}\left[\left((w_k\mathbf{h}_k) \cdot (w_k\mathbf{h}_k^{\mathrm{T}})\right) \circ \mathbf{R}_{\mathbf{xx}}\right] = w_k^2 A_k$ and $\tilde{B}_k = \|w_k\mathbf{g}_k\|_2^2 = w_k^2 B_k$. The above manipulation results in the following expressions:

$$\begin{aligned}
C_{g,\mathbf{W}}(\hat{\mathbf{C}}) &= 10 \cdot \log_{10}\left[\prod_{k=0}^{N-1} \frac{1}{\left(\tilde{A}_k \cdot \tilde{B}_k\right)^{1/N}}\right] \quad \text{(dB)} \\
&= 10 \cdot \log_{10}\left[\left(\prod_{k=0}^{N-1} \frac{1}{w_k^4}\right)\left(\prod_{k=0}^{N-1} \frac{1}{(A_k \cdot B_k)^{1/N}}\right)\right] \quad \text{(dB)} \\
&= 10 \cdot \log_{10}\left[\prod_{k=0}^{N-1} \frac{1}{(A_k \cdot B_k)^{1/N}}\right] - 40 \cdot \log_{10}\left[\prod_{k=0}^{N-1} \frac{1}{w_k}\right] \quad \text{(dB)} \\
&= C_g\left(\hat{\mathbf{C}}\right) - 40 \cdot \log_{10}\left[\prod_{k=0}^{N-1} \frac{1}{w_k}\right] \quad \text{(dB)}.
\end{aligned} \tag{8.12}$$

The final expression for $C_{g,\mathbf{W}}$ is simply the original unified coding gain C_g added by a constant quantity (assuming that the weights are fixed). Therefore, the resulting expression is not fundamentally different from the original unified coding gain. Because of that, we keep the usual coding gain for evaluating transforms.

8.4.4 The transform efficiency

The transform efficiency is a complementary way to evaluate the coding performance of a given DCT approximation [2]. Let $\mathbf{R_{XX}} = \hat{\mathbf{C}} \cdot \mathbf{R_{xx}} \cdot \hat{\mathbf{C}}^{\mathrm{T}}$ be the covariance matrix of the signal $\mathbf{X}$ transformed by $\hat{\mathbf{C}}$. The usual transform efficiency is defined as the ratio $tr(|\mathbf{R_{XX}}|)/\mathrm{su}(|\mathbf{R_{XX}}|)$. It measures how much a particular transform $\hat{\mathbf{C}}$ decorrelates the input signal $\mathbf{X}$. If the input signal can be decorrelated by the transform $\hat{\mathbf{C}}$, then its energy is concentrated into few coefficients in the main diagonal of $\mathbf{R_{XX}}$, resulting in higher ratio $\mathrm{tr}(|\mathbf{R_{XX}}|)/\mathrm{su}(|\mathbf{R_{XX}}|)$. However, this metric does not fully capture the importance of the rows (basis vectors) associated with the low-frequency components. In order to quantify the importance of each row in the transform efficiency metric, we weight the transform matrix $\hat{\mathbf{C}}$ by the square of the weight matrix $\mathbf{W}$. The weighted transform efficiency of $\hat{\mathbf{C}}$ is then given by

$$\eta_{\mathbf{W}}(\hat{\mathbf{C}}) = \frac{\mathrm{tr}\left(|\sqrt{\mathbf{W}} \cdot \mathbf{R_{XX}} \cdot \sqrt{\mathbf{W}}|\right)}{\mathrm{su}\left(|\sqrt{\mathbf{W}} \cdot \mathbf{R_{XX}} \cdot \sqrt{\mathbf{W}}|\right)}. \tag{8.13}$$

Notice that the weighted transform efficiency $\eta_{\mathbf{W}}$ collapses to the usual transform efficiency when the weights are equal to $1/N$.

8.4.5 Specification of the weighting matrix W

The matrix $\mathbf{W}$ is used to adjust the error and efficiency-related metrics in the evaluation of DCT approximations. The coefficients of $\mathbf{W}$ need to reflect the importance of each of the rows of a DCT approximation when applied in the scenario of still image and video coding. In order to determine the real quantities $w_0, w_1, \ldots, w_{N-1}$ that are the diagonal of $\mathbf{W}$, we look into the model adopted to describe images.

The first-order autoregressive model is the most adopted structure for describing natural images [2,4]. Notice also that the DCT itself can be derived from the solution of a transcendental equation obtained with the assumption that the image follows such a model [2].

The DCT matrix rows are the eigenvectors associated with the autocorrelation matrix for a first-order autoregressive model. The autocorrelation matrix coefficients of an autoregressive model of order one are given by (8.9). The eigenvalues of a given matrix, on the other hand, represent the amount of energy associated with its corresponding eigenvector. Its magnitude represents the contribution of the corresponding eigenvector to the whole original matrix. A possible choice for the coefficients $w_0, w_1, \ldots, w_{N-1}$ would be the normalized eigenvalues of the autocorrelation

Table 8.7 Performance of selected approximations in the literature

Transform	$\delta(\mathrm{T}\cdot\mathrm{T}^{\mathrm{T}})$	ϵ_w	MSE_w	C_g	η_w	$A(\alpha)$	$S(\alpha)$
$\mathbf{T}_{\mathrm{H}_5}$	0.000	0.053	0.001	8.53	89.37	28	8
$\mathbf{T}_{\mathrm{RAZ}}$	0.000	0.023	0.001	8.53	89.37	24	6
$\mathbf{T}_{\mathrm{BAS}_1}$	0.177	0.074	0.002	6.27	83.17	21	0
$\mathbf{T}_{\mathrm{BAS}_2}$	0.000	0.132	0.004	8.12	86.86	18	2
$\mathbf{T}_{\mathrm{BAS}_3}$	0.000	0.150	0.005	7.91	85.38	18	0
$\mathbf{T}_{\mathrm{BAS}_4}$	0.000	0.115	0.004	8.33	88.22	24	4
$\mathbf{T}_{\mathrm{BAS}_5}$	0.000	0.361	0.009	7.91	85.38	18	0
$\mathbf{T}_{\mathrm{BAS}_6}$	0.000	0.361	0.009	7.91	85.64	16	0
$\mathbf{T}_{\mathrm{BAS}_7}$	0.000	0.345	0.009	8.12	86.86	18	2
$\mathbf{T}_{\mathrm{BAS}_8}$	0.000	0.136	0.004	7.95	85.31	24	0
$\mathbf{T}_{\mathrm{POT}}$	0.000	0.438	0.022	7.33	80.9	14	0

matrix $\mathbf{R_{XX}}$ of the first-order autoregressive model. The eigenvalues of $\mathbf{R_{XX}}$ in [2, p. 54] are then obtained by using the following expression:

$$w_n = \frac{1-\rho^2}{1-2\rho\cos(n\pi/N)+\rho^2},$$

where $n = 0, 1, \ldots, N-1$. Considering $\rho = 0.95$ [2] and $N = 8$, we have

$$\begin{aligned} w_0 &= 9.37\cdot 10^{-1}, & w_1 &= 1.65\cdot 10^{-2},\\ w_2 &= 4.35\cdot 10^{-3}, & w_3 &= 2.07\cdot 10^{-3},\\ w_4 &= 1.27\cdot 10^{-3}, & w_5 &= 9.25\cdot 10^{-4},\\ w_6 &= 7.49\cdot 10^{-4}, & w_7 &= 6.65\cdot 10^{-4}. \end{aligned} \tag{8.14}$$

Notice the metrics obtained from some of the transforms considered from the literature in Table 8.7. Observe that the weighted metrics are very low in magnitude. This is because emphasis is put on the first rows of the DCT approximation under consideration. It is noteworthy that all considered approximations present null error in its first row. This is because in order to correctly approximate the DCT, any candidate approximation must only have a constant row corresponding to the DC level. The decay on the weights shown in (8.14) associated with null error on the first row results in very low errors that are below 64-bit machine precision used to select and analyze the mentioned transforms.

8.5 Efficient transforms

8.5.1 Multicriteria optimization

In this section, we formalize an optimization problem for obtaining the optimal transforms in the literature [17,19–24,26,27,30,31]. We find the optimal transforms by considering error- and coding-related metrics separately. Several matrices in the literature are considered to well approximate the DCT in terms of error-related metrics, for instance we mention the approximations proposed in [19,27,31]. Although we are interested in matrices that approximate the DCT in the compression domain, matrices that are considerably close to the DCT as measured in a Euclidean distance sense may also effect good performance in compression contexts [28,31].

Let $\mathcal{M}(\mathbf{C}_8)$ be the set of all DCT approximations considered in this work for $N = 8$, which encompass the class of Feig–Winograd approximations [31], approximations based on general integer functions [30], the series of BAS transforms [20–24,26], the SDCT [17], the RDCT [27], LO approximation [19], and the approximation proposed in [43]. We then solve the following two nonlinear multicriteria optimization problems [44,45] for 8×8 matrices over the set $\mathcal{M}(\mathbf{C}_8)$:

$$\min_{\hat{\mathbf{C}} \in \mathcal{M}(\mathbf{C}_8)} \left(\varepsilon_{\mathbf{W}}(\hat{\mathbf{C}}), \mathrm{MSE}_{\mathbf{W}}(\hat{\mathbf{C}}), A(\boldsymbol{\alpha})\right), \tag{8.15}$$

and

$$\min_{\hat{\mathbf{C}} \in \mathcal{M}(\mathbf{C}_8)} \left(-C_g(\hat{\mathbf{C}}), -\eta_{\mathbf{W}}(\hat{\mathbf{C}}), A(\boldsymbol{\alpha})\right), \tag{8.16}$$

both subject to

1. the existence of the inverse transformation and
2. the admissibility of orthogonality or near-orthogonality.

Notice that in the second problem, (8.16), the quantities $C_g(\hat{\mathbf{C}})$ and $\eta_{\mathbf{W}}(\hat{\mathbf{C}})$ are in negative form to comply with the minimization requirement. Also, the objective functions of the multicriteria optimization problems (8.15) and (8.16) are $\mathcal{F}_{\text{error}} = \{\varepsilon_{\mathbf{W}}(\cdot), MSE_{\mathbf{W}}(\cdot), A(\cdot)\}$ and $\mathcal{F}_{\text{coding}} = \{-C_g(\cdot), -\eta_{\mathbf{W}}(\cdot), A(\cdot)\}$, respectively. It is important to highlight that both the problems in (8.15) and (8.16) are discrete and finite as there is a finitely countable number of possible values that the objective functions can assume in the set $\mathcal{M}(\mathbf{C}_8)$. Moreover, the objective functions of both problems in (8.15) and (8.16) are nonlinear in nature. This makes (8.15) and (8.16) nonlinear and discrete problems, which are not suitable to be solved by conventional analytic methods. Because of that, we employ exhaustive search to solve the problems (8.15) and (8.16). The discussed multicriteria problems require the identification of the set of *Pareto* efficient solutions [44], which is given by

$$\begin{aligned}\{\hat{\mathbf{C}} \in \mathcal{M}(\mathbf{C}_8) :\ & \text{there is no } \hat{\mathbf{C}} \in \mathcal{M}(\mathbf{C}_8) \text{ such that} \\ & f(\hat{\mathbf{C}}) \le f\left(\hat{\mathbf{C}}^*\right) \quad \text{for all f} \in \mathcal{F} \quad \text{and} \\ & f_0(\hat{\mathbf{C}}) < f_0\left(\hat{\mathbf{C}}^*\right) \quad \text{for some } \mathrm{f}_0 \in \mathcal{F}\},\end{aligned} \tag{8.17}$$

where $\mathcal{F}$ is either $\mathcal{F}_{\text{error}}$ for problem (8.15) or $\mathcal{F}_{\text{coding}}$ for problem (8.16).

8.5.2 Efficient solutions

The efficient solutions were found through exhaustive search in a search algorithm implemented in Julia programming language [46]. Tables 8.8 and 8.9 display the efficient solutions for the problem (8.16), and Tables 8.10 and 8.11 show the efficient solutions for the problem (8.15).

Table 8.8 Efficient transforms for coding-related optimization (Part 1)

Transform matrix	Orthogonal?	Description
$\mathbf{T}_1^{\text{cod}} = \begin{bmatrix} \frac{1}{2} & \frac{1}{2} & \frac{1}{2} & \frac{1}{2} & \frac{1}{2} & \frac{1}{2} & \frac{1}{2} & \frac{1}{2} \\ 0 & 0 & 0 & \frac{1}{2} & -\frac{1}{2} & 0 & 0 & 0 \\ 0 & \frac{1}{2} & -\frac{1}{2} & 0 & 0 & -\frac{1}{2} & \frac{1}{2} & 0 \\ 0 & -\frac{1}{2} & 0 & 0 & 0 & 0 & \frac{1}{2} & 0 \\ \frac{1}{2} & -\frac{1}{2} & -\frac{1}{2} & \frac{1}{2} & \frac{1}{2} & -\frac{1}{2} & -\frac{1}{2} & \frac{1}{2} \\ 0 & 0 & \frac{1}{2} & 0 & 0 & -\frac{1}{2} & 0 & 0 \\ \frac{1}{2} & 0 & 0 & -\frac{1}{2} & -\frac{1}{2} & 0 & 0 & \frac{1}{2} \\ \frac{1}{2} & 0 & 0 & 0 & 0 & 0 & 0 & -\frac{1}{2} \end{bmatrix}$	Yes	New and in Feig–Winograd class [31]
$\mathbf{T}_2^{\text{cod}} = \begin{bmatrix} \frac{1}{2} & \frac{1}{2} & \frac{1}{2} & \frac{1}{2} & \frac{1}{2} & \frac{1}{2} & \frac{1}{2} & \frac{1}{2} \\ 0 & 0 & \frac{1}{2} & 0 & 0 & -\frac{1}{2} & 0 & 0 \\ 0 & \frac{1}{2} & -\frac{1}{2} & 0 & 0 & -\frac{1}{2} & \frac{1}{2} & 0 \\ 0 & 0 & 0 & -\frac{1}{2} & \frac{1}{2} & 0 & 0 & 0 \\ \frac{1}{2} & -\frac{1}{2} & -\frac{1}{2} & \frac{1}{2} & \frac{1}{2} & -\frac{1}{2} & -\frac{1}{2} & \frac{1}{2} \\ \frac{1}{2} & 0 & 0 & 0 & 0 & 0 & 0 & -\frac{1}{2} \\ \frac{1}{2} & 0 & 0 & -\frac{1}{2} & -\frac{1}{2} & 0 & 0 & \frac{1}{2} \\ 0 & -\frac{1}{2} & 0 & 0 & 0 & 0 & \frac{1}{2} & 0 \end{bmatrix}$	Yes	New and in Feig–Winograd class [31]
$\mathbf{T}_3^{\text{cod}} = \begin{bmatrix} \frac{1}{2} & \frac{1}{2} & \frac{1}{2} & \frac{1}{2} & \frac{1}{2} & \frac{1}{2} & \frac{1}{2} & \frac{1}{2} \\ 0 & \frac{1}{2} & 0 & 0 & 0 & 0 & -\frac{1}{2} & 0 \\ 0 & \frac{1}{2} & -\frac{1}{2} & 0 & 0 & -\frac{1}{2} & \frac{1}{2} & 0 \\ \frac{1}{2} & 0 & 0 & 0 & 0 & 0 & 0 & -\frac{1}{2} \\ \frac{1}{2} & -\frac{1}{2} & -\frac{1}{2} & \frac{1}{2} & \frac{1}{2} & -\frac{1}{2} & -\frac{1}{2} & \frac{1}{2} \\ 0 & 0 & 0 & \frac{1}{2} & -\frac{1}{2} & 0 & 0 & 0 \\ \frac{1}{2} & 0 & 0 & -\frac{1}{2} & -\frac{1}{2} & 0 & 0 & \frac{1}{2} \\ 0 & 0 & \frac{1}{2} & 0 & 0 & -\frac{1}{2} & 0 & 0 \end{bmatrix}$	Yes	New and in Feig–Winograd class [31]

(Continues)

Table 8.8 (Continued)

Transform matrix	Orthogonal?	Description
$\mathbf{T}_4^{\text{cod}} = \begin{bmatrix} \frac{1}{2} & \frac{1}{2} & \frac{1}{2} & \frac{1}{2} & \frac{1}{2} & \frac{1}{2} & \frac{1}{2} & \frac{1}{2} \\ 0 & 0 & 0 & \frac{1}{2} & -\frac{1}{2} & 0 & 0 & 0 \\ \frac{1}{2} & 0 & 0 & -\frac{1}{2} & -\frac{1}{2} & 0 & 0 & \frac{1}{2} \\ 0 & -\frac{1}{2} & 0 & 0 & 0 & 0 & \frac{1}{2} & 0 \\ \frac{1}{2} & -\frac{1}{2} & -\frac{1}{2} & \frac{1}{2} & \frac{1}{2} & -\frac{1}{2} & -\frac{1}{2} & \frac{1}{2} \\ 0 & 0 & \frac{1}{2} & 0 & 0 & -\frac{1}{2} & 0 & 0 \\ 0 & -\frac{1}{2} & \frac{1}{2} & 0 & 0 & \frac{1}{2} & -\frac{1}{2} & 0 \\ \frac{1}{2} & 0 & 0 & 0 & 0 & 0 & 0 & -\frac{1}{2} \end{bmatrix}$	Yes	New and in Feig–Winograd class [31]
$\mathbf{T}_5^{\text{cod}} = \begin{bmatrix} \frac{1}{2} & \frac{1}{2} & \frac{1}{2} & \frac{1}{2} & \frac{1}{2} & \frac{1}{2} & \frac{1}{2} & \frac{1}{2} \\ 0 & 0 & \frac{1}{2} & 0 & 0 & -\frac{1}{2} & 0 & 0 \\ \frac{1}{2} & 0 & 0 & -\frac{1}{2} & -\frac{1}{2} & 0 & 0 & \frac{1}{2} \\ 0 & 0 & 0 & -\frac{1}{2} & \frac{1}{2} & 0 & 0 & 0 \\ \frac{1}{2} & -\frac{1}{2} & -\frac{1}{2} & \frac{1}{2} & \frac{1}{2} & -\frac{1}{2} & -\frac{1}{2} & \frac{1}{2} \\ \frac{1}{2} & 0 & 0 & 0 & 0 & 0 & 0 & -\frac{1}{2} \\ 0 & -\frac{1}{2} & \frac{1}{2} & 0 & 0 & \frac{1}{2} & -\frac{1}{2} & 0 \\ 0 & -\frac{1}{2} & 0 & 0 & 0 & 0 & \frac{1}{2} & 0 \end{bmatrix}$	Yes	New and in Feig–Winograd class [31]
$\mathbf{T}_6^{\text{cod}} = \begin{bmatrix} \frac{1}{2} & \frac{1}{2} & \frac{1}{2} & \frac{1}{2} & \frac{1}{2} & \frac{1}{2} & \frac{1}{2} & \frac{1}{2} \\ 0 & \frac{1}{2} & 0 & 0 & 0 & 0 & -\frac{1}{2} & 0 \\ \frac{1}{2} & 0 & 0 & -\frac{1}{2} & -\frac{1}{2} & 0 & 0 & \frac{1}{2} \\ \frac{1}{2} & 0 & 0 & 0 & 0 & 0 & 0 & -\frac{1}{2} \\ \frac{1}{2} & -\frac{1}{2} & -\frac{1}{2} & \frac{1}{2} & \frac{1}{2} & -\frac{1}{2} & -\frac{1}{2} & \frac{1}{2} \\ 0 & 0 & 0 & \frac{1}{2} & -\frac{1}{2} & 0 & 0 & 0 \\ 0 & -\frac{1}{2} & \frac{1}{2} & 0 & 0 & \frac{1}{2} & -\frac{1}{2} & 0 \\ 0 & 0 & \frac{1}{2} & 0 & 0 & -\frac{1}{2} & 0 & 0 \end{bmatrix}$	Yes	New and in Feig–Winograd class [31]

Table 8.9 Efficient transforms for coding-related optimization (Part 2)

Transform matrix	Orthogonal?	Description
$\mathbf{T}_7^{\text{cod}} = \begin{bmatrix} \frac{1}{2} & \frac{1}{2} & \frac{1}{2} & \frac{1}{2} & \frac{1}{2} & \frac{1}{2} & \frac{1}{2} & \frac{1}{2} \\ \frac{1}{2} & 0 & 0 & 0 & 0 & 0 & 0 & -\frac{1}{2} \\ 0 & \frac{1}{2} & -\frac{1}{2} & 0 & 0 & -\frac{1}{2} & \frac{1}{2} & 0 \\ 0 & 0 & -\frac{1}{2} & 0 & 0 & \frac{1}{2} & 0 & 0 \\ \frac{1}{2} & -\frac{1}{2} & -\frac{1}{2} & \frac{1}{2} & \frac{1}{2} & -\frac{1}{2} & -\frac{1}{2} & \frac{1}{2} \\ 0 & -\frac{1}{2} & 0 & 0 & 0 & 0 & \frac{1}{2} & 0 \\ \frac{1}{2} & 0 & 0 & -\frac{1}{2} & -\frac{1}{2} & 0 & 0 & \frac{1}{2} \\ 0 & 0 & 0 & -\frac{1}{2} & \frac{1}{2} & 0 & 0 & 0 \end{bmatrix}$	Yes	New and in Feig–Winograd class [31]
$\mathbf{T}_8^{\text{cod}} = \begin{bmatrix} \frac{1}{2} & \frac{1}{2} & \frac{1}{2} & \frac{1}{2} & \frac{1}{2} & \frac{1}{2} & \frac{1}{2} & \frac{1}{2} \\ \frac{1}{2} & \frac{1}{2} & \frac{1}{2} & 0 & 0 & -\frac{1}{2} & -\frac{1}{2} & -\frac{1}{2} \\ 0 & \frac{1}{2} & -\frac{1}{2} & 0 & 0 & -\frac{1}{2} & \frac{1}{2} & 0 \\ \frac{1}{2} & 0 & -\frac{1}{2} & -\frac{1}{2} & \frac{1}{2} & \frac{1}{2} & 0 & -\frac{1}{2} \\ \frac{1}{2} & -\frac{1}{2} & -\frac{1}{2} & \frac{1}{2} & \frac{1}{2} & -\frac{1}{2} & -\frac{1}{2} & \frac{1}{2} \\ \frac{1}{2} & -\frac{1}{2} & 0 & \frac{1}{2} & -\frac{1}{2} & 0 & \frac{1}{2} & -\frac{1}{2} \\ \frac{1}{2} & 0 & 0 & -\frac{1}{2} & -\frac{1}{2} & 0 & 0 & \frac{1}{2} \\ 0 & -\frac{1}{2} & \frac{1}{2} & -\frac{1}{2} & \frac{1}{2} & -\frac{1}{2} & \frac{1}{2} & 0 \end{bmatrix}$	Yes	New and in Feig–Winograd class [31]
$\mathbf{T}_9^{\text{cod}} = \begin{bmatrix} \frac{1}{2} & \frac{1}{2} & \frac{1}{2} & \frac{1}{2} & \frac{1}{2} & \frac{1}{2} & \frac{1}{2} & \frac{1}{2} \\ \frac{1}{2} & 0 & 0 & 0 & 0 & 0 & 0 & -\frac{1}{2} \\ \frac{1}{2} & 0 & 0 & -\frac{1}{2} & -\frac{1}{2} & 0 & 0 & \frac{1}{2} \\ 0 & 0 & -\frac{1}{2} & 0 & 0 & \frac{1}{2} & 0 & 0 \\ \frac{1}{2} & -\frac{1}{2} & -\frac{1}{2} & \frac{1}{2} & \frac{1}{2} & -\frac{1}{2} & -\frac{1}{2} & \frac{1}{2} \\ 0 & -\frac{1}{2} & 0 & 0 & 0 & 0 & \frac{1}{2} & 0 \\ 0 & -\frac{1}{2} & \frac{1}{2} & 0 & 0 & \frac{1}{2} & -\frac{1}{2} & 0 \\ 0 & 0 & 0 & -\frac{1}{2} & \frac{1}{2} & 0 & 0 & 0 \end{bmatrix}$	Yes	New and in Feig–Winograd class [31]

(Continues)

Table 8.9 (*Continued*)

Transform matrix	**Orthogonal?**	**Description**
$\mathbf{T}_{10}^{\text{cod}} = \begin{bmatrix} \frac{1}{2} & \frac{1}{2} & \frac{1}{2} & \frac{1}{2} & \frac{1}{2} & \frac{1}{2} & \frac{1}{2} & \frac{1}{2} \\ \frac{1}{2} & \frac{1}{2} & \frac{1}{2} & 0 & 0 & -\frac{1}{2} & -\frac{1}{2} & -\frac{1}{2} \\ \frac{1}{2} & 0 & 0 & -\frac{1}{2} & -\frac{1}{2} & 0 & 0 & \frac{1}{2} \\ \frac{1}{2} & 0 & -\frac{1}{2} & -\frac{1}{2} & \frac{1}{2} & \frac{1}{2} & 0 & -\frac{1}{2} \\ \frac{1}{2} & -\frac{1}{2} & -\frac{1}{2} & \frac{1}{2} & \frac{1}{2} & -\frac{1}{2} & -\frac{1}{2} & \frac{1}{2} \\ \frac{1}{2} & -\frac{1}{2} & 0 & \frac{1}{2} & -\frac{1}{2} & 0 & \frac{1}{2} & -\frac{1}{2} \\ 0 & -\frac{1}{2} & \frac{1}{2} & 0 & 0 & \frac{1}{2} & -\frac{1}{2} & 0 \\ 0 & -\frac{1}{2} & \frac{1}{2} & -\frac{1}{2} & \frac{1}{2} & -\frac{1}{2} & \frac{1}{2} & 0 \end{bmatrix}$	Yes	New and in Feig–Winograd class [31]
$\mathbf{T}_{11}^{\text{cod}} = \begin{bmatrix} 2 & 2 & 2 & 2 & 2 & 2 & 2 & 2 \\ 3 & 2 & 2 & 0 & 0 & -2 & -2 & -3 \\ 3 & 1 & -1 & -3 & -3 & -1 & 1 & 3 \\ 2 & 0 & -3 & -2 & 2 & 3 & 0 & -2 \\ 2 & -2 & -2 & 2 & 2 & -2 & -2 & 2 \\ 2 & -3 & 0 & 2 & -2 & 0 & 3 & -2 \\ 1 & -3 & 3 & -1 & -1 & 3 & -3 & 1 \\ 0 & -2 & 2 & -3 & 3 & -2 & 2 & 0 \end{bmatrix}$	Yes	Introduced in [30]
$\mathbf{T}_{12}^{\text{cod}} = \begin{bmatrix} 1 & 1 & 1 & 1 & 1 & 1 & 1 & 1 \\ 2 & 2 & 1 & 0 & 0 & -1 & -2 & -2 \\ 2 & 1 & -1 & -2 & -2 & -1 & 1 & 2 \\ 1 & 0 & -2 & -2 & 2 & 2 & 0 & -1 \\ 1 & -1 & -1 & 1 & 1 & -1 & -1 & 1 \\ 2 & -2 & 0 & 1 & -1 & 0 & 2 & -2 \\ 1 & -2 & 2 & -1 & -1 & 2 & -2 & 1 \\ 0 & -1 & 2 & -2 & 2 & -2 & 1 & 0 \end{bmatrix}$	Yes	Introduced in [43]
$\mathbf{T}_{13}^{\text{cod}} = \begin{bmatrix} 1 & 1 & 1 & 1 & 1 & 1 & 1 & 1 \\ 1 & 1 & 0 & 0 & 0 & 0 & -1 & -1 \\ 1 & \frac{1}{2} & -\frac{1}{2} & -1 & -1 & -\frac{1}{2} & \frac{1}{2} & 1 \\ 0 & 0 & -1 & 0 & 0 & 1 & 0 & 0 \\ 1 & -1 & -1 & 1 & 1 & -1 & -1 & 1 \\ 1 & -1 & 0 & 0 & 0 & 0 & 1 & -1 \\ \frac{1}{2} & -\frac{1}{2} & \frac{1}{2} & -\frac{1}{2} & -\frac{1}{2} & 1 & -1 & \frac{1}{2} \\ 0 & 0 & 0 & -1 & 1 & 0 & 0 & 0 \end{bmatrix}$	Yes	Introduced in [21]

(Continues)

Table 8.9 (Continued)

Transform matrix	Orthogonal?	Description
$\mathbf{T}_{14}^{\text{cod}} = \begin{bmatrix} 1 & 1 & 1 & 1 & 1 & 1 & 1 & 1 \\ 1 & 1 & 0 & 0 & 0 & 0 & -1 & -1 \\ 1 & 0 & 0 & -1 & -1 & 0 & 0 & 1 \\ 0 & 0 & 1 & 0 & 0 & -1 & 0 & 0 \\ 1 & -1 & -1 & 1 & 1 & -1 & -1 & 1 \\ 0 & 0 & 0 & 1 & -1 & 0 & 0 & 0 \\ 1 & -1 & 0 & 0 & 0 & 0 & 1 & -1 \\ 0 & -1 & 1 & 0 & 0 & 1 & -1 & 0 \end{bmatrix}$	Yes	Introduced in [24]
$\mathbf{T}_{15}^{\text{cod}} = \begin{bmatrix} 1 & 1 & 1 & 1 & 1 & 1 & 1 & 1 \\ 1 & 1 & 0 & 0 & 0 & 0 & -1 & -1 \\ 1 & \frac{1}{2} & -\frac{1}{2} & -1 & -1 & -\frac{1}{2} & \frac{1}{2} & 1 \\ 0 & 0 & 1 & 0 & 0 & -1 & 0 & 0 \\ 1 & -1 & -1 & 1 & 1 & -1 & -1 & 1 \\ 0 & 0 & 0 & 1 & -1 & 0 & 0 & 0 \\ 1 & -1 & 0 & 0 & 0 & 0 & 1 & -1 \\ \frac{1}{2} & -1 & 1 & -\frac{1}{2} & -\frac{1}{2} & 1 & -1 & \frac{1}{2} \end{bmatrix}$	Yes	Introduced in [24]
$\mathbf{T}_{16}^{\text{cod}} = \begin{bmatrix} 1 & 1 & 1 & 1 & 1 & 1 & 1 & 1 \\ 0 & 1 & 0 & 0 & 0 & 0 & -1 & 0 \\ 1 & 0 & 0 & -1 & -1 & 0 & 0 & 1 \\ 1 & 0 & 0 & 0 & 0 & 0 & 0 & -1 \\ 1 & -1 & -1 & 1 & 1 & -1 & -1 & 1 \\ 0 & 0 & 0 & 1 & -1 & 0 & 0 & 0 \\ 0 & -1 & 1 & 0 & 0 & 1 & -1 & 0 \\ 0 & 0 & 1 & 0 & 0 & -1 & 0 & 0 \end{bmatrix}$	Yes	Introduced in [47]

The metrics for the efficient transforms according to the coding and error-based optimization are shown in Tables 8.12 and 8.13.

Notice that several obtained transforms exhibit a very low additive cost. This shows that the problems (8.16) and (8.15) favored low-complexity approximations that were not considered as suitable, according to the multicriteria problem encompassing simultaneously coding- and error-related metrics [31]. This is the intended purpose of the new weighted metrics since the errors in different DCT approximations rows have different influence on its performance. Some of the efficient solutions are transforms present in the literature. The exceptions are the new transforms that belong to the Feig–Winograd class of DCT approximations [31]. Such transforms have not been identified as optimal using the unified optimization

Table 8.10 Efficient transforms for error-related optimization (Part 1)

Transform matrix	**Orthogonal?**	**Description**
$T_1^{err} = \begin{bmatrix} \frac{1}{2} & \frac{1}{2} & \frac{1}{2} & \frac{1}{2} & \frac{1}{2} & \frac{1}{2} & \frac{1}{2} & \frac{1}{2} \\ \frac{1}{2} & 0 & 0 & 0 & 0 & 0 & 0 & \frac{1}{2} \\ \frac{1}{2} & 0 & 0 & -\frac{1}{2} & -\frac{1}{2} & 0 & 0 & \frac{1}{2} \\ 0 & 0 & -\frac{1}{2} & 0 & 0 & \frac{1}{2} & 0 & 0 \\ \frac{1}{2} & -\frac{1}{2} & -\frac{1}{2} & \frac{1}{2} & \frac{1}{2} & -\frac{1}{2} & -\frac{1}{2} & \frac{1}{2} \\ 0 & -\frac{1}{2} & 0 & 0 & 0 & 0 & \frac{1}{2} & 0 \\ 0 & -\frac{1}{2} & \frac{1}{2} & 0 & 0 & \frac{1}{2} & -\frac{1}{2} & 0 \\ 0 & 0 & 0 & -\frac{1}{2} & \frac{1}{2} & 0 & 0 & 0 \end{bmatrix}$	Yes	The same as $\mathbf{T}_9^{cod}$
$\mathbf{T}_2^{err} = \begin{bmatrix} \frac{1}{2} & \frac{1}{2} & \frac{1}{2} & \frac{1}{2} & \frac{1}{2} & \frac{1}{2} & \frac{1}{2} & \frac{1}{2} \\ \frac{1}{2} & \frac{1}{2} & 0 & 0 & 0 & 0 & -\frac{1}{2} & -\frac{1}{2} \\ \frac{1}{2} & 0 & 0 & -\frac{1}{2} & -\frac{1}{2} & 0 & 0 & \frac{1}{2} \\ \frac{1}{2} & 0 & -\frac{1}{2} & 0 & 0 & \frac{1}{2} & 0 & -\frac{1}{2} \\ \frac{1}{2} & -\frac{1}{2} & -\frac{1}{2} & \frac{1}{2} & \frac{1}{2} & -\frac{1}{2} & -\frac{1}{2} & \frac{1}{2} \\ 0 & -\frac{1}{2} & 0 & \frac{1}{2} & -\frac{1}{2} & 0 & \frac{1}{2} & 0 \\ 0 & -\frac{1}{2} & \frac{1}{2} & 0 & 0 & \frac{1}{2} & -\frac{1}{2} & 0 \\ 0 & 0 & \frac{1}{2} & -\frac{1}{2} & \frac{1}{2} & -\frac{1}{2} & 0 & 0 \end{bmatrix}$	No	New and in Feig–Winograd class [31]
$\mathbf{T}_3^{err} = \begin{bmatrix} \frac{1}{2} & \frac{1}{2} & \frac{1}{2} & \frac{1}{2} & \frac{1}{2} & \frac{1}{2} & \frac{1}{2} & \frac{1}{2} \\ \frac{1}{2} & \frac{1}{2} & \frac{1}{2} & 0 & 0 & -\frac{1}{2} & -\frac{1}{2} & -\frac{1}{2} \\ \frac{1}{2} & 0 & 0 & -\frac{1}{2} & -\frac{1}{2} & 0 & 0 & \frac{1}{2} \\ \frac{1}{2} & 0 & -\frac{1}{2} & -\frac{1}{2} & \frac{1}{2} & \frac{1}{2} & 0 & -\frac{1}{2} \\ \frac{1}{2} & -\frac{1}{2} & -\frac{1}{2} & \frac{1}{2} & \frac{1}{2} & -\frac{1}{2} & -\frac{1}{2} & \frac{1}{2} \\ \frac{1}{2} & -\frac{1}{2} & 0 & \frac{1}{2} & -\frac{1}{2} & 0 & \frac{1}{2} & -\frac{1}{2} \\ 0 & -\frac{1}{2} & \frac{1}{2} & 0 & 0 & \frac{1}{2} & -\frac{1}{2} & 0 \\ 0 & -\frac{1}{2} & \frac{1}{2} & -\frac{1}{2} & \frac{1}{2} & -\frac{1}{2} & \frac{1}{2} & 0 \end{bmatrix}$	Yes	New and in Feig–Winograd class [31]

(Continues)

Table 8.10 (*Continued*)

Transform matrix	**Orthogonal?**	**Description**
$\mathbf{T}_4^{\text{err}} = \begin{bmatrix} \frac{1}{2} & \frac{1}{2} & \frac{1}{2} & \frac{1}{2} & \frac{1}{2} & \frac{1}{2} & \frac{1}{2} & \frac{1}{2} \\ \frac{1}{2} & 0 & 0 & 0 & 0 & 0 & 0 & -\frac{1}{2} \\ 1 & \frac{1}{2} & -\frac{1}{2} & -1 & -1 & -\frac{1}{2} & \frac{1}{2} & 1 \\ 0 & 0 & -\frac{1}{2} & 0 & 0 & \frac{1}{2} & 0 & 0 \\ \frac{1}{2} & -\frac{1}{2} & -\frac{1}{2} & \frac{1}{2} & \frac{1}{2} & -\frac{1}{2} & -\frac{1}{2} & \frac{1}{2} \\ 0 & -\frac{1}{2} & 0 & 0 & 0 & 0 & \frac{1}{2} & 0 \\ \frac{1}{2} & -1 & 1 & -\frac{1}{2} & -\frac{1}{2} & 1 & -1 & \frac{1}{2} \\ 0 & 0 & 0 & -\frac{1}{2} & \frac{1}{2} & 0 & 0 & 0 \end{bmatrix}$	Yes	New and in Feig–Winograd class [31]
$\mathbf{T}_5^{\text{err}} = \begin{bmatrix} \frac{1}{2} & \frac{1}{2} & \frac{1}{2} & \frac{1}{2} & \frac{1}{2} & \frac{1}{2} & \frac{1}{2} & \frac{1}{2} \\ 1 & \frac{1}{2} & \frac{1}{2} & 0 & 0 & -\frac{1}{2} & -\frac{1}{2} & -1 \\ \frac{1}{2} & 0 & 0 & -\frac{1}{2} & -\frac{1}{2} & 0 & 0 & \frac{1}{2} \\ \frac{1}{2} & 0 & -1 & -\frac{1}{2} & \frac{1}{2} & 1 & 0 & -\frac{1}{2} \\ \frac{1}{2} & -\frac{1}{2} & -\frac{1}{2} & \frac{1}{2} & \frac{1}{2} & -\frac{1}{2} & -\frac{1}{2} & \frac{1}{2} \\ \frac{1}{2} & -1 & 0 & \frac{1}{2} & -\frac{1}{2} & 0 & 1 & -\frac{1}{2} \\ 0 & -\frac{1}{2} & \frac{1}{2} & 0 & 0 & \frac{1}{2} & -\frac{1}{2} & 0 \\ 0 & -\frac{1}{2} & \frac{1}{2} & -1 & 1 & -\frac{1}{2} & \frac{1}{2} & 0 \end{bmatrix}$	Yes	New and in Feig–Winograd class [31]

problem considering both coding- and error-related metrics in [31]. The formalism introduced in this work by splitting the optimization problem into two smaller problems allows for the identification of new transforms as optimal.

One can notice that the transforms $\mathbf{T}_9^{\text{cod}}$, $\mathbf{T}_{11}^{\text{cod}}$, and $\mathbf{T}_{12}^{\text{cod}}$ are in the efficient solution set of both problems 8.15 and 8.16. This suggests that those transforms are good approximations for the DCT both in contexts of low error compared to the exact DCT and of image/video coding. Therefore, we select $\mathbf{T}_9^{\text{cod}}$, $\mathbf{T}_{11}^{\text{cod}}$, and $\mathbf{T}_{12}^{\text{cod}}$ and submit them for hardware implementation.

Table 8.11 Efficient transforms for error-related optimization (Part 2)

Transform matrix	Orthogonal?	Description
$\mathbf{T}_6^{\mathrm{err}} = \begin{bmatrix} 2 & 2 & 2 & 2 & 2 & 2 & 2 & 2 \\ 3 & 2 & 2 & 0 & 0 & -2 & -2 & -3 \\ 3 & 1 & -1 & -3 & -3 & -1 & 1 & 3 \\ 2 & 0 & -3 & -2 & 2 & 3 & 0 & -2 \\ 2 & -2 & -2 & 2 & 2 & -2 & -2 & 2 \\ 2 & -3 & 0 & 2 & -2 & 0 & 3 & -2 \\ 1 & -3 & 3 & -1 & -1 & 3 & -3 & 1 \\ 0 & -2 & 2 & -3 & 3 & -2 & 2 & 0 \end{bmatrix}$	Yes	Introduced in [30]
$\mathbf{T}_7^{\mathrm{err}} = \begin{bmatrix} 1 & 1 & 1 & 1 & 1 & 1 & 1 & 1 \\ 2 & 2 & 1 & 0 & 0 & -1 & -2 & -2 \\ 2 & 1 & -1 & -2 & -2 & -1 & 1 & 2 \\ 1 & 0 & -2 & -2 & 2 & 2 & 0 & -1 \\ 1 & -1 & -1 & 1 & 1 & -1 & -1 & 1 \\ 2 & -2 & 0 & 1 & -1 & 0 & 2 & -2 \\ 1 & -2 & 2 & -1 & -1 & 2 & -2 & 1 \\ 0 & -1 & 2 & -2 & 2 & -2 & 1 & 0 \end{bmatrix}$	Yes	Introduced in [43]
$\mathbf{T}_8^{\mathrm{err}} = \begin{bmatrix} 1 & 1 & 1 & 1 & & 1 & 1 & 1 & 1 \\ 1 & 1 & 1 & 0 & & 0 & -1 & -1 & -1 \\ 1 & 1 & -1 & -1 & \vdots & -1 & -1 & 1 & 1 \\ 1 & 0 & -1 & 0 & & 0 & 1 & 0 & -1 \\ 1 & -1 & -1 & 1 & & 1 & -1 & -1 & 1 \\ 1 & -1 & 1 & 0 & & 0 & -1 & 1 & -1 \\ 1 & -1 & 1 & -1 & \vdots & -1 & 1 & -1 & 1 \\ 1 & -1 & 1 & -1 & & 1 & -1 & 1 & -1 \end{bmatrix}$	No	Introduced in [20]

Table 8.12 Performance of efficient approximations for coding-related optimization

Transform	$\delta(\mathbf{T}\cdot\mathbf{T}^{\mathrm{T}})$	ϵ_w	MSE_w	C_g	η_w	$A(\alpha)$	$S(\alpha)$
$\mathbf{T}_1^{\mathrm{cod}}$	0.000	1.034	0.073	7.33	80.9	14	8
$\mathbf{T}_2^{\mathrm{codd}}$	0.000	0.775	0.042	7..33	80.9	14	8
$\mathbf{T}_3^{\mathrm{cod}}$	0.000	0.577	0.026	7.33	80.9	14	8
$\mathbf{T}_4^{\mathrm{cod}}$	0.000	0.895	0.069	7.33	80.9	14	8
$\mathbf{T}_5^{\mathrm{cod}}$	0.000	0.636	0.039	7.33	80.9	14	8
$\mathbf{T}_6^{\mathrm{cod}}$	0.000	0.438	0.022	7.33	80.9	14	8
$\mathbf{T}_7^{\mathrm{cod}}$	0.000	0.47	0.020	7.33	80.9	14	8
$\mathbf{T}_8^{\mathrm{cod}}$	0.000	0.192	0.005	8.18	87.43	22	16
$\mathbf{T}_9^{\mathrm{cod}}$	0.000	0.331	0.016	7.33	80.9	14	8
$\mathbf{T}_{10}^{\mathrm{cod}}$	0.000	0.053	0.001	8.18	87.43	22	16
$\mathbf{T}_{11}^{\mathrm{cod}}$	0.000	0.021	0.000	8.50	91.01	24	10
$\mathbf{T}_{12}^{\mathrm{cod}}$	0.000	0.023	0.001	8.53	89.37	24	6
$\mathbf{T}_{13}^{\mathrm{cod}}$	0.000	0.132	0.004	8.12	86.86	18	2
$\mathbf{T}_{14}^{\mathrm{cod}}$	0.000	0.361	0.009	7.91	85.64	16	0
$\mathbf{T}_{15}^{\mathrm{cod}}$	0.000	0.345	0.009	8.12	86.86	18	2
$\mathbf{T}_{16}^{\mathrm{codd}}$	0.000	0.438	0.022	7.33	80.9	14	0

Table 8.13 Performance of efficient approximations for error-related optimization

Transform	$\delta(\mathbf{T}\cdot\mathbf{T}^{\mathrm{T}})$	ϵ_w	MSE_w	C_g	η_w	$A(\alpha)$	$S(\alpha)$
$\mathbf{T}_1^{\mathrm{err}}$	0.000	0.331	0.016	7.33	80.9	14	8
$\mathbf{T}_2^{\mathrm{err}}$	0.125	0.115	0.004	6.05	83.08	18	12
$\mathbf{T}_3^{\mathrm{err}}$	0.000	0.053	0.001	8.18	87.43	22	16
$\mathbf{T}_4^{\mathrm{err}}$	0.000	0.312	0.016	7.54	81.99	16	8
$\mathbf{T}_5^{\mathrm{err}}$	0.000	0.053	0.001	8.14	86.81	22	12
$\mathbf{T}_6^{\mathrm{err}}$	0.000	0.021	0.000	8.50	91.01	24	10
$\mathbf{T}_7^{\mathrm{err}}$	0.177	0.074	0.002	6.27	83.17	21	0
$\mathbf{T}_8^{\mathrm{err}}$	0.000	0.361	0.009	7.91	85.64	16	0

Let

$$\mathbf{P} = \begin{bmatrix} 1 & 0 & 0 & 0 & 0 & 0 & 0 & 0 \\ 0 & 0 & 0 & 0 & -1 & 0 & 0 & 0 \\ 0 & 0 & 1 & 0 & 0 & 0 & 0 & 0 \\ 0 & 0 & 0 & 0 & 0 & -1 & 0 & 0 \\ 0 & 1 & 0 & 0 & 0 & 0 & 0 & 0 \\ 0 & 0 & 0 & 0 & 0 & 0 & 0 & -1 \\ 0 & 0 & 0 & 1 & 0 & 0 & 0 & 0 \\ 0 & 0 & 0 & 0 & 0 & 0 & 1 & 0 \end{bmatrix}, \tag{8.18}$$

and $\mathbf{D} = \text{diag}([1 \quad 2 \quad 1 \quad 2 \quad 1 \quad 2 \quad 1 \quad 2])$. Let also

$$\mathbf{B}_1 = \begin{bmatrix} 1 & 1 & 0 & 0 & 0 & 0 & 0 & 0 \\ 1 & -1 & 0 & 0 & 0 & 0 & 0 & 0 \\ 0 & 0 & 1 & 0 & 0 & 0 & 0 & 0 \\ 0 & 0 & 0 & 1 & 0 & 0 & 0 & 0 \\ 0 & 0 & 0 & 0 & 1 & 0 & 0 & 0 \\ 0 & 0 & 0 & 0 & 0 & 1 & 0 & 0 \\ 0 & 0 & 0 & 0 & 0 & 0 & 1 & 0 \\ 0 & 0 & 0 & 0 & 0 & 0 & 0 & 1 \end{bmatrix}, \tag{8.19}$$

$$\mathbf{B}_2 = \begin{bmatrix} 1 & 0 & 0 & 1 & 0 & 0 & 0 & 0 \\ 0 & 1 & 1 & 0 & 0 & 0 & 0 & 0 \\ 0 & 1 & -1 & 0 & 0 & 0 & 0 & 0 \\ 1 & 0 & 0 & -1 & 0 & 0 & 0 & 0 \\ 0 & 0 & 0 & 0 & 1 & 0 & 0 & 0 \\ 0 & 0 & 0 & 0 & 0 & 1 & 0 & 0 \\ 0 & 0 & 0 & 0 & 0 & 0 & 1 & 0 \\ 0 & 0 & 0 & 0 & 0 & 0 & 0 & 1 \end{bmatrix}, \tag{8.20}$$

and

$$\mathbf{B}_3 = \begin{bmatrix} 1 & 0 & 0 & 0 & 0 & 0 & 0 & 1 \\ 0 & 1 & 0 & 0 & 0 & 0 & 1 & 0 \\ 0 & 0 & 1 & 0 & 0 & 1 & 0 & 0 \\ 0 & 0 & 0 & 1 & 1 & 0 & 0 & 0 \\ 0 & 0 & 0 & 1 & -1 & 0 & 0 & 0 \\ 0 & 0 & 1 & 0 & 0 & -1 & 0 & 0 \\ 0 & 1 & 0 & 0 & 0 & 0 & -1 & 0 \\ 1 & 0 & 0 & 0 & 0 & 0 & 0 & -1 \end{bmatrix}. \tag{8.21}$$

The fast algorithms for $\mathbf{T}_9^{\text{cod}}$, $\mathbf{T}_{11}^{\text{cod}}$, and $\mathbf{T}_{12}^{\text{cod}}$ share a similar structure according

$$\mathbf{T}_9^{\text{cod}} = \mathbf{P} \cdot \mathbf{K}_9 \cdot \mathbf{B}_1 \cdot \mathbf{B}_2 \cdot \mathbf{B}_3, \tag{8.22}$$

$$\mathbf{T}_{11}^{\text{cod}} = \mathbf{P} \cdot \mathbf{K}_{11} \cdot \mathbf{B}_1 \cdot \mathbf{B}_2 \cdot \mathbf{B}_3, \tag{8.23}$$

and

$$\mathbf{T}_{12}^{\text{cod}} = \mathbf{D} \cdot \mathbf{K}_{12} \cdot \mathbf{B}_1 \cdot \mathbf{B}_2 \cdot \mathbf{B}_3, \tag{8.24}$$

where

$$\mathbf{K}_9 = \begin{bmatrix} \frac{1}{2} & 0 & 0 & 0 & 0 & 0 & 0 & 0 \\ 0 & \frac{1}{2} & 0 & 0 & 0 & 0 & 0 & 0 \\ 0 & 0 & 0 & \frac{1}{2} & 0 & 0 & 0 & 0 \\ 0 & 0 & -\frac{1}{2} & 0 & 0 & 0 & 0 & 0 \\ 0 & 0 & 0 & 0 & 0 & 0 & 0 & -\frac{1}{2} \\ 0 & 0 & 0 & 0 & 0 & \frac{1}{2} & 0 & 0 \\ 0 & 0 & 0 & 0 & -\frac{1}{2} & 0 & 0 & 0 \\ 0 & 0 & 0 & 0 & 0 & 0 & \frac{1}{2} & 0 \end{bmatrix}, \tag{8.25}$$

$$\mathbf{K}_{11} = \begin{bmatrix} 2 & 0 & 0 & 0 & 0 & 0 & 0 & 0 \\ 0 & 2 & 0 & 0 & 0 & 0 & 0 & 0 \\ 0 & 0 & 1 & 3 & 0 & 0 & 0 & 0 \\ 0 & 0 & -3 & 1 & 0 & 0 & 0 & 0 \\ 0 & 0 & 0 & 0 & 0 & -2 & -2 & -3 \\ 0 & 0 & 0 & 0 & 2 & 3 & 0 & -2 \\ 0 & 0 & 0 & 0 & -3 & 2 & -2 & 0 \\ 0 & 0 & 0 & 0 & -2 & 0 & 3 & -2 \end{bmatrix}, \tag{8.26}$$

$$\mathbf{K}_{12} = \begin{bmatrix} 1 & 0 & 0 & 0 & 0 & 0 & 0 & 0 \\ 0 & 0 & 0 & 0 & 0 & \frac{1}{2} & 1 & 1 \\ 0 & 0 & 1 & 2 & 0 & 0 & 0 & 0 \\ 0 & 0 & 0 & 0 & -1 & -1 & 0 & \frac{1}{2} \\ 0 & 1 & 0 & 0 & 0 & 0 & 0 & 0 \\ 0 & 0 & 0 & 0 & \frac{1}{2} & 0 & -1 & 1 \\ 0 & 0 & -2 & 1 & 0 & 0 & 0 & 0 \\ 0 & 0 & 0 & 0 & -1 & 1 & -\frac{1}{2} & 0 \end{bmatrix}. \tag{8.27}$$

The fast algorithms for those transforms require, respectively, 14, 24, and 24 additions. The same transforms require, respectively, eight, eight, and six bit-shifting operations.

8.6 Video coding experiments

The selected transforms were embedded into a public-available HEVC reference software [48] for assessing their applicability in the video-coding context. Unlike older codecs—such as MPEG [49], H.261 [50], H.263 [51], H.264 [52]—the HEVC employs not only an 8-point integer DCT (IDCT) but also transforms of size 4, 16, and 32 [53]. Different transform sizes affect a series of optimization routines which greatly improve the processing of low-textured or big smooth areas [54].

Aiming at the derivation of large blocklength transforms for HEVC embedding, the selected matrices $\mathbf{T}_9^{\text{cod}}$, $\mathbf{T}_{11}^{\text{cod}}$, and $\mathbf{T}_{12}^{\text{cod}}$ were submitted to the JAM-scaling algorithm [36]. This method is capable of generating 16- and 32-point output matrices based on an input 8-point transform. Such scalable method is then suitable for furnishing all the necessary matrices for further video-coding experiments.

For the coding experiments, the original 8-, 16-, and 32-point integer transforms of HEVC were substituted by $\mathbf{T}_9^{\text{cod}}$, $\mathbf{T}_{11}^{\text{cod}}$, and $\mathbf{T}_{12}^{\text{cod}}$ and their scalable counterparts derived from (8.4). The original 4-point IDCT was kept unaltered since it is already a low-complexity transformation. One video sequence of each A–F class described in the common test conditions (CTC) documentation [55] was considered for testing. Classes A and B contain high-resolution media for broadcasting and video on demand services; Classes C and D are lower resolution videos for the same applications; Class E presents typical examples of video conferencing data; and Class F provides computer-edited image sequences [56].

Namely, the first 100 frames of the following 8-bit video sequences were considered: `PeopleOnStreet` (2,560×1,600 at 30 fps), `BasketballDrive` (1,920×1,080 at 50 fps), `RaceHorses` (832×480 at 30 fps), `BlowingBubbles` (416×240 at 50 fps), `KristenAndSara` (1,280×720 at 60 fps), and `BasketballDrillText` (832×480 at 50 fps). Configurations `All Intra` (AI), `Random Access` (RA), `Low Delay B` (LD-B), and `Low Delay P` (LD-P) in the Main profile with default values were tested.

The frame-by-frame MSE and peak signal-to-noise ratio (PSNR) [53] for each YUV color channel were selected as figures of merit. Then, for all the considered videos, the rate distortion (RD) curve with recommended quantization parameter (*QP*) values, i.e., 22, 27, 32, and 37 [55], were computed. The resulting RD curves are illustrated in Figure 8.1. The present analysis focuses on AI configuration because it is more adequate for assessing novel intra-coding tools [55]. The Bjøntegaard's delta PSNR (BD-PSNR) [57,58] for the modified versions of the HEVC software was also calculated. Those measurements are shown in Table 8.14. Notice that the transform $\mathbf{T}_{12}^{\text{cod}}$ and its scalable versions were the best performing. Replacing the IDCT by this DCT approximation results in a quality loss not greater than 0.47 dB for the AI configuration. Worst performance for $\mathbf{T}_{12}^{\text{cod}}$ in the other configurations—RA, LD-B, and LD-P—arises for the RaceHorses video sequence, where approximately 0.55 dB is lost when compared to the original HEVC implementation. Results for transform $\mathbf{T}_{11}^{\text{cod}}$ are quite similar to those of $\mathbf{T}_{12}^{\text{cod}}$, and consistently superior of those produced by $\mathbf{T}_9^{\text{cod}}$.

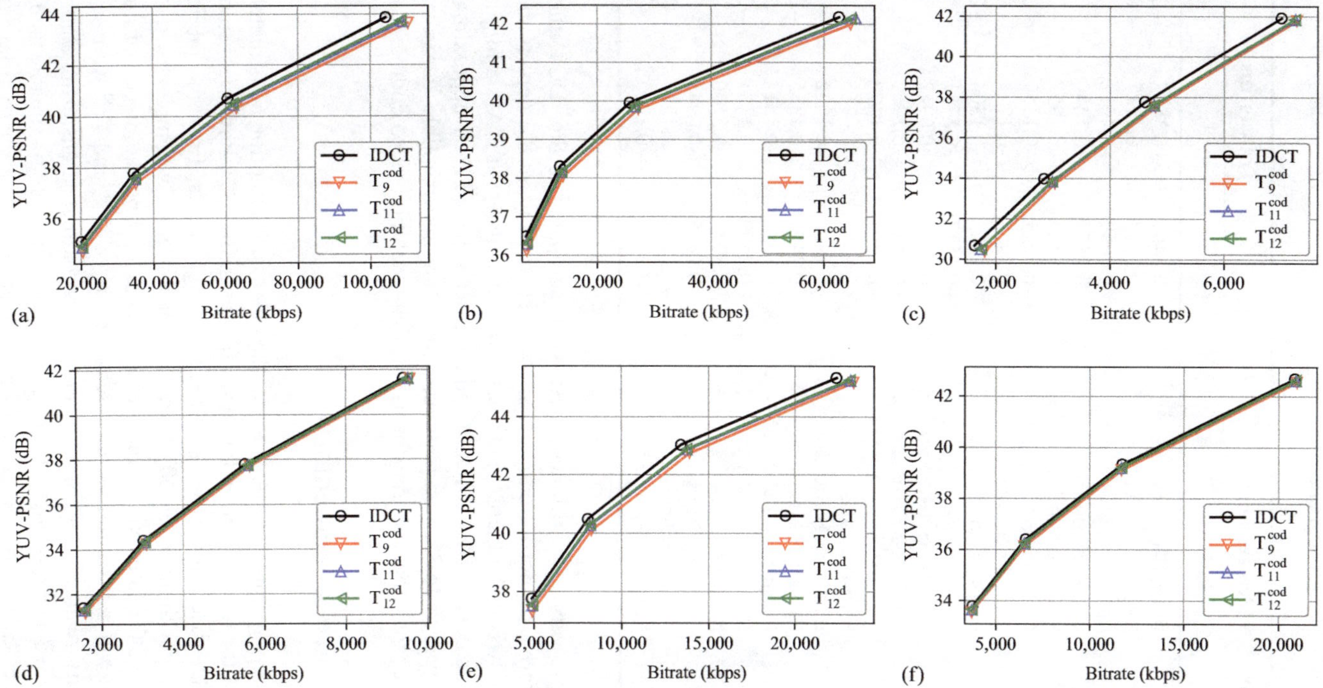

Figure 8.1 *Rate distortion curves of the modified HEVC software in AI configuration for the considered videos: (a) PeopleOnStreet, (b) BasketballDrive, (c) RaceHorses, (d) BlowingBubbles, (e) KristenAndSara, and (f) BasketballDrillText*

Table 8.14 Average BD-PSNR (dB) of the modified HEVC reference software for tested video sequences

Configuration	Video sequence	Transforms		
		$\mathbf{T}_9^{\text{cod}}$	$\mathbf{T}_{11}^{\text{cod}}$	$\mathbf{T}_{12}^{\text{cod}}$
AI	`PeopleOnStreet`	−0.5409	−0.3587	−0.2999
	`BasketballDrive`	−0.3313	−0.1917	−0.1692
	`RaceHorses`	−0.6699	−0.4972	−0.4714
	`BlowingBubbles`	−0.2571	−0.1440	−0.0839
	`KristenAndSara`	−0.4723	−0.2957	−0.2682
	`BasketballDrillText`	−0.2027	−0.1159	−0.1036
RA	`PeopleOnStreet`	−0.2904	−0.1723	−0.1467
	`BasketballDrive`	−0.2533	−0.1506	−0.1412
	`RaceHorses`	−0.9517	−0.5751	−0.5521
	`BlowingBubbles`	−0.1988	−0.0942	−0.0821
	`BaskeballDrillText`	−0.2532	−0.1460	−0.1372
LD-B	`BasketballDrive`	−0.2391	−0.1372	−0.1271
	`RaceHorses`	−0.9267	−0.5765	−0.5460
	`BlowingBubbles`	−0.1992	−0.1022	−0.0806
	`KristenAndSara`	−0.2409	−0.1335	−0.1230
	`BasketballDrillText`	−0.2943	−0.1814	−0.1748
LD-P	`BasketballDrive`	−0.2399	−0.1381	−0.1276
	`RaceHorses`	−0.8921	−0.5576	−0.5344
	`BlowingBubbles`	−0.1865	−0.0945	−0.0813
	`KristenAndSara`	−0.2188	−0.1218	−0.1118
	`BasketballDrillText`	−0.2701	−0.1700	−0.1646

Despite the very low arithmetic cost when compared to the IDCT, the considered transforms do not introduce significant errors. Figure 8.2 depicts the first frame of the KristenAndSara video sequence encoded with AI configuration according to the default HEVC IDCT and the three selected DCT approximations. The QP value was set to 32. Compression artifacts are imperceptible, demonstrating the suitability of the proposed transforms for high-resolution video coding.

8.7 Hardware implementation

With the increasing chip transistors count, the power consumption constraints have become one of the top challenges for hardware designing [59]. Power-consumption optimization has been addressed at different design levels, from the fabrication process to the algorithmic level [60]. In the current section, low-complexity hardware architecture for DCT approximations are addressed. Among the discussed approximations, the novel transformations represented by $\mathbf{T}_9^{\text{cod}}$, $\mathbf{T}_{11}^{\text{cod}}$, and $\mathbf{T}_{12}^{\text{cod}}$ were selected to be physically realized. The aforementioned approximate transforms algorithms were implemented as digital architecture employing a field-programmable gate array (FPGA) chip. FPGAs allow digital hardware design without a physical layout process, providing a cheap, flexible, and fast solution for prototyping circuits which

(a) MSE-Y = 39.5389, MSE-U = 43.5331, MSE-V = 44.5911, PSNR-Y = 7.2308, PSNR-U = 2.8825, PSNR-V = 2.2593

(b) MSE-Y = 39.1467, MSE-U = 43.2443, MSE-V = 44.1933, PSNR-Y = 7.9143, PSNR-U = 3.0807, PSNR-V = 2.4760

(c) MSE-Y = 39.2998, MSE-U = 43.3973, MSE-V = 44.3515, PSNR-Y = 7.6401, PSNR-U = 2.9741, PSNR-V = 2.3874

(d) MSE-Y = 39.3436, MSE-U = 43.3028, MSE-V = 44.3352, PSNR-Y = 7.5634, PSNR-U = 3.0395, PSNR-V = 2.3964

Figure 8.2 Compression of the first frame of KristenAndSara, with $QP = 32$ in AI coding configuration, using the following embedded transforms: (a) the IDCT, (b) $\boldsymbol{T}_9^{cod}$, (c) $\boldsymbol{T}_{11}^{cod}$, and (d) $\boldsymbol{T}_{12}^{cod}$

eventually can be implemented in application-specific integrated circuits [61, p. 258]. Nevertheless, FPGAs have been also considered as final product platforms in industrial applications [62–64]. The selected DCT approximations were implemented in the ML605 platform, which is equipped with a Xilinx *Virtex-6 XC6VLX240T-1FFG1156* FPGA chip, developed through the Xilinx System Generator library for Simulink™ [65].

In general, several metrics are considered to evaluate the area, speed, and energy consumption of a hardware piece. The Xilinx FPGAs basically consists of (i) configurable logic blocks (CLBs), which contain combinational logic and registers; (ii) input and output blocks, representing the interface between the FPGA and the external world elements; (iii) programmable interconnections; and (iv) other resources, such as RAM blocks, three-state buffers, and global clock buffers [61]. The basic building block of the CLB is the *slice*, which contains essentially look-up tables (LUTs) and storage elements that can be configured as latches or flip-flops (FFs). The system latency is defined in terms of the number of clock cycles required for a single set of inputs to be processed [61, p. 244], while the critical path delay T_{cd} is given by the delay of the longest combinational path and it determines the maximum clock frequency $f_{\max} = T_{cd}^{-1}$ [66, p. 61].

The power consumption can be classified into (i) static or quiescent power (P_q), which is due to leakage current flowing through the transistors and is not related to the circuit switching activities; and (ii) dynamic power (P_d), a dissipation that is related to switching activities and capacitance charge and discharge due to logical operations [60,61,67]. The dynamic power consumption tends to be higher when the circuit-operation frequency is increased [67]. The area-time metric is a common measure that takes into account the circuit area and the processing time [68,69].

In the analysis of the hardware resources, we take into account the number of slices, LUTs, and FFs as figures of merit. It is also considered the critical path delay T_{cd}, in ns, and its related maximum operational frequency $f_{\max}$, in MHz, as well as the circuit latency T_{lat}, in ns. The static power P_q and dynamic power and P_d were estimated through the *Xilinx XPower Analyzer* tool. As energy consumption metrics, we consider the static power, given in watts (W), and the energy per clock cycle (E_d) dissipated by the dynamic power, given by $P_d/f_{\max}$, in nJ. The area-time is considered as the product of the number of slices and the critical path delay.

The digital architecture for the considered 8-point transforms were constructed according to the fast algorithms represented by (8.22)–(8.24). The fast algorithm was pipelined in order to provide maximum throughput. The architecture structure consists in four stages, where the first three are additive blocks and the last one consists of low-complexity multiplications—which is implemented with additions and bit-shifting operations. The pipeline latency depends on the latency of each stage. The first three stages are the same for the three transforms and present latency of one clock cycle each, while the fourth stage varies according to each transform, either $\mathbf{T}_9^{\text{cod}}$, $\mathbf{T}_{11}^{\text{cod}}$, or $\mathbf{T}_{12}^{\text{cod}}$. Eight parallel inputs with wordlength of 8 bit were considered—a typical wordlength for images.

The architecture for the general case is shown in Figure 8.3, in which the additive stages present 14 adders. The designs were tested employing hardware co-simulation via JTAG cable and compared to the theoretical result using 100,000 random input vectors of 8-bit integers.

For the $\mathbf{T}_9^{\text{cod}}$, the multiplicative stage presents only right-shifts operations, which correspond to the divisions by 2 imposed by the matrix $\mathbf{K}_9$ in the factorization (8.22). Bit-shifting operations can be synthesized in FPGA simply with proper rewiring. No extra additions are required since matrix $\mathbf{K}_9$ possesses only one nonzero entry per row as it can be seen in (8.25). Consequently, such stage does not present any latency, and the pipeline essentially consists only in the first three stages shown in Figure 8.3 and present latency of three clock cycles as in Figure 8.4.

In the approximation induced by $\mathbf{T}_{11}^{\text{cod}}$, the multiplicative stage shown in Figure 8.5 is associated with matrix $\mathbf{K}_{11}$ in factorization (8.23). The matrix $\mathbf{K}_{11}$ in (8.26) requires some additions and multiplications by 2, which is implemented with left shift, and multiplication by 3, which is implemented with an adder and a left shift. Consequently, the multiplier stage is decomposed in two internal stages associated with the adders of the multiplier by 3 and the inherent adders imposed by matrix $\mathbf{K}_{11}$ due to two or more nonzero elements in the same row. The full pipeline is, then, composed of the five stages—the three adder stages represented by the

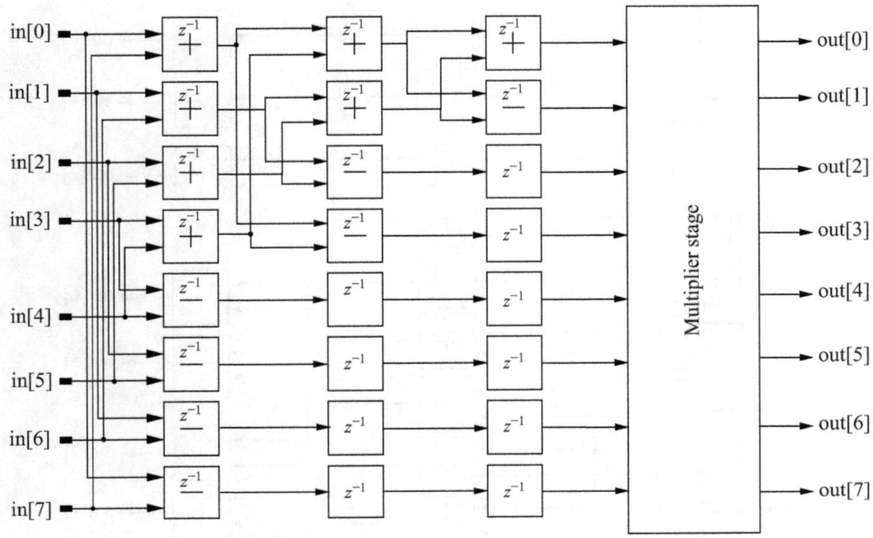

Figure 8.3 General architecture for the considered 8-point DCT approximations

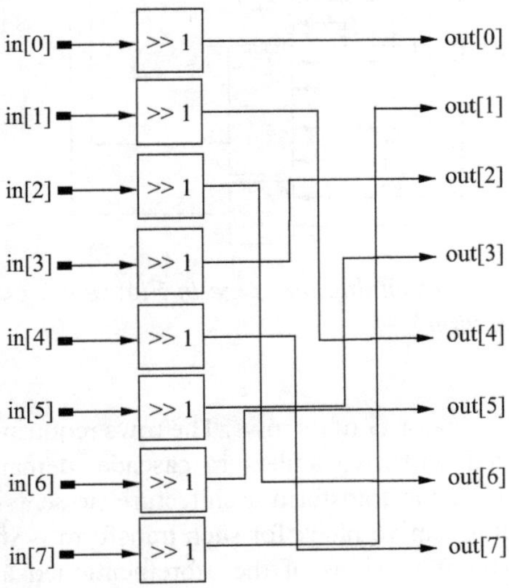

Figure 8.4 Low-complexity multiplier stage in Figure 8.3 for the $\mathbf{T}_9^{\text{cod}}$ DCT approximation

matrices $\mathbf{B}_1$, $\mathbf{B}_2$, and $\mathbf{B}_3$ plus the two internal stages in the multiplier block representing $\mathbf{K}_{11}$. This results in a latency of five clock cycles.

The transform $\mathbf{T}_{12}^{\text{cod}}$ presents the multiplicative stage according to matrix $\mathbf{K}_{12}$ in the factorization (8.24) as in Figure 8.6. Such matrix in (8.27) possesses three

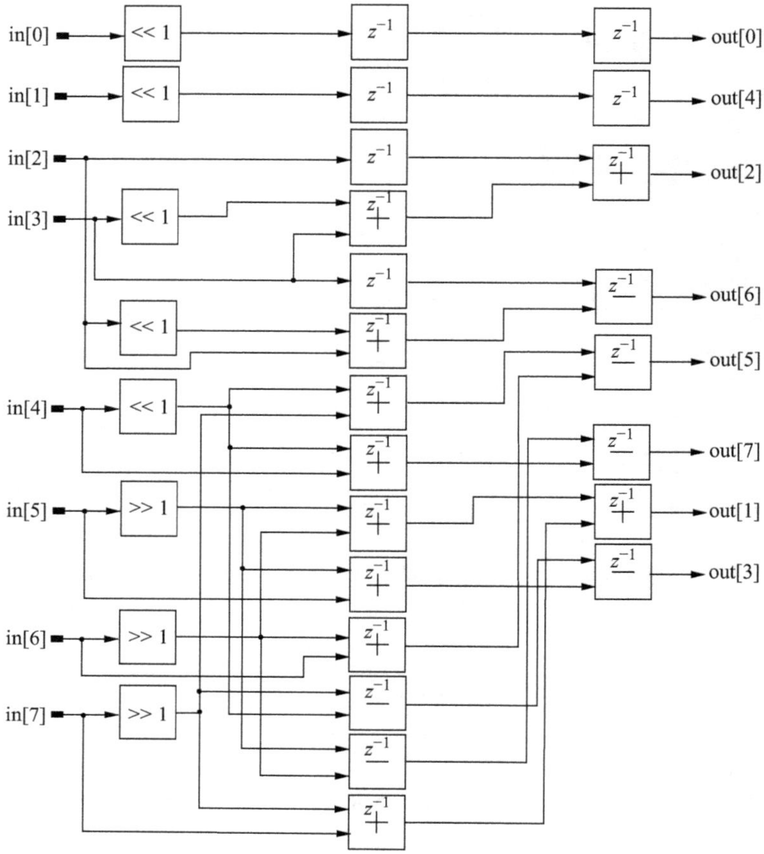

Figure 8.5 Low-complexity multiplier stage in Figure 8.3 for the $\mathbf{T}_{11}^{\text{cod}}$ *DCT approximation*

nonzero elements in the majority of its rows. The rows requiring three simultaneous inputs are implemented with two adders in cascade, demanding the two-clock cycles. Similarly to $\mathbf{T}_{11}^{\text{cod}}$, the transform architecture presents latency of the five-clock cycles. The multiplicative block for such transform is shown in Figure 8.6.

The 16- and 32-point versions of the aforementioned approximations were obtained through the JAM method [36] and described in Section 8.2.2. Such method employs two instantiations of the N-point transform to derive a $2N$-point transform including an extra additive stage. This implies that the 16-point versions of $\mathbf{T}_{9}^{\text{cod}}$, $\mathbf{T}_{11}^{\text{cod}}$, and $\mathbf{T}_{12}^{\text{cod}}$ possess four, six and six clock cycles of latency, respectively. Similarly, the 32-point version of $\mathbf{T}_{9}^{\text{cod}}$, $\mathbf{T}_{11}^{\text{cod}}$, and $\mathbf{T}_{12}^{\text{cod}}$ demand five, seven and seven clock cycles of latency, respectively.

The 8-, 16-, and 32-point IDCTs employed in the HEVC open-reference software [48] were also implemented for reference. We considered integer multipliers

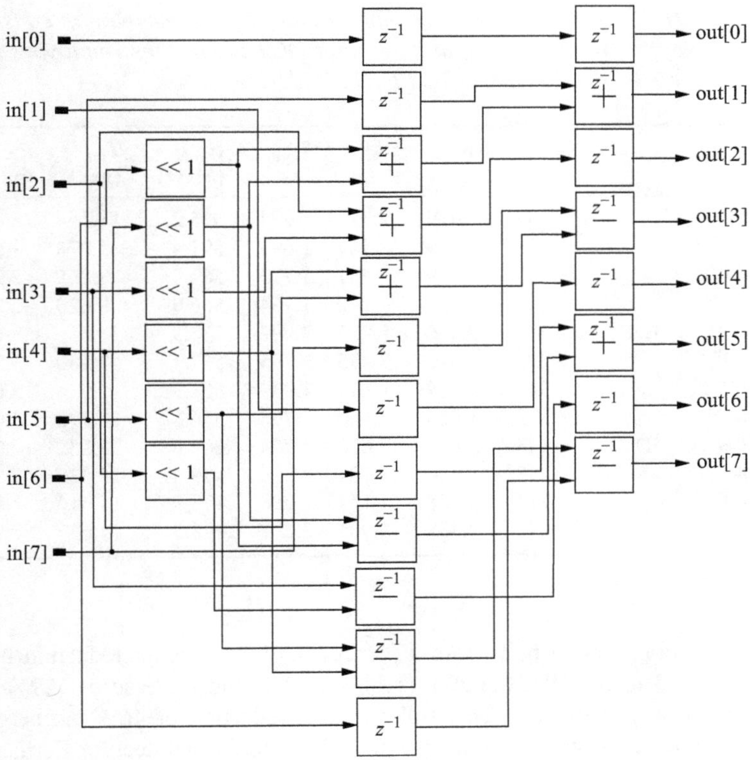

Figure 8.6 Low-complexity multiplier stage in Figure 8.3 for the $\mathbf{T}_{12}^{\text{cod}}$ *DCT approximation*

with 8-bit constants. The hardware usage for all the synthesized DCT algorithms aforementioned is shown in Table 8.15. The improvement provided by the approximate DCT architecture in terms of chip area, speed, and energy consumption is noticeable. All the three selected transforms and its 16- and 32-point versions outperform the HEVC IDCT architecture in terms of hardware resources with small video quality degradation (cf. Figure 8.1).

The transform $\mathbf{T}_9^{\text{cod}}$ presents the smallest hardware resource consumption among the implemented transforms. It requires approximately one-fifth of the FFs and one-eighth of the area-time required by the HEVC IDCT architecture. The transform $\mathbf{T}_{11}^{\text{cod}}$ has less than half of the latency time compared to HEVC IDCT architecture. The transform $\mathbf{T}_{12}^{\text{cod}}$ requires almost one-tenth of the total LUTs required by the HECV IDCT architecture.

The advantage of the selected transforms over the HEVC IDCT architecture in terms of hardware resources consumption is even more prominent on the 16- and 32-point versions. All the 16-point versions of $\mathbf{T}_9^{\text{cod}}$, $\mathbf{T}_{11}^{\text{cod}}$, and $\mathbf{T}_{12}^{\text{cod}}$ require less than one-tenth of area-time metric compared to the 16-point version of the HEVC IDCT architecture.

Table 8.15 Hardware-resource consumption and power consumption considered approximate 8-, 16-, and 32-point DCT algorithms employing Xilinx Virtex-6 XC6VLX240T 1FFG1156 device

Transform	**Slices**	**FF**	**LUT**	T_{cpd} **(ns)**	$F_{\max}$ **(MHz)**	T_{lat} **(ns)**	E_d **(nJ/cycle)**
8-Point HEVC IDCT	687	2,470	2,371	3.094	323.2	18.564	3.706
8-Point-$\mathbf{T}_9^{\text{cod}}$	86	219	141	1.691	591.4	5.073	1.551
8-Point-$\text{T}_{11}^{\text{cod}}$	144	459	341	1.773	564.0	8.865	2.237
8-Point-$\mathbf{T}_{12}^{\text{cod}}$	154	435	270	1.764	566.9	8.820	2.247
16-Point HEVC IDCT	2,597	9,406	8,827	3.097	322.9	21.68	8.634
16-Point-$\mathbf{T}_9^{\text{cod}}$	225	615	475	1.795	557.1	7.180	3.744
16-Point-$\mathbf{T}_{11}^{\text{cod}}$	341	1,104	911	1.989	502.8	11.934	5.074
16-Point-$\text{T}_{12}^{\text{cod}}$	323	1,046	760	1.796	556.8	10.776	5.023
32-Point HEVC IDCT	10,148	35,992	33,678	4.794	208.6	38.352	14.271
32-Point-T_9^{cod}	488	1,424	1,291	2.086	479.4	10.430	7.720
32-Point-$\mathbf{T}_{11}^{\text{cod}}$	713	2,528	2,154	2.088	478.9	14.616	7.778
32-Point-$\mathbf{T}_{12}^{\text{cod}}$	680	2,408	1,901	2.091	478.2	14.637	7.931

The 32-point version built from $\mathbf{T}_9^{\text{cod}}$ presents a slice-count reduction of 95% when compared to the HEVC IDCT architecture and an increase of 129% in the maximum-allowed frequency. This still allows a reduction of 46% in energy dissipated per clock cycle when compared to the HEVC IDCT architecture. Furthermore, the processing latency is also reduced in 73% and the area-time measure is 98% lower. The static power is not considerably affected, suggesting that the stand-by leakage by the FPGA chip dominates the quiescent energy consumption.

8.8 Conclusions

In this chapter, we reported the implementation of hardware-friendly DCT approximations obtained through a discrete optimization problem. The nonlinear discrete optimization problem considered the traditional and the proposed metrics for DCT approximations in the context of image/video compression. New metrics for DCT approximations assessment were proposed by the introduction of a weighting that discriminates the contribution of each of the transform basis vectors according to its effect in the transform performance in image/video-compression scenarios.

Three of the optimal solutions were selected for further analysis on software-simulation environment with an HEVC codec. The selected three transforms and their 16- and 32-point JAM scalable versions were embedded into the HEVC framework resulting in a modified HEVC-based coding scheme. The original and modified methods were compared and their performance assessed to the standard transforms used in the HEVC. Those transforms were implemented in hardware on the Xilinx *Virtex-6 XC6VLX240T-1FFG1156* FPGA chip, developed through the

Xilinx System Generator library for Simulink [65]. The transforms were implemented with pipelined stages for higher throughput. The three selected transforms outperformed the HEVC architecture in all the hardware-consumption-related metrics, achieving higher maximum operating frequency, and smaller consumptions of area and memory. The new transform $\mathbf{T}_9^{\text{cod}}$, not previously reported in the literature, exhibited the smallest hardware resource consumption compared with all the transforms. At the same time, it still could furnish images of high-quality compression according to the software experiments.

Acknowledgements

Dr. Renato J. Cintra, Dr. Fábio Mariano Bayer, Dr. Vítor de Andrade Coutinho, and Thiago Lopes Trugillo da Silveira thank Conselho Nacional de Pesquisa (CNPq) for partial support. Thiago Lopes Trugillo da Silveira also thanks Coordenação de Aperfeiçoamento de Pessoal de Nível Superior (CAPES) for partial support.

References

[1] N. Ahmed and K. R. Rao, *Orthogonal Transforms for Digital Signal Processing*. New York, NY: Springer-Verlag, 1975.

[2] V. Britanak, P. Yip, and K. R. Rao, *Discrete Cosine and Sine Transforms*. Academic Press, 2007.

[3] A. V. Oppenheim and R. W. Schafer, *Discrete-Time Signal Processing*, 3rd ed. Upper Saddle River, NJ: Prentice Hall, 2009.

[4] R. C. Gonzalez and R. E. Woods, *Digital Image Processing*, 2nd ed. Upper Saddle River, NJ: Prentice-Hall, Inc., 2001.

[5] V. Patil and R. Kumar, "A fast inverse motion compensation algorithm for DCT-domain video transcoder," *IEEE Transactions on Circuits and Systems for Video Technology*, vol. 18, no. 3, pp. 394–399, 2008.

[6] L. K. Ray and K. Ut-Va, "Motion compensation on DCT domain," *EURASIP Journal on Advances in Signal Processing*, vol. 2001, no. 3, pp. 1687–6180, 2001.

[7] E. Feig and S. Winograd, "Fast algorithms for the discrete cosine transform," *IEEE Transactions on Signal Processing*, vol. 40, no. 9, pp. 2174–2193, 1992.

[8] C. Loeffler, A. Ligtenberg, and G. S. Moschytz, "A Practical Fast 1-D DCT Algorithms with 11 Multiplications," in *International Conference on Acoustics, Speech, and Signal Processing*, vol. 2, 1989, pp. 988–991.

[9] H. W.-H. Chen, C. Smith, and S. Fralick, "A fast computational algorithm for the discrete cosine transform," *IEEE Transactions on Communications*, vol. 25, no. 9, pp. 1004–1009, 1977.

[10] Y. Arai, T. Agui, and M. Nakajima, "A fast DCT-SQ scheme for images," *IEICE Transactions*, vol. E71, no. 11, pp. 1095–1097, 1988.

[11] P. Duhamel and H. Hollmann, "Split radix FFT algorithm," *Electronics Letters*, vol. 20, no. 1, pp. 14–16, 1984.

[12] M. T. Heideman and C. S. Burrus, *Multiplicative Complexity, Convolution, and the DFT*. New York, NY: Springer-Verlag, 1988, originally presented as the author's thesis (Ph. D. – Rice University) under title: Applications of multiplicative complexity theory to convolution and the discrete Fourier transform.

[13] E. Simonotto, M. Riani, C. Seife, M. Roberts, J. Twitty, and F. Moss, "Visual perception of stochastic resonance," *Physical Review Letters*, vol. 78, pp. 1186–1189, 1997.

[14] J. Liu, A. Harris, and N. Kanwisher, "Perception of face parts and face configurations: An fMRI study," *Journal of Cognitive Neuroscience*, vol. 22, no. 1, pp. 203–211, 2010.

[15] R. P. N. Rao, B. A. Olshausen, and M. S. Lewicki, Eds., *Probabilistic Models of the Brain: Perception and Neural Function*. MIT Press, Mar. 2002.

[16] V. S. Ramachandran and R. L. Gregory, "Perceptual filling in of artificially induced scotomas in human vision," *Nature*, vol. 350, pp. 699–702, 1991.

[17] T. I. Haweel, "A new square wave transform based on the DCT," *Signal Processing*, vol. 81, no. 11, pp. 2309–2319, 2001.

[18] A. K. Jain, *Fundamentals of Digital Image Processing*. Upper Saddle River, NJ, USA: Prentice-Hall, Inc., 1989.

[19] K. Lengwehasatit and A. Ortega, "Scalable variable complexity approximate forward DCT," *IEEE Transactions on Circuits and Systems and Systems for Video Technology*, vol. 14, no. 11, pp. 1236–1248, 2004.

[20] S. Bouguezel, M. O. Ahmad, and M. N. S. Swamy, "A multiplication-free transform for image compression," in *2nd International Conference on Signals, Circuits and Systems*, Monastir, TN, 2008, pp. 1–4.

[21] S. Bouguezel, M. O. Ahmad, and M. N. S. Swamy, "Low-complexity 8×8 transform for image compression," *Electronics Letters*, vol. 44, no. 21, pp. 1249–1250, 2008.

[22] S. Bouguezel, M. O. Ahmad, and M. N. S. Swamy,, "A novel transform for image compression," in *53rd IEEE International Midwest Symposium on Circuits and Systems*, Aug. 2010, pp. 509–512.

[23] S. Bouguezel, M. O. Ahmad, and M. N. S. Swamy, "Binary discrete cosine and Hartley transforms," *IEEE Transactions on Circuits and Systems I*, vol. 60, no. 4, pp. 989–1002, 2013.

[24] S. Bouguezel, M. O. Ahmad, and M. N. S. Swamy, "A low-complexity parametric transform for image compression," in *IEEE International Symposium on Circuits and Systems*, Rio de Janeiro, BR, May 2011, pp. 2145–2148.

[25] D. Zhang, S. Lin, Y. Zhang, and L. Yu, "Complexity controllable DCT for real-time H.264 encoder," *Journal of Visual Communication and Image Representation*, vol. 18, no. 1, pp. 59–67, 2007.

[26] S. Bouguezel, M. O. Ahmad, and M. N. S. Swamy, "A fast 8×8 transform for image compression," in *International Conference on Microelectronics*, Dec. 2009, pp. 74–77.

[27] R. J. Cintra and F. M. Bayer, "A DCT approximation for image compression," *IEEE Signal Processing Letters*, vol. 18, no. 10, pp. 579–582, 2011.

[28] R. J. Cintra, "An integer approximation method for discrete sinusoidal transforms," *Circuits, Systems, and Signal Processing*, vol. 30, no. 6, pp. 1481–1501, 2011.

[29] F. M. Bayer and R. J. Cintra, "DCT-like transform for image compression requires 14 additions only," *Electronics Letters*, vol. 48, no. 15, pp. 919–921, 2012.

[30] R. J. Cintra, F. M. Bayer, and C. J. Tablada, "Low-complexity 8-point DCT approximations based on integer functions," *Signal Processing*, vol. 99, pp. 201–214, 2014.

[31] C. J. Tablada, F. M. Bayer, and R. J. Cintra, "A class of DCT approximations based on the Feig–Winograd algorithm," *Signal Processing*, vol. 113, pp. 38–51, 2015.

[32] A. Edirisuriya, A. Madanayake, R. J. Cintra, and F. M. Bayer, "A multiplication-free digital architecture for 16×16 2-D DCT/DST transform for HEVC," in *2012 IEEE 27th Convention of Electrical and Electronics Engineers in Israel*, Nov. 2012, pp. 1–5.

[33] P. K. Meher, S. Y. Park, B. K. Mohanty, K. S. Lim, and C. Yeo, "Efficient integer DCT architectures for HEVC," *IEEE Transactions on Circuits and Systems for Video Technology*, vol. 24, no. 1, pp. 168–178, 2014.

[34] T. L. T. da Silveira, F. M. Bayer, R. J. Cintra, S. Kulasekera, A. Madanayake, and A. J. Kozakevicius, "An orthogonal 16-point approximate DCT for image and video compression," *Multidimensional Systems and Signal Processing*, vol. 27, no. 1, pp. 87–104, 2016.

[35] T. L. T. da Silveira, R. S. Oliveira, F. M. Bayer, R. J. Cintra, and A. Madanayake, "Multiplierless 16-point DCT approximation for low-complexity image and video coding," *Signal, Image and Video Processing*, vol. 11, no. 2, pp. 227–233, 2017.

[36] M. Jridi, A. Alfalou, and P. K. Meher, "A generalized algorithm and reconfigurable architecture for efficient and scalable orthogonal approximation of DCT," *IEEE Transactions on Circuits and Systems I*, vol. 62, no. 2, pp. 449–457, 2015.

[37] M. Jridi and P. K. Meher, "Scalable approximate DCT architectures for efficient HEVC-compliant video coding," *IEEE Transactions on Circuits and Systems for Video Technology*, vol. 27, no. 8, pp. 1815–1825, 2017.

[38] R. J. Cintra, F. M. Bayer, and C. J. Tablada, "Low-complexity 8-point DCT approximations based on integer functions," *Signal Processing*, vol. 99, pp. 201–214, 2014.

[39] G. H. Golub and C. F. Van Loan, *Matrix Computations*, 3rd ed. Baltimore, MD: Johns Hopkins University Press, 1996.

[40] J. Katto and Y. Yasuda, "Performance evaluation of subband coding and optimization of its filter coefficients," *Journal of Visual Communication and Image Representation*, vol. 2, no. 4, pp. 303–313, 1991.

[41] M. J. Liang and T. D. Tran, "Fast multiplierless approximations of the DCT with the lifting scheme," *IEEE Transactions on Signal Processing*, vol. 49, no. 12, pp. 3032–3044, 2001.

[42] O. Egger, W. Li, and M. Kunt, "High compression image coding using an adaptive morphological subband decomposition," *Proceedings of the IEEE*, vol. 83, no. 2, pp. 272–287, 1995.

[43] R. S. Oliveira, R. J. Cintra, F. M. Bayer, T. L. T. da Silveira, A. Madanayake, and A. Leite, "Low-complexity 8-point DCT approximation based on angle similarity for image and video coding," *Multidimensional Systems and Signal Processing*, 2018.

[44] M. Ehrgott, *Multicriteria Optimization*, 2nd ed. Auckland, New Zealand, Springer, 2005.

[45] V. Barichard, M. Ehrgott, X. Gandibleux, and V. T'Kindt, *Multiobjective Programming and Goal Programming: Theoretical Results and Practical Applications*, 1st ed. Springer, 2009.

[46] J. Bezanson, A. Edelman, S. Karpinski, and V. B. Shah, "Julia: A fresh approach to numerical computing," *Society of Industrial and Applied Mathematics Review*, vol. 59, pp. 65–98, 2017.

[47] U. S. Potluri, A. Madanayake, R. J. Cintra, F. M. Bayer, S. Kulasekera, and A. Edirisuriya, "Improved 8-point approximate DCT for image and video compression requiring only 14 additions," *IEEE Transactions on Circuits and Systems I*, vol. 61, no. 6, pp. 1727–1740, 2014.

[48] Joint Collaborative Team on Video Coding (JCT-VC), HEVC References Software Documentation. Fraunhofer Heinrich Hertz Institute, 2013.

[49] D. J. L. Gall, "The MPEG video compression algorithm," *Signal Processing: Image Communication*, vol. 4, no. 2, pp. 129–140, 1992.

[50] International Telecommunication Union, "ITU-T recommendation H.261 version 1: Video codec for audiovisual services at $p \times 64$ kbits," ITU-T, Tech. Rep. 1990.

[51] International Telecommunication Union, "ITU-T recommendation H.263 version 1: Video coding for low bit rate communication," ITU-T, Tech. Rep. 1995.

[52] A. Luthra, G. J. Sullivan, and T. Wiegand, "Introduction to the special issue on the H.264/AVC video coding standard," *IEEE Transactions on Circuits and Systems for Video Technology*, vol. 13, no. 7, pp. 557–559, 2003.

[53] J. Ohm, G. J. Sullivan, H. Schwarz, T. K. Tan, and T. Wiegand, "Comparison of the coding efficiency of video coding standards—Including high efficiency video coding (HEVC)," *IEEE Transactions on Circuits and Systems for Video Technology*, vol. 22, no. 12, pp. 1669–1684, 2012.

[54] M. T. Pourazad, C. Doutre, M. Azimi, and P. Nasiopoulos, "HEVC: The new gold standard for video compression: How does HEVC compare with H.264/AVC?," *IEEE Consumer Electronics Magazine*, vol. 1, no. 3, pp. 36–46, 2012.

[55] F. Bossen, "Common test conditions and software reference configurations," Document JCTVC-L1100, Joint Collaborative Team on Video Coding (JCT-VC) of ITU-T SG16 WP3 and ISO/IEC JTC1/SC29/WG11, 2013.

[56] M. Naccari and M. Mrak, "Chapter 5 – Perceptually optimized video compression," in *Academic Press Library in Signal Processing Image and Video Compression and Multimedia*, ser. Academic Press Library in Signal Processing, S. Theodoridis and R. Chellappa, Eds. Elsevier, 2014, vol. 5, pp. 155–196.

[57] G. Bjøntegaard, "Calculation of average PSNR differences between RD-curves," in *13th VCEG Meeting*, Austin, TX, USA, Apr. 2001, document VCEG-M33.

[58] P. Hanhart and T. Ebrahimi, "Calculation of average coding efficiency based on subjective quality scores," *Journal of Visual Communication and Image Representation*, vol. 25, no. 3, pp. 555–564, 2014, QoE in 2D/3D Video Systems.

[59] X. Chen and N. K. Jha, "A 3-D CPU-FPGA-DRAM hybrid architecture for low-power computation," *IEEE Transactions on Very Large Scale Integration (VLSI) Systems*, vol. 24, no. 5, pp. 1649–1662, 2016.

[60] F. Sun, H. Wang, F. Fu, and X. Li, "Survey of FPGA low power design," in *International Conference on Intelligent Control and Information Processing (ICICIP)*. IEEE, 2010, pp. 547–550.

[61] J.-P. Deschamps, G. J. Bioul, and G. D. Sutter, *Synthesis of Arithmetic Circuits: FPGA, ASIC and Embedded Systems*. John Wiley & Sons, 2006.

[62] F. Salewski and S. Kowalewski, "Hardware/software design considerations for automotive embedded systems," *IEEE Transactions on Industrial Informatics*, vol. 4, no. 3, pp. 156–163, 2008.

[63] E. Monmasson and M. N. Cirstea, "FPGA design methodology for industrial control systems—A review," *IEEE Transactions on Industrial Electronics*, vol. 54, no. 4, pp. 1824–1842, 2007.

[64] E. Monmasson, L. Idkhajine, M. N. Cirstea, I. Bahri, A. Tisan, and M. W. Naouar, "FPGAs in industrial control applications," *IEEE Transactions on Industrial Informatics*, vol. 7, no. 2, pp. 224–243, 2011.

[65] Xilinx System Generator for DSP. https://www.xilinx.com/products/design-tools/vivado/integration/sysgen. Last accessed on Jun 17, 2019.

[66] P. Athanas, D. Pnevmatikatos, and N. Sklavos, *Embedded Systems Design with FPGAs*. Springer Science & Business Media, 2012.

[67] L. Shang, A. S. Kaviani, and K. Bathala, "Dynamic power consumption in virtex™-II FPGA family," in *Proceedings of the 2002 ACM/SIGDA Tenth International Symposium on Field-programmable Gate Arrays*. ACM, 2002, pp. 157–164.

[68] B. Sugla and D. A. Carlson, "Extreme area-time tradeoffs in VLSI," *IEEE Transactions on Computers*, vol. 39, no. 2, pp. 251–257, 1990.

[69] R. P. Brent and H. Kung, "The area-time complexity of binary multiplication," *Journal of the ACM (JACM)*, vol. 28, no. 3, pp. 521–534, 1981.

Chapter 9

Reconfigurable and approximate computing for video coding

Francesca Palumbo[1] *and Carlo Sau*[2]

9.1 Constraints and needs of video-coding systems

Systems perception capabilities are fundamental to make modern interconnected systems smart and, in this context, video coding is certainly a key technology. Video coding standards have been traditionally based on monolithic specifications coupled to a fixed set of profiles, capable of implementing different subset of functionalities of the given reference standard. Standards and profiles typically *offer* different features and enable different trade-offs that make them suitable for different applications and scenarios. Nevertheless, keeping the pace with constantly evolving scenarios and user's needs, for example, in terms of offered quality or consumed power, is not that simple and has led to continuous releases/updates of standards and profiles.

Ideally, designers should be allowed, at design time, to select and combine coding tools and profiles to optimally match the given requirements, while guaranteeing that customized applications are still interoperable. Nevertheless, monolithic specifications tend to hide parallelism and the dataflow nature of video-coding algorithms that, on the contrary, can be successfully exploited to guarantee efficient implementations and to play with different trade-offs [1].

> To address these issues, related to codecs design and customization, designers have been called to conceive efficient design-time methodologies capable of managing the adaptation of video-coding technologies along time.

Nowadays, systems are required to be flexible and versatile, capable of supporting multiple codecs and profiles simultaneously, and to enable dynamic reconfiguration among them [2]. Fortunately, both standards and profiles, even migrating from one generation to another, tend to present commonalities that are

[1]Intelligent system DEsign and Applications (IDEA) group, University of Sassari, Italy
[2]Microelectronics and Bioengineering Lab (EOLAB), University of Cagliari, Italy

exploitable both at specification level and at the implementation level to facilitate (re-)configurability.

To address these issues, related to codecs execution, designers have been called to conceive efficient run-time methodologies capable of addressing these flexibility needs.

In [3], the concept of dynamic parameter adjustment is expressed. As depicted in Figure 9.1, tuning current processing to the content variation and changing user/system requirements is made possible mastering the run-time variation of different parameters, which are stated as follows:

- Functional parameters—Allow for tuning output of the computation, they may include, for example, filter and transform lengths, and quantization levels.
- Architectural parameters—Allow for tuning the guaranteed performance and energy consumption, without modifying the output bits of the computing mechanism. An example can be the computation parallelism that may affect the given throughput positively/negatively affecting the energy consumption.

Both functional and architectural parameters can be tuned to optimize codecs before deployment and to also make it flexible to the changing conditions at run time. The aim of this chapter is describing the different design-time and run-time technologies that can be adopted for these purposes.

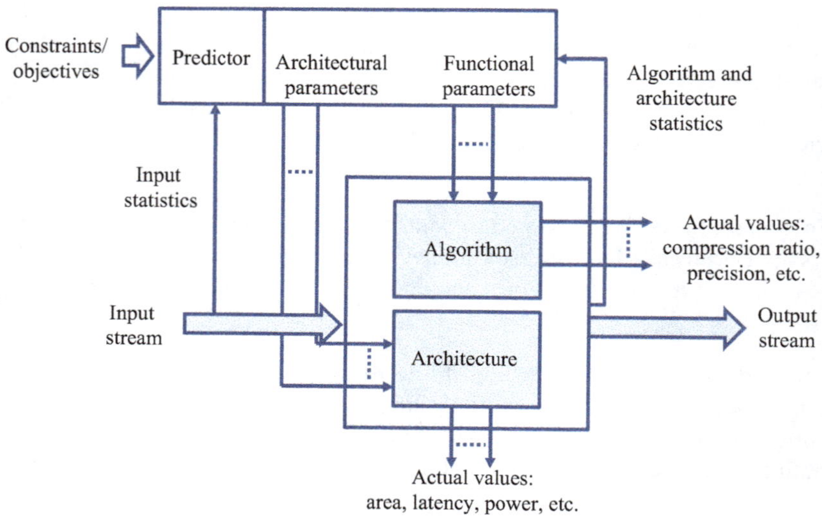

Figure 9.1 Dynamic parameter adjustment [3]

9.1.1 Chapter organization

The chapter is organized as follows:

- Section 9.2: The lack in flexibility of traditional monolithic codec specifications, not suitable to model commonalities among codecs and foster reusability among successive codec generations/updates, was the main trigger for the development of a new standard initiative within the ISO/IEC MPEG committee, called reconfigurable video coding (RVC). MPEG-RVC framework exploits the dataflow nature behind video coding to foster flexible and reconfigurable codec design, as well as to support dynamic reconfiguration.
- Section 9.3: The inherent resiliency of various functional blocks (like motion estimation in the high-efficiency video coding, HEVC) and the varying levels of user perception make video coding suitable to apply approximate computing techniques. Approximate computing, if properly supported at design time, allows achieving run-time trade-offs, representing a new direction in hardware–software codesign research. The main assumption behind approximate computing, exploited within video coding, is that the degree of accuracy (in this case during codec execution) is not required to be the same all the time.
- Section 9.4: This section will try to put together the concepts addressed in the chapter and will remark on which are, on our opinion, some interesting research directions provided the open challenges.

As a final guideline in the study of this chapter, please note that for each discussed techniques emphasis on design-time and run-time support available to the designers is provided. Moreover, the main objective for each discussed technique would be the (potentially) achievable run-time flexibility, which will consent them to address highly variable constraints and workloads.

9.2 Reconfigurable video coding

The MPEG-RVC [4], an ISO/IEC standard, aims at providing a framework for dynamic development, implementation, and adoption of standardized video coding solutions with features of higher flexibility and reusability. The intent of the MPEG-RVC framework [1] is to enable a smoother transition from one standard to another, being able to maintain an effective support of the previous ones, by exploiting the commonalities among the standards. Indeed, traditional monolithic schemes and specification formalism do not allow highlighting commonalities among codecs (either at the specification level or at the implementation one). To overcome this issue, MPEG-RVC provides a high-level specification formalism, based on RVC-CAL that is a subset [1] of the Caltrop Actor Language (CAL) normalized as a part of the RVC standard,[1] constituting a starting point model for the direct software and hardware synthesis. Basically, RVC does not define a codec

[1]CAL is a dataflow and actor oriented language specified within the Ptolemy project at the University of California at Berkeley [5].

but provides a way to describe codecs in a modular way (combining together functional units from the modular video tool library, VTL), along with a set of instruments to facilitate their analysis, customization, implementation and optimization [6–16]. Low-level details of C/C++ codec implementations are abstracted away by describing a codec as an "abstract" RVC-CAL network of actors taken from the standard library. Modularity is the key feature that enables the possibility of designing multi-standard video-decoding applications and devices by allowing software [17,18] and hardware [19–21] reuse across video standards.

9.2.1 Overview of the standard and its features

The MPEG-RVC framework defines two standards:

- ISO/IEC 23001-4 (or MPEG-B part 4)—Framework and standard languages used to describe the components of the framework.
- ISO/IEC 23002-4 (or MPEG-C part 4)—VTL.

To maximize reconfiguration, a lot of effort in the MPEG-RVC community has been spent on choosing the appropriate granularity level for the components constituting the standard library. Reuse is possible if the actors specifying codecs are not too coarse but, at the same time, they should not be too fine to lead to an explosion in terms of modules in the library to make the high-level reconfiguration process unfeasible.

The RVC-CAL features that foster its adoption in reconfigurable implementation, as reported in [4], are as follows:

- Natural capability of expressing parallelism—Codec specifications in RVC-CAL are networks of actors that encapsulate their own state and execute concurrently. This feature enables easier parallel implementations on multicore infrastructures [17].
- Modularity—Changing the implementation of an actor does not affect the rest of the network thanks to encapsulation. The dataflow form of the standard RVC specification guarantees that any reconfiguration of the user-defined proprietary library would not affect the implementation, which will remain consistent with the abstract RVC decoder model specified with the standard library.
- Facilitated scheduling—Since control and processing are decoupled. This not only allows for more facilitated scheduling but also allows for more flexibility. Various scheduling schemes, depending on optimization criteria, can be adopted.
- Adaptivity—Atomic firing of an action triggers the actor in a well-defined, and always known, state. One of the key features of a dataflow programming in general lays in its ability to be reconfigured dynamically.

Figure 9.2 illustrates both normative and proprietary components and how codecs can be built in an MPEG-RVC compliant manner.

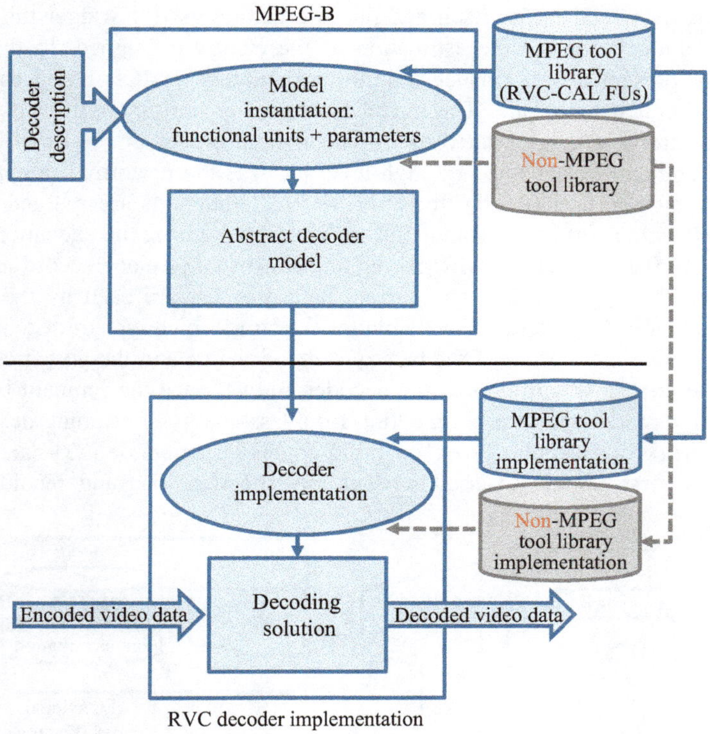

Figure 9.2 Illustration of the RVC framework, showing the dataflow-based normative specification as well as the decoder implementation process [17]

9.2.2 Dynamic reconfiguration exploiting MPEG-RVC

In terms of implementation, MPEG-RVC proved to be extremely flexible, and many different targets have been used for implementation along time. The work in [17] presents the automatic synthesis (C/C++ code have been synthesized leveraging on [6]) of an MPEG-4 simple profile (SP) decoder onto a dual-core platform. This work, despite not specifically reporting on a reconfigurable implementation, is particularly useful to understand the methodology for developing RVC dataflow codec and their implementation on multicore infrastructures.

In [22], Gorin *et al.* propose the Just-In-Time Adaptive Decoder Engine (Jade) that is responsible of managing the run-time reconfiguration of a decoder, by partially/completely switching the executing network leveraging on a low-level virtual machine. Reconfiguration, as discussed in Section 9.1, can take place for both functional and extra-functional (i.e., changes in requirements) reasons, and the work in [18] presents a combination of Jade with on-line energy estimations to decide upon optimal decoder implementation, where the functional parameters adjusted is the decoder type itself and the information used to trigger this kind of reconfiguration is the on-line estimation of the energy consumed. In that work, Ren *et al.* present a performance monitoring counters (PMCs) based estimation model, integrated with Jade. The model is driven by run-time-retrieved counters values monitoring energy-related events. In modern processors, a set of PMCs is provided and can be accessed by high-level tools as the performance application programming interface (PAPI). In [18], the PAPI functions are merged into the RVC-CAL framework, and estimations are performed during the execution to feed Jade. Figure 9.3 illustrates how dynamic adjustment takes place according to this technique. PMCs, constraints and current battery status are used by the *energy-aware manager*; this latter is responsible for selecting which type of decoder has to be used. Online monitoring of the battery state of charge and the power consumption of the current description of the decoder allows, once the remnant battery is lower than a predefined threshold, opting for a less-energy-consuming description.

The work of Beaumin *et al.* on multi-context accelerator [23] can be considered the first attempt to combine dataflow specifications and reconfigurable

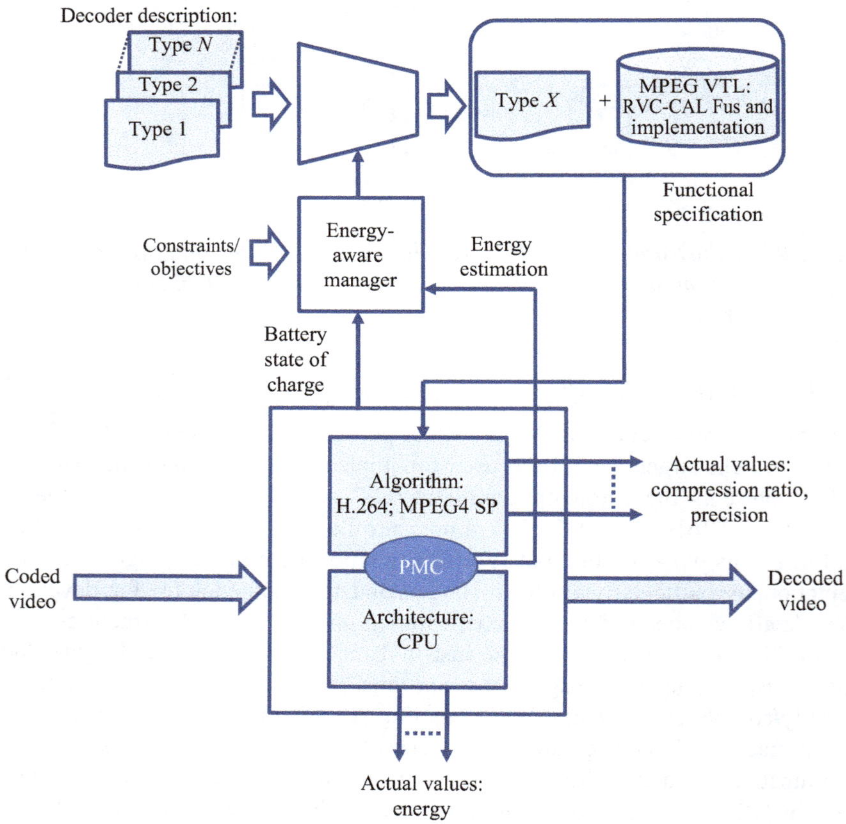

Figure 9.3 Energy-aware decoder reconfiguration

computing, to define a coprocessing unit. This latter exploits RVC-CAL principles to build the reconfigurable coprocessor, which is in charge of executing different dataflow specifications. The design of the reconfigurable coprocessor is based on a set of heterogeneous processing elements (PEs), called *network units* (NUs). Each NU executes a different dataflow network instantiating as many *processing units* (PUs) as needed to implement the CAL actors in the corresponding dataflow. Communication among PUs is FIFO based, a full mesh infrastructure is made available in each NU and it is configurable to process any input CAL specification. An example of NU is depicted in Figure 9.4. NUs are customized at design time on the basis of the input specifications. In [23], authors present the implementation of a coprocessing unit for the MPEG-4-SP implemented using the MPEG-RVC standard. The definition of the coprocessing unit, with NUs and PEs, is performed at design time. At run time, FIFO management is supported (as depicted in Figure 9.5): if the *FIFO controller* detects that the allocated memory for a FIFO is not sufficient, a complementary set of unused allocated FIFOs is activated. The required reconfiguration time is not affecting computation, since this type of reconfiguration acts at a coarse level, involving just the interconnect, and not at the level of the PUs execution. Dynamic management allows to better fit to workload variability.

In [20,21], MPEG-RVC is exploited for hardware reconfiguration purposes: optimized multifunctional coarse-grained virtual reconfigurable platforms are derived from disjoined specifications, exploiting the combination of three different MPEG-RVC compliant tools, namely, TURNUS [11,24], Xronos [9] and MDC [12,25]. Authors showed how coarse-grained virtual reconfiguration can be exploited to achieve functional adaptivity by implementing on a Xilinx Virtex 5 FPGA:

- An MPEG-4 multi-dataflow decoder [20,21] capable of switching among the MPEG-4-SP intra-only decoder and the MPEG-4-SP inter-decoder.
- JPEG codec prototype [21] capable of performing both coding and decoding over the same substrate.

Moreover, the same coarse-grained hardware-reconfiguration approach is demonstrated to be effective also for requirements-oriented adaptivity purposes. In [21],

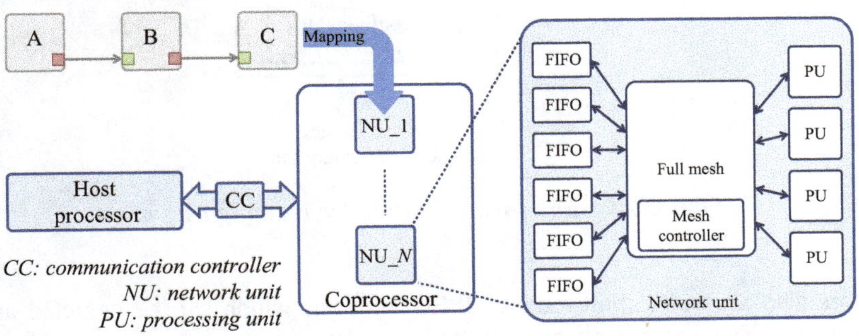

Figure 9.4 Dataflow process network based coprocessor [23]

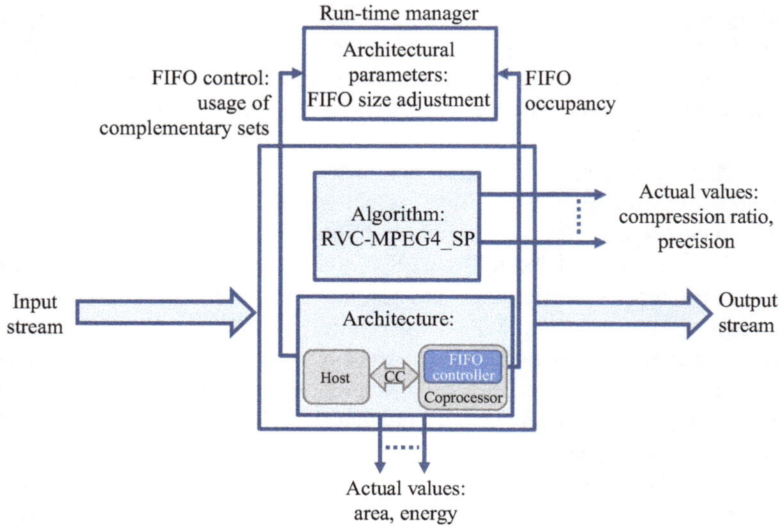

Figure 9.5 MPEG-RVC reconfigurable MPEG-4-SP processor to coprocessor system

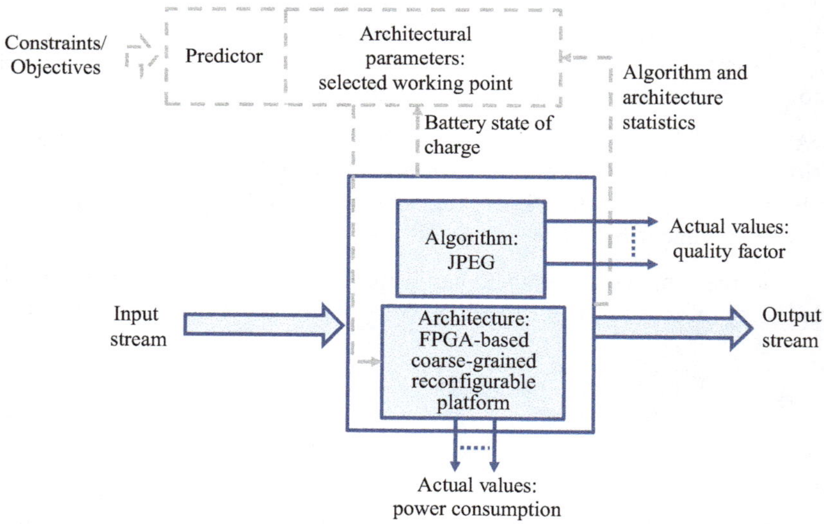

Figure 9.6 Energy-aware decoder reconfiguration

authors also present a multi-quality JPEG encoder, which can be depicted as in Figure 9.6. The computing part of this decoder, leveraging on coarse-grained virtual reconfiguration, enables surfing among different quality factors for the compressed images at the price of power consumption. In fact, dynamic power tends to increase going from a certain working point to better quality one, since

Table 9.1 Summary of run-time reconfigurable RVC-based approaches

Work	Algorithm	Architecture	Functional param.	Architectural param.
[18]	H.264 MPEG-4-SP	CPU	Decoder type	–
[23]	MPEG-4-SP	Host + custom coprocessor	–	FIFO depth
[20,21]	MPEG-4-SP	Custom HW	Execution profile	Profile ID
[21]	JPEG		Offered quality	

more actors are active in the computation. Using such an approach, user-triggered adaption along with power vs. performance (self-)adaptation is both feasible. The dashed parts of the figure indicate that, despite run-time reconfiguration is physically possible, the loop of dynamic (self-)adaptation was not supported at the time of that publication.

9.2.3 MPEG-RVC key remarks

This section recaps the goals and achievements of MPEG-RVC with respect to flexibility support in the video-coding domain. MPEG-RVC provides a standard to define codec specification in a modular way, favoring reconfiguration in different ways as summarized hereafter:

- Modular descriptions allow limiting the effort of passing from one profile to another or from one version of the standard to the successive one. Indeed, variations can be confined within atomic actors, while keeping the rest of the network unchanged.
- Several different tools have been proposed at the state of the art to improve design productivity, ranging from high-level design space exploration down to hardware synthesis.
- Reconfiguration is also enabled at run-time and performance/requirements-aware dynamic-system management is enabled. Table 9.1 provides a summary of the analyzed approaches.

9.3 Approximate video-coding systems

Human-perceptual tolerance makes video coding particularly suitable to the usage of approximate computing techniques, due to the fact that it can tolerate a varying degree of errors in the output visual quality after the coding process. Moreover, video-coding applications approximate by construction a video-coding sequence exploiting information redundancies, such as perceptual, statistical, spatial and temporal ones. For instance, temporal redundancy is exploited in motion estimation that is one of the most computationally intensive steps in the HEVC standard [26]. Motion estimation exhibits a high degree of resilience for little arithmetical errors. Indeed, it performs a search for the already processed image block (reference block) that is mostly similar to the currently processed block (current block).

Thus, if the chosen reference block is not the best one, the coding algorithm result will not be inconsistent. Actually, the general approach of fast motion estimation algorithms tends to reduce motion estimation complexity, leading to nonoptimal results and causing minor degradation to the coding efficiency. Motion estimation is then a perfect candidate for the adoption of inexact operators as those available within the approximate computing field, and this is just one of the many examples.

Approximate computing can be defined as "*the idea that computer systems can let applications trade-off accuracy for efficiency*" [27] by accepting, under certain applicability conditions, inaccurate results rather than guaranteed accurate ones. To do that "*techniques where the system intentionally exposes incorrectness to the application layer in return for conserving some resource*" are exploited [27]. Generally speaking, the emphasis is on codesigning the architecture, responsible for exposing accuracy-efficiency trade-offs, and the programming layer abstraction, which makes available the trade-offs to the programmers. Approximate solutions, for example, can be used to reduce computational complexity and energy consumption by exploring the trade-off between energy and quality of the system output. Given this consideration, approximation found good applicability in the video-coding scenario.

According to textbook definitions [28], approximate computing provides three degrees of freedom by acting at different levels:

- *Data level*—It consists in reducing the quality of the application in a controlled way by processing either less up-to-date data (temporal decimation), less input data (spatial decimation [29]), less accurate data (word-length optimization [30]) or even corrupted data.
- *Hardware level*—The exactness of computation can be relaxed by acting on the underlying circuitry functionally, by adopting inexact operators [31], and technologically, by playing with system parameters such as voltage supply [32,33].
- *Computation level*—It corresponds to computation and algorithm modifications. It aims at approximating the processing to decrease the computational complexity [29,34,35].

This classification is not so rigid, some works tend to aggregate different techniques, as in [28,29] where they apply computational approximation (given by loop perforation [36]) in combination with data approximation [37], to achieve larger benefits. Moreover, other taxonomy and classifications are available in the literature [27,28,39]. In [39], for example, classification of approximate computing techniques is based on features,[2] as explained hereafter:

- *Compute* vs. *data*—In compute-based techniques, approximation takes place while running the program (e.g., in low supply voltage static random access memory (SRAM) approximation occurs while accessing data, either for read or

[2]Classification of approximate computing techniques in details is beyond the scope of this chapter. Please refer to Table 9.1 of [39] for more details, where the authors reported a detailed and up-to-date taxonomy of approximate computing techniques referring to more than 40 state of the art works.

write). In data-based techniques, approximation takes place even when the program is not running and acts on storage or representation of data values (e.g., in low refresh dynamic random access memory (DRAM) approximation may take place at any point in the lifetime of the data when stored).

- *Deterministic* vs. *nondeterministic*—In the former case, for each and every input, the same error is always provided (e.g., code perforation). In the latter one, given a set of inputs, the occurrence of an error can only be probabilistically evaluated (e.g., synchronization elision).
- *Coarse-grained* vs. *fine-grained*—These techniques have different scopes and range of applicability. In fact, in coarse-grained techniques, arbitrary errors are induced while approximating an entire function or while reducing the overall data footprint. In this case, the information loss increases as more data and more instructions are omitted. On the contrary, using fine-grained techniques, errors are confined to the execution of a single instruction or on a single information.

To quantify the error and define the tolerance of approximation, for each approximate technique, the *quality of results* (QoR) has to be specified to define the level of acceptance of the computed results, according to user specifications and scenarios constraints. QoR is exploitable to guide compiler and run time, in choosing the optimal degree of available approximation and/or for controlling the optimal approximate execution engine (determining the software and hardware approximation mechanisms to be used) [40].

Approximate computing techniques can be clearly implemented either in software, at the code or data level, or directly in hardware, using, for example, inexact operators. Nevertheless, it is important to understand that in the definition of flexible solutions, implementing dynamic parameters adjustment, programmability and reconfigurability become fundamental whatever the chosen target platform is. In hardware, dynamism and flexibility are not straightforward and tuning approximation at run time may be challenging due to the fact that execution efficiency (corresponding to system specialization) and flexibility (corresponding to system generality) are normally colliding features. To overcome this issue, as discussed in many examples in the following sections, in many cases, run-time dynamism is achieved by adopting different flavors of the reconfigurable computing paradigm.[3]

9.3.1 Data-level approximation

Data are a natural source of approximation. The usage of less or less precise data induces errors in the computation, which can still be tolerated according to the given scenario.

Palomino *et al.* in [29] present a thermal optimization technique that employs a varying degree of approximations to reduce the temperature associated with the HEVC process, while maintaining good-quality results. The long-term goal here is

[3]Discussions and classification on reconfigurable approaches are beyond the scope of this chapter, details can be found in [41].

to avoid the side effects of the increased high-power density of modern system on chip, where thermal hot spots negatively affect the reliability and lifetime of devices aggravating most of the aging effects. Being video coding highly data dependent, authors adopt a content-driven approximate computing technique by analyzing at run time the tolerance to imprecise computations of each frame, to adjust the execution profile used accordingly, exploring the trade-offs between on-chip temperatures, computational complexity and visual quality. The different regions of a video sequence are classified, according to texture and motion information, to determine their potential error resilience; the higher is their tolerance supposed to be, the more aggressive is the approximation (selectively eliminating operations/computations at the algorithm and data levels) without compromising the output-video quality. This work considers HEVC, which encoding process is based upon a recursive quad-tree partitioning [42] of the basic coding unit (CU—64 × 64 pixel), which can be subdivided iteratively till reaching 8 × 8 sub-blocks (see Figure 9.7(a)). Blocks are analyzed to determine the optimal compression mode. Reaching the deepest 8 × 8 division turned out not to be always needed: in case of homogeneous blocks (with little to none motion), 64 × 64 provides good coding efficiency, on the contrary highly textured and high-motion regions require a finer splitting granularity. Spatial and temporal approximation are then used to code the sequence, leveraging on intra- (based on matching functions like sum of absolute difference (SAD)) and inter- (based on finite input response (FIR) filters) predictions that are composed of primitive pixel-level operations (i.e., addition, subtraction and multiplication), which can be approximated at various granularity achieving obviously different degradation degrees. Data-level approximation can be applied as in Figure 9.7(b), where for each coding granularity level is presented a different data-approximation degree. Figure 9.7(c) depicts an example of how variable degrees of data approximation can be used while encoding; the more aggressive is the approximation degree, the more coding results may be affected. As depicted in Figure 9.8, authors in [29] present a lightweight *approximate mode selection heuristic algorithm* based on *error tolerance*, which computes how far data approximation can be pushed, while preserving the requested quality. Such algorithm determines at the CU level the *data-approximation mode*, which is a functional parameter of quad-tree partitioning and can range from 0 (1:1, no approximation) to 3 (1:8, given 8 pixel only 1 is used in prediction calculations).

Another clear example of data approximation, acting on both data precision and amount, is the work proposed by Paim *et al.* [43] on approximating the transformation step that is typical of video-coding applications. Transform operations are pervasive in video-coding algorithms. They are present throughout the whole processing pipeline, including spatial and temporal predictions. Transformation means porting the input signal, in this case an image block, from the color space domain to the frequency of colors variation domain. It is implemented as a double multiplication of the image block by a given coefficients matrix, typically coming

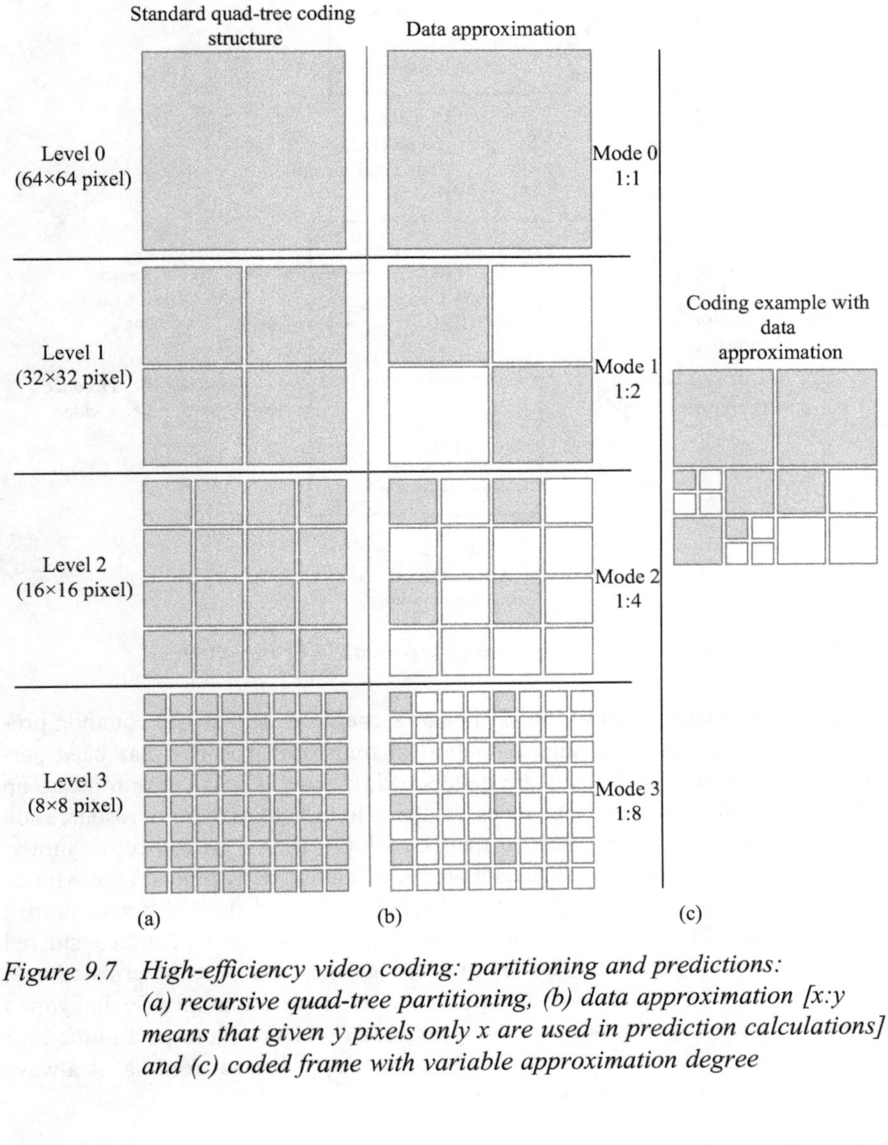

Figure 9.7 High-efficiency video coding: partitioning and predictions: (a) recursive quad-tree partitioning, (b) data approximation [x:y means that given y pixels only x are used in prediction calculations] and (c) coded frame with variable approximation degree

from a discrete cosine transform (DCT) or discrete sine transform or from alternative solutions such as the discrete Tchebichef transform (DTT). Paim *et al.* [43], on the top of an already approximated (computational level approximation that will be discussed later on) version of the DTT matrix [44], apply truncation and pruning in order to simplify the computation and the system (a hardware accelerator) itself, resulting in area, power and compression efficiency improvements. First, a

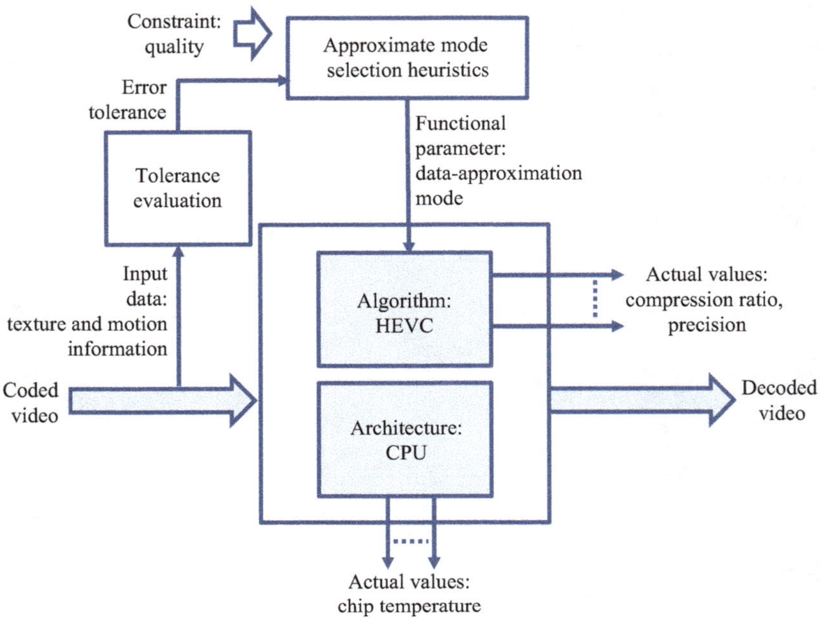

Figure 9.8 Run-time approximate CU encoding

quality-driven quantization of the DTT matrix coefficients, that is a common process in computational level approximation of transform matrices, has been performed. The chosen coefficients are quite small, making it possible to truncate up to 4 bit of the intermediate image block data and, in turn, to save lot of resources on the intermediate storage and second matrix multiplication with respect to similar literature works. Besides acting on data precision, authors also applied approximate computing on the processed amount of data: they derived three different pruned versions of the DTT accelerator by removing up to three rows of the considered DTT matrix (see Figure 9.9). This led to additional improvements in terms of area, power (up to 39% and to 43%, respectively, in application-specific integrated circuit (ASIC) and field-programmable gate array (FPGA) with respect to the case without pruning) and throughput, in front of a quality degradation that is always less than 2 dB in terms of PSNR.

9.3.2 Hardware level approximation

Simplifications on the physical platform capable of performing the computation are referred to as hardware level approximation. One possibility is to simplify, or rather relax, the technological parameters of the platform: voltage supply can be lowered [32] or even removed [33] on some or all parts of the design, always or only when approximation can be afforded. In particular, in [33], the power gating applied during approximated operations is also causing data-level approximation due to the resulting truncation of the operands. Another option for hardware level

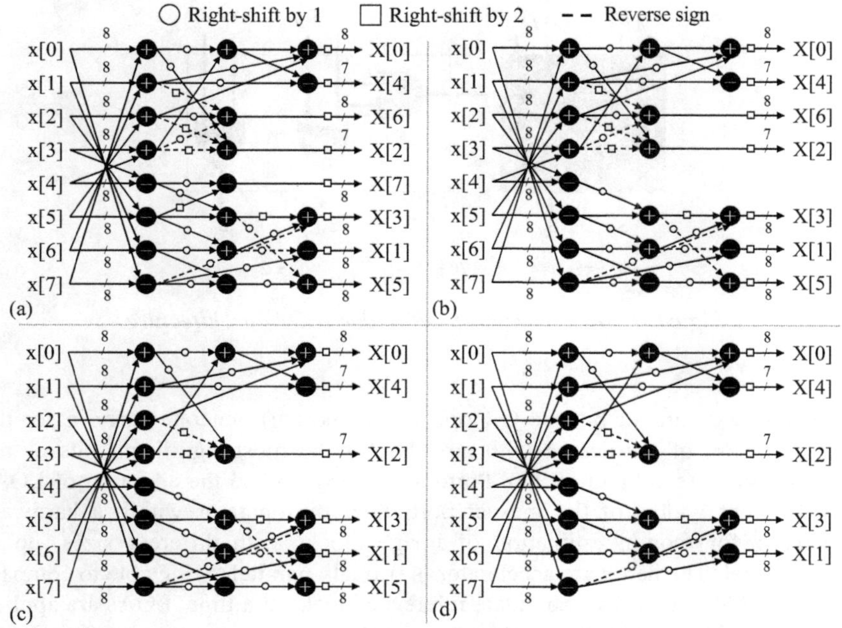

Figure 9.9 Approximate one-dimensional DTT architecture with different numbers of calculated rows: (a) 8 (all), (b) 7, (c) 6 and (d) 5

approximation is to adopt, by purpose, inexact operators. This means that, for instance, additions and multiplications implemented by those operators will lead to results that are not correct, but almost [31].

As already said, video coding is the perfect candidate application for applying approximate computing. Prediction related parts of video-coding algorithms turn out to be the most suitable at this purpose, since they implement, already, approximations and they are usually the most computationally intensive steps, meaning that approximation can be very effective on them. The SADs is the final step of the motion-estimation algorithm, involved in temporal prediction. Motion estimation aims at evaluating which is the best already processed frames blocks (reference block) that is mostly similar to the currently processed frame block (current block). The decision is taken by calculating the differences between pixels of candidate reference blocks and the current one, whereupon the absolute values of the pixel differences are summed up with the SAD step.

Inexact operators, in particular, adders, can be applied on SAD to simplify the process, reducing the complexity and, in turn, latency (frequency) and energy consumption [45]. One alternative to regular, exact ripple-carry adders (RCAs), commonly adopted in SAD, are lower part OR adders (LOAs). LOAs split addition onto two pieces: the most significant bits are processed with a regular, precise sum; the least significant bits instead (i) are not subjected to carry propagation, (ii) are

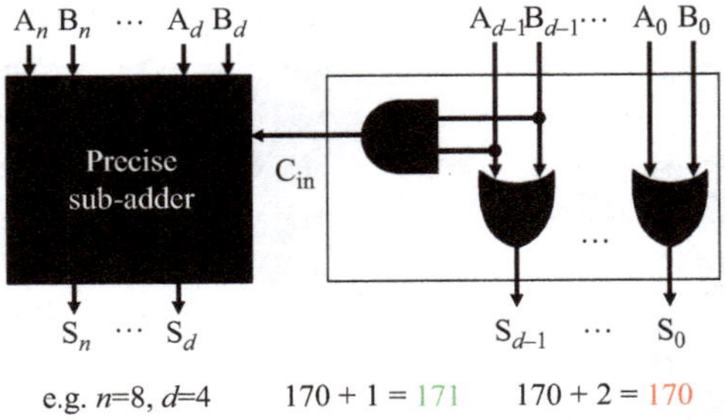

Figure 9.10 Lower part OR adder (LOA) architecture

processed by means of bitwise OR operations and (iii) generate carry-in for the precise portion of the design with an AND of the most significant bits of the imprecise logic (see Figure 9.10). Porto *et al.* [45] studied the adoption of LOAs with different widths of the inexact portion of the operator within a hardware accelerator for motion estimation of image blocks with different sizes, up to 16 × 16 pixel. The hardware accelerator is strongly parallel, being able to compare the current block with 169 candidate reference blocks at a time. LOAs are applied only to the adder operators calculating the difference between pixels of the reference and current block. Overall, the core of the accelerator involves 364 adders, among which 208 have been replaced with imprecise LOAs. Three different amounts of approximated least significant bits, 3, 4 and 5, have been considered for the LOAs, resulting in three different levels of imprecision. Results show that these levels can be quantified in terms of Bjøntegaard delta rate that is between 0.6% and 2.5% for the luma component and between 0.45% and 1.85% for the chroma one, respectively, going from 3 to 5 bit on the approximated part of the adder (all these values make the quality degradation negligible). This is the price that, with respect to a regular full RCA design, has to be paid for achieving a power saving going from 9.7% to 22.1% on the single SAD computing core and from 7% to 11% on the whole accelerator.

El-Harouni *et al.* [46] also investigated the possibility of employing approximate adders on the SAD step of the video-coding motion estimation, going even more deeply throughout a dynamic approximate video codec solution. Again, they adopted LOAs, but they realized four different implementations of the imprecise adders for the least significant bits, starting from the corresponding single bit instance, going from a full precise solution to a trivial only-shortcut-based one (see Figure 9.11). LOAs have been considered to build different SAD computing tiles for several block sizes and with various bit length of the imprecise sub-adder. The SAD computing tiles have also been profiled so that their bitrate, power/energy consumption and quality may be known. An interesting measure here is that, on the overall motion-estimation performance, for a given approximate adder architecture not always taking less bits on the imprecise sub-adder leads to the best results. The bitrate, power/energy and quality information is made available at run

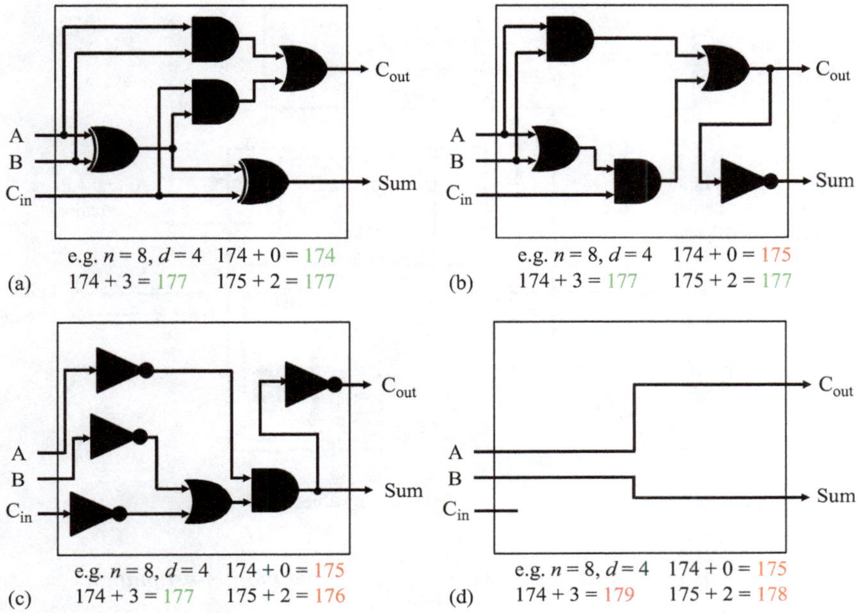

Figure 9.11 Four different implementations of single bit adders: (a) exact or (b), (c) and (d) approximate

time to a selection unit capable to decide, at run time, which SAD tiles have to be adopted according to the current user requirements in terms of tolerable error and required energy reduction. In such a way, the system activates only the SAD tiles necessary to meet the user (quality and energy) and application (block size) constraints while power gating the others (technological hardware-level approximation combined to the functional, operator related one). The resulting architecture is basically capable of tuning at run time two kinds of architectural parameters: the SAD tiles size and their approximation level. In such a way, as can be seen in Figure 9.12, the resulting quality (functional) and energy (architectural) values can be adjusted, according to the user requirements, as well as the specific input stream (block size impacts also on energy since only the properly SAD tiles are used in the current computation, while the others are switched off). The maximum bitrate enhancement for the final architecture is achieved for the maximum bit length of the imprecise sub-adder within the LOAs (6 bit), while in the worst case, the PSNR of the decoded video is decreased by 0.57 dB. In terms of energy, the same behavior is not seen since it is not always true that the lowest consumption is reached for the most approximated LOAs.

Hardware approximation can be achieved also by working on memories; the latter is one of the main sources of power dissipation in multimedia applications due to the number of memory accesses. It is possible to trade precision with

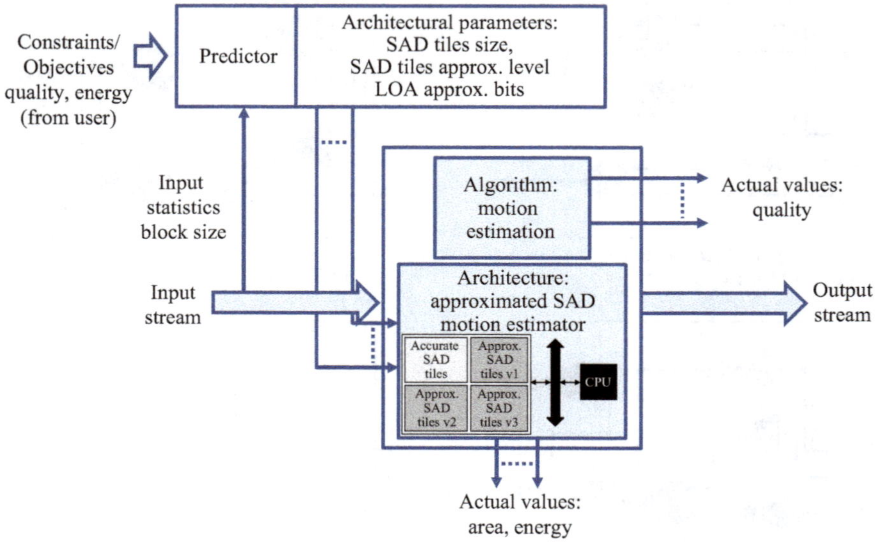

Figure 9.12 Energy/quality/size-aware approximated motion-estimation accelerator

consumption, lowering VDD. Supply voltage scaling is one of the most effective techniques for power reduction, as dynamic power dissipation is quadratically dependent on VDD. Nevertheless, low-voltage operations cause the failure probabilities of SRAM bit-cells to significantly increase. In the video-coding scenario, this is doable, again due to the fact that human perception, which makes our eyes not so much sensitive to small variations. The work in [47] presents 64 kB approximate SRAM architecture for low-power video applications, where 6 transistor memory cells are used. To avoid significant error, this works loads the most significant bits in protected cells, and the less significant ones in unprotected cells. Three different supply voltages are enabled. Protecting the most significant bits allows for preserving a good-image quality, PSNR is still around 30 dB when the VDD is decreased to 0.6 V, in comparison to the standard nonprotected ones, where PSNR is lowered down to 5 dB.

9.3.3 Computation level approximation

Approximate computing can also be adopted at the algorithm level, in order to simplify the processing and benefit from the consequent throughput, latency, consumption improvements. Computation level approximation is basically the main pillar at the basis of video-coding algorithms by nature. However, further approximations on the computational level are still possible at different stages of the algorithm.

As already seen previously, video-coding transformation step offers a certain amount of place for applying approximate computing. Actually, the coefficient matrices adopted in transformations can see themselves as approximations of more precise solutions [48]. One of the most beaten research street is related to the further approximation of the video-coding transform matrices. Indeed, it is very common to replace DCT matrix with a quantized DCT matrix where coefficients, that are decimal numbers lying in the range $[-1,1]$ are rounded to the closest integer value, resulting in a rounded cosine transform. In other cases, the rounding is performed smartly by differentiating highest and lowest values of each matrix row from the remaining coefficients in order to preserve correlation between pixels [49] (see Figure 9.13). Additionally to the coefficients rounding approximation, also similarities between DCT matrices or decomposed (odd–even decomposition) DCT matrices blocks are exploited to reuse parts of the design for smaller or bigger block size transformations [50]. Decomposed DCT matrices, where original DCT coefficients are approximated at algorithm level, have been also coupled to approximate computing at the data level: truncation of the matrix block coefficients in an odd–even decomposed matrix where odd block is used for both odd and even part has been proposed in literature [51]. In other cases, always relying on DCT matrix decomposition, block matrix coefficient simplification has not been driven by mathematical (simply rounding), input data (correlation between pixels) or data precision (truncation) motivations, but it has been driven

$$C_8 = P_8 \begin{pmatrix} E_4 & 0 \\ 0 & O_4 \end{pmatrix} \begin{pmatrix} I_4 & J_4 \\ J_4 & -I_4 \end{pmatrix}$$

$$E_4 = \begin{pmatrix} 64 & 64 & 64 & 64 \\ 83 & 36 & -36 & -83 \\ 64 & -64 & -64 & 64 \\ 36 & -83 & 83 & -36 \end{pmatrix} \approx 32\times \begin{pmatrix} 2 & 2 & 2 & 2 \\ 2 & 1 & -1 & -2 \\ 2 & -2 & -2 & 2 \\ 1 & -2 & 2 & -1 \end{pmatrix} = \widetilde{E}_4$$

$$O_4 = \begin{pmatrix} 89 & 75 & 50 & 18 \\ 75 & -18 & -89 & -50 \\ 50 & -89 & 18 & 75 \\ 18 & -50 & 75 & -89 \end{pmatrix} \approx 32\times \begin{pmatrix} 2 & 2 & 1 & 0 \\ 1 & 0 & -2 & -2 \\ 2 & -2 & 0 & 1 \\ 0 & -1 & 2 & -2 \end{pmatrix} = \widetilde{O}_4$$

Figure 9.13 Example of even–odd decomposed 8 × 8 DCT matrix approximation by quantization: C_8 is the original 8 × 8 DCT matrix, P_8 an 8 × 8 permutation matrix, I_4 and J_4 the 4 × 4 identity and its reflection, respectively, E_4 and O_4 the even–odd 4 × 4 decomposition matrices that are approximated with E_3 and O_4, respectively by hardware implementation. Potluri *et al.* [52] proposed a DCT simplified block matrix where coefficients have been approximated in order to minimize the number of multiplications and additions required for the transformation computation, resulting in a 8 × 8 DCT calculation with only 14 additions.

Hardware implementation-driven approximation has been also the approach followed by Masera *et al.* [53], where authors adopted the Walsh–Hadamard transform (WHT) to calculate the DCT. WHT is basically a decomposition of the DCT mainly requiring, besides the computing of the DCT common butterfly operations, also the calculation of the Givens rotations. If all the Givens rotations are calculated, WHT is capable of providing a complete, exact DCT. Authors proposed to skip some or all Givens rotations calculations in order to simplify the complexity of the algorithm and save power, but introducing errors in the results since WHT with less Givens rotations than expected is only an approximation of DCT. A study on how many Givens rotations have to be skipped depending on the magnitude of the DCT inputs and on the generated quality degradation has been conducted. Two different implementations have been developed, folding or not the DCT calculation circuitry for horizontal and vertical multiplication. The implementations have four different operation modes for the DCT implementation going from 0% (no power saving, no quality degradation) to 100% (maximum power saving, maximum quality degradation), passing by two intermediate points skipping, respectively, 37% and 55% of the Givens rotations. It is also possible to decide if Givens rotation has to be computed for a specific input by setting a certain threshold on its magnitude, thus obtaining an adaptive behavior. Such architecture can then act dynamically and change the operation mode (architectural parameter), that is power vs. quality trade-off, at run time depending on the user needs or, adaptively, on the current input data magnitude (see Figure 9.14).

Generally speaking, to be effective, algorithmic approximations must target computation-intensive blocks. In the HEVC standard, it has been demonstrated that computation is dominated by motion compensation [34,54] that is implemented with fractional pixel interpolation filters and represents between 62% and 80% of a decoder complexity. Basically, motion compensation leverages on temporal redundancy of the video sequences to transmit only motion vectors of the objects (image blocks) in the scene rather than complete, very similar frames. When fractional pixel motion has to be compensated, interpolation is necessary to predict a certain image block from the corresponding reference one. For instance, in HEVC luma and chroma, interpolations are implemented by means of two separable FIR filters, respectively, acting horizontally and vertically.

Computation level approximate computing can be applied in the motion-compensation step of HEVC to reduce filtering complexity: instead of using legacy N taps filters (N is 7 or 8 for luma and 4 for chroma in HEVC), approximated solutions with reduced number of taps can be very effective. Up to 20% of energy can be saved for software HEVC decoder with approximate 7, 5 and 3 taps luma, and 3 and 2 taps chroma motion compensation interpolation filter [34]. Authors combine this kind of approximation with the skipping of in-loop filtering (deblocking and SAO), constituting the final step of motion compensation in the

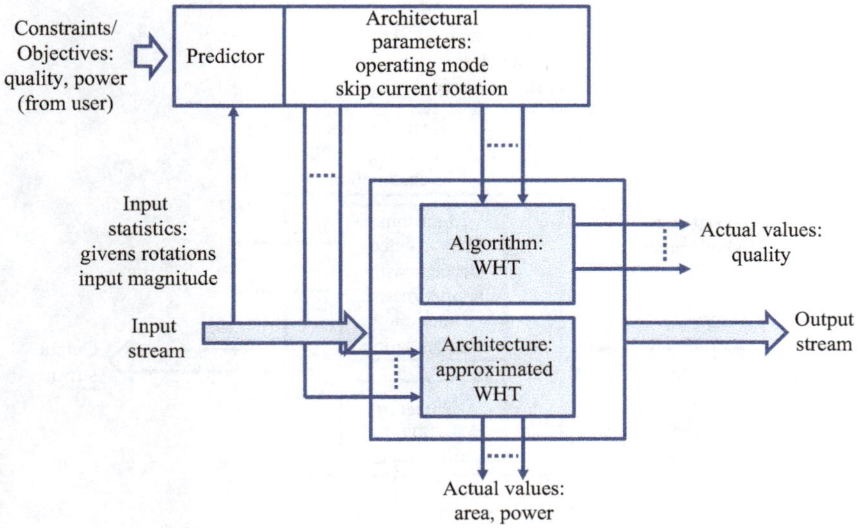

Figure 9.14 Energy/Quality approximated WHT (DCT calculation) accelerator

video decoder. Four different operating modes have been proposed, each one with a different interpolation and in-loop filtering approximation and, in turn, different levels of energy savings up to 40% of the total consumption. However, energy saving does not come for free. A price has to be paid in terms of quality of the decoded images. In particular, in the worst case, the quality degradation in PSNR is about 2 dB. Note that approximation has to be applied selectively in order to avoid drawbacks on the results of the computation while, at the same time, taking advantages from its benefits in terms of energy saving. For instance, in the HEVC fractional interpolation case, approximate filters usage should be avoided on I-frames to limit quality drift between reference frames while achieving substantial energy savings [55]. Authors envisioned also closing the loop and controlling the operating mode (quality vs. energy trade-off) of the decoder according to the instantaneous energy consumption (see Figure 9.15). In an extremely user-oriented application field such as the video coding one, an important metric is also related to the user itself. Thus, besides the numerical, PSNR-based quality degradation evaluation, the perception of the final user represents a key aspect to understand if a certain approximated implementation can be actually applied in the practice. Subjective tests have been performed for approximate HEVC fractional interpolation: the degradation of the perceptual quality seems to be imperceptible for most of the considered sample testers [56].

According to the work proposed by Nogues *et al.* [55], in [57] and [35], the corresponding hardware implementation has been presented. The implementation exploits the full parallelism of the algorithm by pipelining horizontal and vertical filtering by thorough intermediate FIFO buffers. Virtual coarse-grain

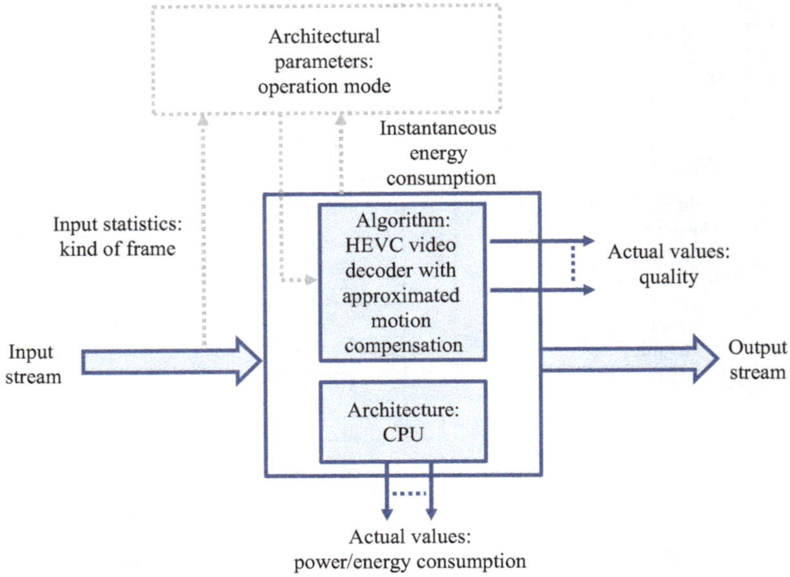

Figure 9.15 Energy/Quality approximated motion compensation for HEVC decoding

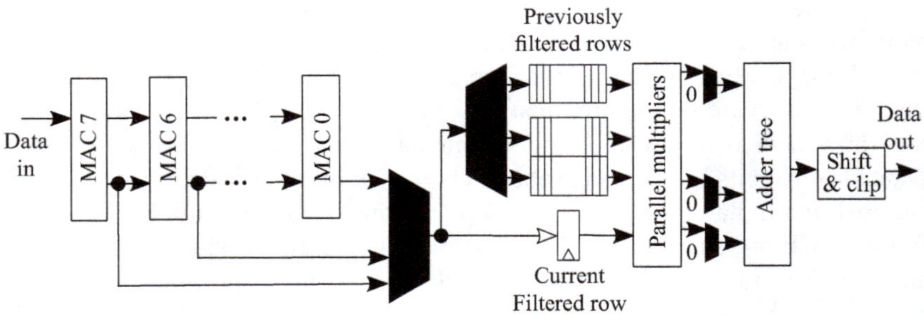

Figure 9.16 Hardware architecture of the reconfigurable approximate HEVC interpolation filter

reconfiguration (like the one used in [12]) is implemented with multiplexers and allows the system to exclude certain filtering stages from the computation upon request at run time, thus achieving approximation, as shown in Figure 9.16. According to the supported configurations, three different operating points are provided: high, *H*, where the legacy filters (8/7-tap *luma* and 4-tap *chroma*) are executed; medium, *M*, with 5-tap *luma* and 3-tap *chroma*; low, *L*, with 3-tap *luma* and 2-tap *chroma*. Considering the legacy *H* mode, approximation allows the saving of up to 19% and 38% energy per group of frame blocks, respectively, for luma and chroma.

At the computational level, approximate computing, due to the reduction of algorithm complexity, can easily lead to an enhancement in terms of throughput. However, the throughput achieved by the legacy (nonapproximated) implementation could yet be enough for the desired behavior. In such cases, one possibility for a further optimization of the system consumption is to apply frequency scaling on the top of approximation and reconfiguration. Thus, the system throughput can be kept above the minimum (corresponding to the throughput of the legacy configuration), while the power consumption is reduced as a consequence of the frequency decrease. Note that, power-consumption reduction through frequency decrease means also execution time enlarging. For this reason, a corresponding energy reduction is not always guaranteed: in [35], the energy saving of underclocked L mode with respect to H mode increases from 19% to 27% for luma but remains unaltered on 38% for chroma.

The fractional interpolator for motion estimation proposed in [35], described as a standalone accelerator without any system level integration, could be used at run time to play with quality and power/energy parameters. Indeed, as depicted in Figure 9.17, it could be put in the loop with user or input-driven requirement in order to switch among possible architectural operating modes dynamically. Power/energy can be driven by the remaining battery level on the device or by PMCs providing some kind of information on the current consumption of the system. Quality can be driven by the user, by purpose or automatically depending on the distance of the screen from the human eye. But the operating mode selection,

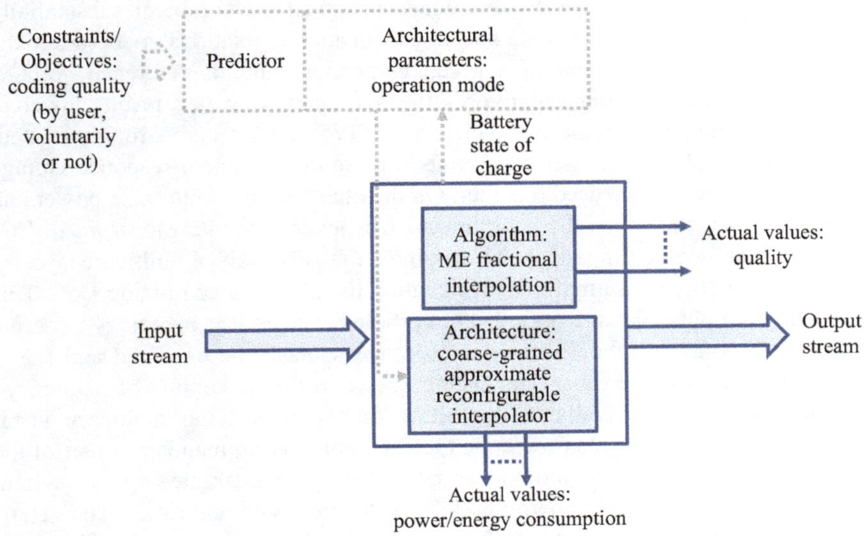

Figure 9.17 Possible dynamic behavior of a reconfigurable approximate interpolator for motion estimation/compensation in HEVC

according to the work of Nogues *et al.* [55], should also consider the current frame nature (I, P or B) in order to limit the quality degradation, thus introducing some kind of adaptivity in the loop.

To understand the potentials of approximate multifrequency approaches in the practice, Table 9.2 reports a comparison of the work proposed in [35] with similar state-of-the-art solutions. The detailed discussion of such results is beyond the scope of this chapter and it is available in [35]. However, from the table, it is possible to appreciate how (i) the power/energy vs. quality trade-off curve exists despite the additional logic required to implement reconfiguration on hardware and (ii) the reconfigurable approximate multifrequency solution is capable of competing and beating most of the literature, even not dynamic (not reconfigurable) approaches in terms of throughput and energy efficiency.

The work in [63] combines computational level approximate computing with *dynamic voltage and frequency scaling* (DVFS). Within the approximate computing domain, an *imprecise computation task* can be defined as that task that can be logically divided into a *mandatory subtask* plus an *optional subtask*. The former has to be completed before the deadline to ensure the baseline quality of service (QoS), while the latter may not be completed before the deadline. When the latter situation occurs, QoS associated to the overall task will not be optimal, and different levels of approximation can be offered staring from the minimal baseline one. DVFS couples *dynamic frequency scaling* (DFS) and *dynamic voltage scaling* (DVS). DFS relates to the possibility of adjusting the execution frequency of the system on the fly to preserve, according to the computational requirements, the consumed power (which dynamic part depends directly from the frequency) and, in turn, to limit heat dissipation. DVS is the possibility of incrementing (over-volting) or decrementing (under-volting) the reference voltage of a component substantially affecting its power consumption, due to the quadratic relationship between the dynamic power consumption of a given component and its reference voltage. Nevertheless, under-volting negatively affects frequency, which results lowered. This last consideration should make clear why DVS and DFS are strongly related. Indeed, when real time or user constraints impose quick system response voltage and clock have to be increased together. On the other hand, to minimize power and heat dissipation, they should be minimized whenever possible. Mo *et al.* in [63] focus specifically on optimizing task mapping on DVFS-enabled multicore systems to minimize energy consumption and execution time, while maximizing QoS. This is basically a multi-objective optimization problem, where it is necessary to define (i) in which processor should a certain task (or subtask) be executed and (ii) at which frequency/voltage values (for each processor different couples of frequency–voltage values can be available depending on the underlying multicore infrastructure). The goal is to maximize the QoS (by completing mandatory part of the computation plus as much optional as possible) while guaranteeing the system constraints (as the available energy budget or the requested real time). The details of this optimization problem are beyond the scope of this chapter, but [63] demonstrates how DVFS can be quite effectively used in combination to those applications where imprecise computation task is affordable, and video coding

Table 9.2 Potentials of reconfigurable approximate multifrequency architecture with respect to other literature works. H, M *and* L *are the system operating modes, while in brackets are reported results achievable when multifrequency is enabled*

		FPGA (nm)	Chroma	LUT	Register	BRAM	Multiplier	Resolution	Frame rate (fps)	dP (mW)	dE × pel (pJ)
[58]		Stratix III, 65	N	5,710	3,861	1	–	3,840 × 2,160	30	379	1,523
[59]		Virtex-6, 40	N	1,890	1,224	2	–	2,560 × 1,600	60	168	684
[60]		Virtex-6, 40	Y	3,929	3,422	6	–	3,840 × 2,160	30	NA	NA
[61]		Virtex-5, 65	Y	5,017	2,550	2	–	2,560 × 1,600	30	89	362
[62]		Virtex-6, 40	N	14,225	9,984	–	–	3,840 × 2,160	45	NA	NA
[35]	*H*	Virtex-5, 65	Y	2,903	522	28	80	3,840 × 2,160	60	107(107)	108(108)
	M									101(90)	101(90)
	L									78(64)	78(64)
	H	Zynq-7000, 28	Y	3,235	515	17	80	3,840 × 2,160	60	64(64)	64(64)
	M									55(50)	55(50)
	L									38(32)	38(32)

is certainly among them. Basically, DVFS is exploited to guarantee enough frequency–voltage levels to those cores running the mandatory parts of the computation, while execution is relaxed for the optional parts according to the achievable compromises among power budget and QoS. In fact, thanks to the various degrees of tolerance behind audio/video streaming, frames with a lower quality are acceptable rather than having any missing frame.

9.3.4 Approximate computing key remarks

In this section, the main achievements, potentials and limits of adopting approximate computing within video-coding applications are resumed. It has to be remarked that the approximate nature of the same video-coding algorithm plays a big role on the achievable results with approximate computing, since it means basically changing the degree of approximation that is already present on the application.

In general, the main considerations that can be derived by observing the state of the art are

- Approximate computing is very effective on the video-coding domain, allowing the optimization of throughput, consumption and memory demand.
- Approximation has to be applied selectively and requires a dynamic behavior according to current user requirements/constraints or to current input data. Table 9.3 resumes the analyzed approximated dynamic solutions.

Table 9.3 Summary of the dynamic (run time) approximate video-coding solutions

Work	Algorithm	Architecture	Functional param.	Architectural param.	Approx.* level	Features**
[29]	HEVC	CPU	Operating mode	–	d	dat
						ndt
						fgr
[46]	ME(HEVC)	ASIC/FPGA	–	Tiles size	d, h	cmp
				Tiles approx.		ndt
				Approx. bits		cgr
[53]	DCT(HEVC)	ASIC	–	Operating mode	c	cmp
				Skip current		ndt
				Rotation		fgr
[55]	HEVC	CPU	Operating mode	–	c	cmp
						ndt
						cgr
[35]	ME(HEVC)	FPGA	–	Operating mode	c	cmp
						ndt
						cgr

*d = data, h = hardware, c = computational.
**cmp = computation, dat = data; det = deterministic, ndt = non-deterministic; cgr = coarse-grained, fgr = fine-grained.

- Frequency–voltage couple, multifrequency and multi-VDD approaches seem to be good allies, both in custom heterogeneous accelerators and in commercial multicore infrastructures, to act on approximation involving precision vs. power budget.
- In most of the cases, there are still lots of places to further approximate the algorithms and the architecture under different levels of approximations and on several algorithms steps and architecture aspects.

9.4 Final remarks

The main focus of this chapter was flexibility, intended as the capability of the system to offer different trade-offs reacting to internal and external constraints and triggers. Adaptivity and reconfigurability concepts in video coding are not new, and MPEG has studied this concept in various ways. To favor long-term adaptivity, that is the capability of better supporting standards evolution, the MPEG-RVC framework has been defined. MPEG-RVC is a dataflow-based codecs specification, the main advantages of which are (i) the possibility of exploiting modularity for parallelization and reusability purposes and (ii) the availability of a wide set of design-time tools helping designers in porting specifications over various targets. RVC-based approaches are used sometimes also at run time to manage variable constraints by allowing simultaneous support of multiple codecs, profiles and trade-offs.

At the state of the art, a recent trend to support various trade-off executions on the same system, which is taking place also in the video coding application field, is related to the possibility of adopting approximate computing techniques. Different chapters of this book address this topic, here the emphasis is mainly on how those techniques can be used at run time, by acting both on functional and architectural system parameters that, by modifying on the fly processing and platform features, enable the possibility of serving variable workload, guaranteeing different requirements and constraints, and of offering different QoS levels. Despite the big promises demonstrated by the presented analysis, approximate computing is not mainstream. In this regard, Ceze and Sampson in [40] have identified different directions and open issues: investing on *common definitions* and *way of measuring* the quality, *defining proper accelerators* that can aggregate and combine different approximation methodologies, and the *lack of tools support*. The lesson learned in the context of MPEG-RVC may turn out to be useful to address them: raising the level of abstraction along with providing proper design instruments facilitated both the possibility of defining reconfigurable codec specifications and their physical implementation. The usage of co-design techniques, having proper hardware–software interfaces (masking within easily programmable modular hardware accelerators the technological details providing the trade-offs) and a solid compiler workflow, would really unlock the potentials of approximate computing.

Acknowledgment

This research was supported in part by the ECSEL project FitOptiVis (#783162).

References

[1] Bhattacharyya SS, Eker J, Janneck JW, *et al.* Overview of the MPEG Reconfigurable Video Coding Framework. Signal Processing Systems. 2011;63(2):251–263.

[2] Bystrom M, Richardson IEG, Kannangara CS, *et al.* Dynamic Replacement of Video Coding Elements. Signal Processing: Image Communication. 2010;25(4):303–313.

[3] Burleson W, Tessier R, Goeckel D, *et al.* Dynamically parameterized algorithms and architectures to exploit signal variations for improved performance and reduced power. In: IEEE International Conference on Acoustics, Speech, and Signal Processing, ICASSP 2001, Salt Palace Convention Center, Salt Lake City, UT, USA, Proceedings, May 7–11, 2001; 2001. p. 901–904.

[4] Lucarz C, Amer I, and Mattavelli M. Reconfigurable video coding: Objectives and technologies. In: Proceedings of the International Conference on Image Processing, ICIP 2009, Cairo, Egypt, November 7–10, 2009; 2009. p. 749–752.

[5] University of California at Berkeley. The Caltrop Project; 2003. Available from: https://ptolemy.berkeley.edu/projects/embedded/caltrop/index.html.

[6] Roquier G, Wipliez M, Raulet M, *et al.* Automatic software synthesis of dataflow program: An MPEG-4 simple profile decoder case study. In: Proceedings of the IEEE Workshop on Signal Processing Systems, SiPS 2008, Washington, DC, USA, October 8–10, 2008; 2008. p. 281–286.

[7] Lucarz C, Piat J, and Mattavelli M. Automatic Synthesis of Parsers and Validation of Bitstreams Within the MPEG Reconfigurable Video Coding Framework. Signal Processing Systems. 2011;63(2):215–225.

[8] Nezan J, Siret N, Wipliez M, *et al.* Multi-purpose systems: A novel dataflow-based generation and mapping strategy. In: 2012 IEEE International Symposium on Circuits and Systems, ISCAS 2012, Seoul, Korea (South), May 20–23, 2012; 2012. p. 3073–3076.

[9] Bezati E, Mattavelli M, and Janneck JW. High-level synthesis of dataflow programs for signal processing systems. In: 8th International Symposium on Image and Signal Processing and Analysis, ISPA 2013, Trieste, Italy, September 4–6, 2013; 2013. p. 750–754.

[10] Palumbo F, Sau C, and Raffo L. DSE and profiling of multi-context coarse-grained reconfigurable systems. In: 8th International Symposium on Image and Signal Processing and Analysis, ISPA 2013, Trieste, Italy, September 4–6, 2013; 2013. p. 744–749.

[11] Brunet SC, Wiszniewska M, Bezati E, *et al.* TURNUS: An open-source design space exploration framework for dynamic stream programs. In: Proceedings of the 2014 Conference on Design and Architectures for Signal and Image Processing, DASIP 2014, Madrid, Spain, October 8–10, 2014; 2014. p. 1–2.

[12] Palumbo F, Carta N, Pani D, *et al.* The Multi-Dataflow Composer Tool: Generation of On-the-Fly Reconfigurable Platforms. Journal of Real-Time Image Processing. 2014;9(1):233–249.
[13] Fanni T, Sau C, Raffo L, *et al.* Automated power gating methodology for dataflow-based reconfigurable systems. In: Proceedings of the 12th ACM International Conference on Computing Frontiers, CF'15, Ischia, Italy, May 18–21, 2015; 2015. p. 61:1–61:6.
[14] Brunet SC, Bezati E, and Mattavelli M. High level synthesis of Smith–Waterman dataflow implementations. In: 2017 IEEE International Conference on Acoustics, Speech and Signal Processing, ICASSP 2017, New Orleans, LA, USA, March 5–9, 2017; 2017. p. 1173–1177.
[15] Bezati E, Brunet SC, Mattavelli M, *et al.* Clock-Gating of Streaming Applications for Energy Efficient Implementations on FPGAs. IEEE Transactions on Computer-Aided Design of Integrated Circuits and Systems. 2017;36(4):699–703.
[16] Suriano L, Madroñal D, Rodriguez A, *et al.* A Unified hardware/software monitoring method for reconfigurable computing architectures using PAPI. In: 13th International Symposium on Reconfigurable Communication-Centric Systems-on-Chip, ReCoSoC 2018, Lille, France, July 9–11, 2018; 2018. p. 1–8.
[17] Amer I, Lucarz C, Mattavelli M, *et al.* Reconfigurable Video Coding on Multicore: An Overview of Its Main Objectives. IEEE Signal Processing Magazine, Special Issue on Signal Processing on Platforms with Multiple Cores, Part 1: Overview and Methodology. 2009;26(6):113–123.
[18] Ren R, Juárez E, Sanz C, *et al.* Energy-Aware Decoder Management: A Case Study on RVC-CAL Specification based on Just-in-Time Adaptive Decoder Engine. IEEE Transactions on Consumer Electronics. 2014;60(3): 499–507.
[19] Sau C, and Palumbo F. Automatic generation of dataflow-based reconfigurable co-processing units. In: Proceedings of the 2014 Conference on Design and Architectures for Signal and Image Processing, DASIP 2014, Madrid, Spain, October 8–10, 2014; 2014. p. 1–8.
[20] Sau C, Raffo L, Palumbo F, *et al.* Automated design flow for coarse-grained reconfigurable platforms: An RVC-CAL multi-standard decoder use-case. In: XIVth International Conference on Embedded Computer Systems: Architectures, Modeling, and Simulation, SAMOS 2014, Agios Konstantinos, Samos, Greece, July 14–17, 2014; 2014. p. 59–66.
[21] Sau C, Meloni P, Raffo L, *et al.* Automated Design Flow for Multi-Functional Dataflow-Based Platforms. Signal Processing Systems. 2016; 85(1):143–165.
[22] Gorin J, Yviquel H, Prêteux FJ, *et al.* Just-in-time adaptive decoder engine: A universal video decoder based on MPEG RVC. In: Proceedings of the 19th International Conference on Multimedia 2011, Scottsdale, AZ, USA, November 28–December 1, 2011; 2011. p. 711–714.

[23] Beaumin C, Sentieys O, Casseau E, *et al.* A coarse-grain reconfigurable hardware architecture for RVC-CAL-based design. In: Design and Architectures for Signal and Image Processing (DASIP), 2010 Conference on; 2010. p. 152–159.

[24] École polytechnique fédérale de Lausanne. TURNUS; 2003. Available from: http://www.turnus.co and http://github.com/turnus.

[25] Palumbo F, Sau C, Fanni T, *et al.* Dataflow-based design of coarse-grained reconfigurable platforms. In: 2016 IEEE International Workshop on Signal Processing Systems, SiPS 2016, Dallas, TX, USA, October 26–28, 2016; 2016. p. 127–129.

[26] Ohm JR, Sullivan GJ, Schwarz H, *et al.* Comparison of the Coding Efficiency of Video Coding Standards Including High Efficiency Video coding (HEVC). IEEE Transactions on Circuits and Systems for Video Technology. 2012;22.

[27] Sampson A. Approximate Computing: An Annotated Bibliography. Available from: http://approximate.computer/approxbib/.

[28] Agrawal A, Choi J, Gopalakrishnan K, *et al.* Approximate computing: Challenges and opportunities. In: IEEE International Conference on Rebooting Computing, ICRC 2016, San Diego, CA, USA, October 17–19, 2016; 2016. p. 1–8.

[29] Palomino D, Shafique M, Susin AA, *et al.* Thermal optimization using adaptive approximate computing for video coding. In: 2016 Design, Automation & Test in Europe Conference & Exhibition, DATE 2016, Dresden, Germany, March 14–18, 2016; 2016. p. 1207–1212.

[30] Hilaire T, Ménard D, and Sentieys O. Bit accurate roundoff noise analysis of fixed-point linear controllers. In: IEEE International Conference on Computer-Aided Control Systems, CACSD 2008, San Antonio, TX, USA, September 3–5, 2008; 2008. p. 607–612.

[31] Shafique M, Hafiz R, Rehman S, *et al.* Invited – Cross-layer approximate computing: From logic to architectures. In: Proceedings of the 53rd Annual Design Automation Conference, DAC 2016, Austin, TX, USA, June 5–9, 2016; 2016. p. 99:1–99:6.

[32] Esmaeilzadeh H, Sampson A, Ceze L, *et al.* Architecture support for disciplined approximate programming. In: Proceedings of the 17th International Conference on Architectural Support for Programming Languages and Operating Systems, ASPLOS 2012, London, UK, March 3–7, 2012; 2012. p. 301–312. Available from: https://doi.org/10.1145/2150976.2151008.

[33] Venkataramani S, Chippa VK, Chakradhar ST, *et al.* Quality programmable vector processors for approximate computing. In: The 46th Annual IEEE/ACM International Symposium on Microarchitecture, MICRO-46, Davis, CA, USA, December 7–11, 2013; 2013. p. 1–12.

[34] Nogues E, Menard D, and Pelcat M. Algorithmic-level Approximate Computing Applied to Energy Efficient HEVC Decoding. IEEE Transactions on Emerging Topics in Computing. 2016.

[35] Sau C, Palumbo F, Pelcat M, *et al.* Challenging the Best HEVC Fractional Pixel FPGA Interpolators With Reconfigurable and Multifrequency Approximate Computing. Embedded Systems Letters. 2017;9(3):65–68.

[36] Sidiroglou-Douskos S, Misailovic S, Hoffmann H, *et al.* Managing performance vs. accuracy trade-offs with loop perforation. In: SIGSOFT/FSE'11 19th ACM SIGSOFT Symposium on the Foundations of Software Engineering (FSE-19) and ESEC'11: 13th European Software Engineering Conference (ESEC-13), Szeged, Hungary, September 5–9, 2011; 2011. p. 124–134.

[37] Shafique M, Zatt B, Walter FL, *et al.* Adaptive power management of on-chip video memory for multiview video coding. In: DAC. ACM; 2012. p. 866–875.

[38] Mittal S. A Survey of Techniques for Approximate Computing. ACM Computing Surveys. 2016;48(4):62:1–62:33.

[39] Moreau T, Miguel JS, Wyse M, *et al.* A Taxonomy of General Purpose Approximate Computing Techniques. Embedded Systems Letters. 2018; 10(1):2–5.

[40] Ceze L, and Sampson A. Approximate Computing: Unlocking Efficiency with Hardware-Software Co-Design. GetMobile. 2016;20(3):12–16.

[41] Compton K, and Hauck S. Reconfigurable Computing: A Survey of Systems and Software. ACM Computing Surveys. 2002;34(2):171–210. Available from: http://doi.acm.org/10.1145/508352.508353.

[42] Finkel RA, and Bentley JL. Quad Trees a Data Structure for Retrieval on Composite Keys. Acta Informatica. 1974;4(1):1–9. Available from: https://doi.org/10.1007/BF00288933.

[43] Paim G, Rocha LMG, Santana GM, *et al.* Power-, Area-, and Compression-Efficient Eight-Point Approximate 2-D Discrete Tchebichef Transform Hardware Design Combining Truncation Pruning and Efficient Transposition Buffers. IEEE Transactions on Circuits and Systems I: Regular Papers. 2018:1–14.

[44] Oliveira PAM, Cintra RJ, Bayer FM, *et al.* Low-complexity Image and Video Coding Based on an Approximate Discrete Tchebichef Transform. CoRR. 2016. abs/1609.07630. Available from: http://arxiv.org/abs/1609.07630.

[45] Porto R, Agostini L, Zatt B, *et al.* Energy-efficient motion estimation with approximate arithmetic. In: 2017 IEEE 19th International Workshop on Multimedia Signal Processing (MMSP); 2017. p. 1–6.

[46] El-Harouni W, Rehman S, Prabakaran BS, *et al.* Embracing approximate computing for energy-efficient motion estimation in high efficiency video coding. In: Design, Automation Test in Europe Conference Exhibition (DATE), 2017; 2017. p. 1384–1389.

[47] Ataei S, and Stine JE. A 64 kB Approximate SRAM Architecture for Low-Power Video Applications. Embedded Systems Letters. 2018;10(1):10–13. Available from: https://doi.org/10.1109/LES.2017.2750140.

[48] Gonzalez R, and Woods R. Digital Image Processing. Pearson Education; 2011.

[49] Jridi M, Alfalou A, and Meher PK. Efficient approximate core transform and its reconfigurable architectures for HEVC. Journal of Real-Time Image Processing. 2018; Available from: https://doi.org/10.1007/s11554-018-0768-x.

[50] Jridi M, and Meher PK. Scalable Approximate DCT Architectures for Efficient HEVC-Compliant Video Coding. IEEE Transactions on Circuits and Systems for Video Technology. 2017;27(8):1815–1825.

[51] Renda G, Masera M, Martina M, *et al*. Approximate Arai DCT architecture for HEVC. In: 2017 New Generation of CAS (NGCAS); 2017. p. 133–136.

[52] Potluri US, Madanayake A, Cintra RJ, *et al*. Improved 8-Point Approximate DCT for Image and Video Compression Requiring Only 14 Additions. IEEE Transactions on Circuits and Systems I: Regular Papers. 2014;61(6): 1727–1740.

[53] Masera M, Martina M, and Masera G. Adaptive Approximated DCT Architectures for HEVC. IEEE Transactions on Circuits and Systems for Video Technology. 2017;27(12):2714–2725.

[54] Bossen F, Bross B, Sühring K, *et al*. HEVC Complexity and Implementation Analysis. IEEE Transactions on Circuits and Systems for Video Technology. 2012;22(12):1685–1696.

[55] Nogues E, Raffin E, Pelcat M, *et al*. A modified HEVC decoder for low power decoding. In: Proc. of the ACM CF Conf. ACM; 2015. p. 60.

[56] Sidaty N, Heulot J, Nogues E, *et al*. Reducing computational complexity in HEVC decoder for mobile energy saving. In: Subm. to EUSIPCO Conf.; 2017.

[57] Palumbo F, Sau C, Evangelista D, *et al*. Runtime energy versus quality tuning in motion compensation filters for HEVC. In: Proc. of the PDeS Conf.; 2016.

[58] Afonso V, Maich H, Agostini L, *et al*. Low cost and high throughput FME interpolation for the HEVC emerging video coding standard. In: Proc. of the IEEE LASCAS Conf.; 2013.

[59] Kalali E, Adibelli Y, and Hamzaoglu I. A reconfigurable HEVC sub-pixel interpolation hardware. In: Proc. of the IEEE ICCE Conf.; 2013.

[60] Kalali E, and Hamzaoglu I. A low energy HEVC sub-pixel interpolation hardware. In: Proc. of the IEEE ICIP Conf.; 2014. p. 1218–1222.

[61] Diniz CM, Shafique M, Bampi S, *et al*. A Reconfigurable Hardware Architecture for Fractional Pixel Interpolation in High Efficiency Video Coding. IEEE Transactions on Computer-Aided Design of Integrated Circuits and Systems. 2015;34(2):238–251.

[62] Ghani FA, Kalali E, and Hamzaoglu I. FPGA implementations of HEVC sub-pixel interpolation using high-level synthesis. In: Proc of the IEEE DTIS Conf.; 2016.

[63] Mo L, Kritikakou A, and Sentieys O. Energy-Quality-Time Optimized Task Mapping on DVFS-Enabled Multicores. IEEE Transactions on Computer-Aided Design of Integrated Circuits and Systems. 2018;37(11):2428–2439. Available from: https://doi.org/10.1109/TCAD.2018.2857300.

Chapter 10

Future video coding: new tools and algorithms

Roberto Iacoviello[1] *and Paola Sunna*[2]

10.1 Context and international bodies

In the recent years, there has been a real revolution in the world of film and television, thanks to the advent of new digital formats that has involved the entire chain of production of multimedia products. This revolution has impacted the multimedia industry, consumer electronics and communication networks, opening new opportunities for convergence.

Video quality has grown exponentially, aiming to emulate the chromatic richness, dynamics and rendering of details typical of human vision, posing major challenges with respect to bandwidth in transmission channels and media storage, making necessary to have new and more performing standards of compression of the video signal that leaves the quality unchanged.

Moreover, video content itself has experienced important changes related to the quality delivered to the users and also in the way they consume it. From one point of view, HD and beyond-HD resolutions have become increasingly popular and from the other point of view video-on-demand, mobile television services, stereo and multiview capture and display are some examples of how the video content is evolving nowadays. All these services demand efficient solutions to store huge amounts of data and to deliver the same video content at different resolutions. Although communication networks have also evolved to provide higher capacities, these new requirements concerning video content require to compress the video signal very efficiently in order to store it and stream it reliably.

The international bodies that work on compression standards are ISO/IEC MPEG (International Organization for Standardization/International Electrotechnical Commission–Moving Picture Experts Group) and ITU-T (International Telecommunication Union—Telecommunication Standardization Sector), which have created standards that have obtained universal consensus with respect to adoption by different product sectors (e.g., the compression of the human genome) and at the level of geographical coverage.

[1]Rai Centre for Research, Technological Innovation and Experimentation (CRITS), Turin, Italy
[2]European Broadcasting Union, Geneva, Switzerland

ISO/IEC MPEG, formally ISO/IEC JTC 1/SC 29/WG11, is a joint technical committee of the international organizations ISO and IEC with the aim of defining standards for the representation and compression of audio and video signals in digital format.

The ISO is a federation with government that embraces more than 130 standards bodies from as many countries worldwide. The task of the ISO is to promote the development and unification of standards to allow the smooth exchange of goods and services. It also coordinates the scientific, technological and economic environment and establishes references for a variety of sectors such as information technology, mechanics, electrical engineering through the publication of International Standards.

The IEC is an international organization that has a mandate to define standards for electricity, electronics and related technologies. The IEC was founded in 1906 and has the participation of 85 countries.

The ITU-T is the United Nations agency that brings together experts from all over the world in order to develop international standards, known as ITU-T Recommendations, which have as their object Information and Communication Technologies (ICT).

The Joint Video Exploration Team (JVET) has been recently established by MPEG and ITU/VCEG (ITU Video Coding Expert Group) to explore a future video-coding technology with a compression capability that significantly exceeds the one of the current standard (High Efficiency Video Coding, HEVC) (see Figure 10.1).

Since video-coding technology has been researched from the 1970s, various coding techniques were explored and some of them were included in compression standards. In particular, the standard for MPEG-2 [1] was published in 1994, when it represented the state of the art in video coding, and it still continues to be used today in some segments of the digital video market.

An important period in the history of video coding was 1998 and 1999, when H.263 second and third version [2], MPEG-4 visual [3] and H.26L [4] were published. At that time, the three video codecs were influenced by each other and were finally harmonized into advanced video coding (AVC)/H.264 [5]. AVC/H.264 defined various categories of algorithm related to video coding: for example, coding structure, intra-prediction and picture management. Such categories of coding tool are known as common coding tools and are still being used up to now. Since then, the video compression problem has been commonly addressed using a block-based

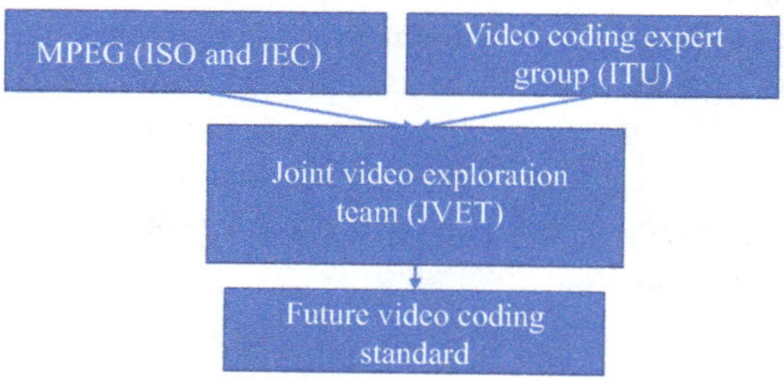

Figure 10.1 JVET structure

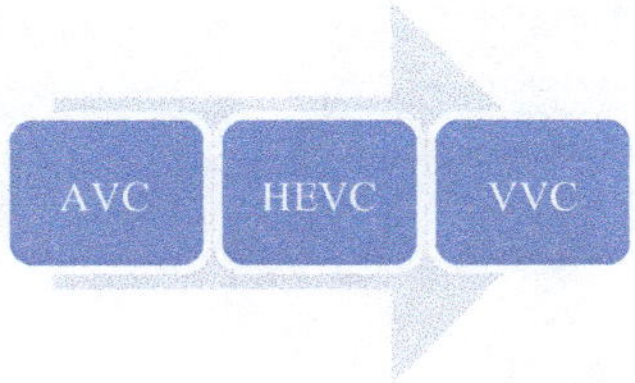

Figure 10.2 JVET standards evolution

hybrid video codec (encoder+decoder), which basically uses the following blocks: a prediction stage to take advantage of spatial and temporal redundancy in the video signal; a discrete cosine transform to represent the prediction residual in a more convenient transformed domain; a quantization process that aims to maximize the zero run-lengths and an entropy coder to efficiently represent these runs.

Even though, in 2013, high-efficiency video coding (HEVC) [6] introduced several new coding tools in addition to AVC/ H.264 (e.g., quad-tree based splitting) most algorithms were relying on increasing precision or adaptation of existing coding tools without significant variations, taking advantage of the increased computational resource due to the growth of semiconductor technologies [7]. Recently, a further video coding standard named versatile video coding (VVC) was started [8], and it is expected that a significant improvement of coding gain over HEVC can be achieved, but it is again evolving through similar design methodology (i.e.: using increasing precision and adaptation relative to HEVC), (Figure 10.2).

10.2 Video coding

This paragraph provides an overview of the fundamentals principles of the video coding, including the tool chain, from acquisition of the video sequence via coding and transmission to display. The focus is on the aspects that concern the encoding and decoding process that include the representation format of video sequences and the representation of color. Moreover, the fundamental concept and the main building blocks of hybrid video coding and how they interact are presented.

10.2.1 Video-coding chain

The building blocks of the video-coding system are summarized below (Figure 10.3):

- Source acquisition: Source of the video sequence which is the output in a digital form.
- Preprocessing: Operations on the raw uncompressed video source material, such as trimming, color format conversion, color correction or de-noising.
- Encoding: Transformation of the input video sequence into a coded bitstream with the purpose of generating a compact representation of the input video sequence which is suitable for the transmission method in the given application scenario.

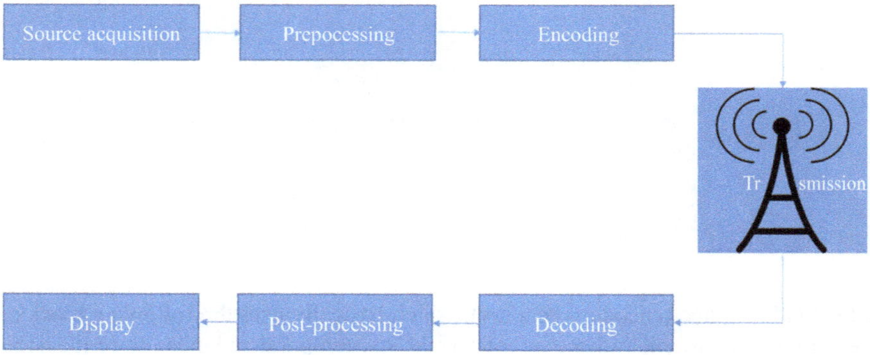

Figure 10.3 Video-coding system block scheme

- Transmission: Packaging the bitstream into an appropriate format and transmission over the channel. Transmission includes sending and delivery of the video data to the receiver side, as well as potential methods for loss protection and loss recovery.
- Decoding: Transformation of the received bitstream into a reconstructed video sequence. Since video encoding usually requires lossy compression to achieve the target transmission bitrate constraints, the decoded video constitutes an approximation of the original source video. If unrecoverable transmission losses have occurred, the decoder applies concealment strategies to recover the corrupted video sequence.
- Post-processing: Operations on the reconstructed video sequence for enhancement or for adaptation of the sequence for display. These operations can, e.g., include color correction, trimming or resampling.
- Display: Presentation of the video sequence for viewing. The video sequence needs to be transferred into the appropriate color format, and it needs to be adapted to the target resolution.

10.2.1.1 Source acquisition

The sources of the video sequence can be very different: capture of natural scenes, scanning of photographed film material or synthetically generated video material like computer-generated animation. In a simple signal model, an acquired video sequence can be described as an undistorted video signal which is additively distorted by a noise component. Additive noise components, which interfere with the content, can be: camera noise, film grain, motion blur, or artifacts from digitized analogue or physical source material. Furthermore, nonadditive artifacts in the video sequence may, for example, be related to geometrical distortions or color alignment issues of the camera lens. When natural scenes are captured, camera properties like the construction and the resolution of the sensor chip or the shutter speed play an important role for the resulting video-sequence properties and the picture quality.

10.2.1.2 Preprocessing

In the preprocessing part, the raw video signal undergoes different signal processing and enhancement operations in preparation for the next steps in the video chain. In particular, specific preprocessing may be applied to adapt the signal to the requirements of the target application.

Usually, the target video properties of the preprocessing step are: frame rate, picture resolution, color format.

Additionally, dedicated preprocessing can be applied to the source material to allow the encoder to improve compression efficiency (i.e. local lowpass filtering or denoising filtering).

10.2.1.3 Encoding

The encoder transforms the preprocessed video sequence into its bitstream-coded representation. The bitstream in contemporary standards has a packet structure in which the encoded information is hierarchically organized. The information contained in the bitstream includes the selection of the tools used for the encoding, constraints on decoder-side buffer requirements and the resulting bit rate for transmission.

Optimization of encoding parameters is a crucial aspect of encoder design: the selection of encoding tools as well as the bit allocation over picture's regions has to be considered.

An important variable is the time available for encoding: in real-time applications, the encoder can only use past information of the video it is encoding; in storage or media distribution applications, offline video processing enables better encoder optimization and parameters tuning.

In preparation for transmission, encoded bits are encapsulated into a transport format according to the application requirements.

10.2.1.4 Transmission

The transmission is classified according to the delay imposed by the application requirements. In telepresence applications, the transmission may be in real time and bidirectional. In live broadcast applications, i.e., sports, a certain delay may be tolerable, and there is usually no feedback between receiver and sender. In video on-demand applications, transmission is decoupled from the production process.

One of the aspects of the design of a transmission system includes the accessibility to the signal. For example, in multichannel applications such as television, it is important to switch between different channels; in telepresence applications, where more than two participants are connected in a video call, the organization and redirection of video signals to the correct receivers is required; in streaming applications, features like fast-forward and backward search are desired.

Another important aspect is the robustness of the signal transmission against distortion and data losses. In a packet-oriented transmission scenario, a protection mechanism against losses and potentially the recovery of lost packets are required.

10.2.1.5 Decoding

The decoder stores the received bitstream in a buffer and reconstructs the encoded data in a video sequence in the format decided by the encoder. The reconstructed signal, at the output of the decoder, corresponds exactly (if there are no losses) to the reference video available inside the loop at the encoder side. In case of transmission losses, the decoder must be able to integrate the missing data: in this case error resilience is required and the decoder must be able to apply error-concealment strategies.

In a standard video coding system, the standard itself defines only the behaviour of the decoding process and the various manufacturers build their products (i.e.: decoders) according to these specifications. The encoder design, on the other hand, is left entirely to the manufacturers but has to be compatible with the target decoder.

An important aspect of the decoder is the buffer management for the decoded pictures. The decoded pictures inside the buffer may be used as a reference in the decoding process.

10.2.1.6 Post-processing

The picture output from the decoder may be inserted into a post-processing stage where image-enhancement operations, display adaptation and error-concealment operations may be applied.

If the format of the decoded video does not correspond to the format required for display, operations similar to those of the preprocessing stage are applied (e.g. color format, picture resolution, frame rate...).

10.2.1.7 Display

The display device is the interface to the human-vision system and the final block of the diagram in Figure 10.3.

The video format, the color space of the source material and other information, called Video Usability Information, can be useful to prepare the video for proper viewing and this information can be signalled in the bitstream. Although they are not part of the normative decoding process, it is useful to convey the intended parameters for display to the decoding side.

10.3 Inside a video

A video sequence consists of a series of pictures that are presented with some constant time intervals. To obtain the impression of motion, a picture rate of at least 24 pictures per second is required. The unit of measurement of the picture rate is Hz with $1\ \text{Hz} = 1\ \text{s}^{-1}$. The acronym fps is also used. For high definition, HD, 50–60 Hz is used while for Ultra-HD, UHD, the specifications define a picture rate up to 120 Hz [9].

A picture is an array of samples with intensity values (Figure 10.4). A sample in such array is also denoted as pixel.

In the case of a monochrome image, i.e., with only one single color component, the picture consists of a single sample array. For the representation of color pictures, three color components are used (Figure 10.5) and correspondingly three intensity arrays.

In the context of video coding, an input video sequence is encoded in the bitstream format that contains all the information and data that are necessary to

Figure 10.4 Picture as an array of samples

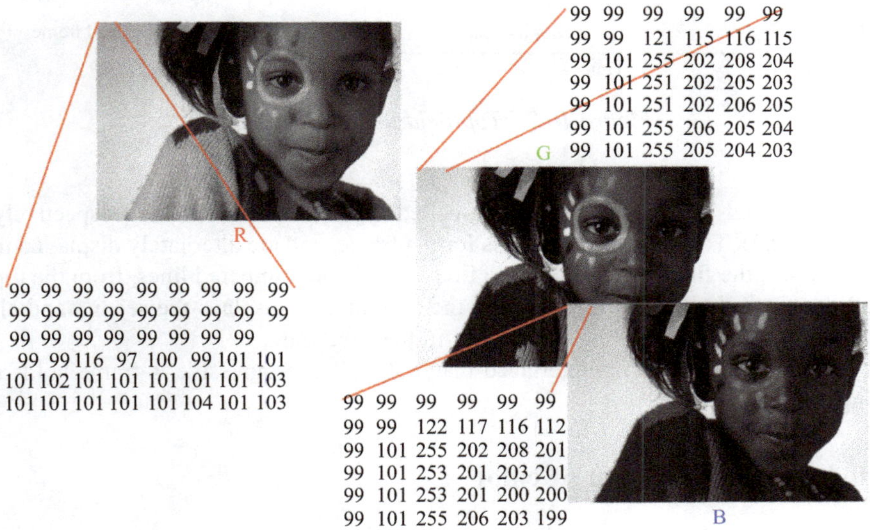

Figure 10.5 Picture split in three intensity arrays

decode and reconstruct the bitstream and display the output video sequence at the receiver side. In most video-encoding applications, encoding a video sequence in a bitstream format corresponds to a lossy compression, i.e., the reconstructed sequence at the decoder is an approximation of the original input sequence.

Historically, the video format for capturing, distributing and displaying television programs was interlaced (Figure 10.6), a format built for cathode-ray tube screens with a relatively slow-fading characteristics. In interlaced video, the even

Figure 10.6 Interlaced frame

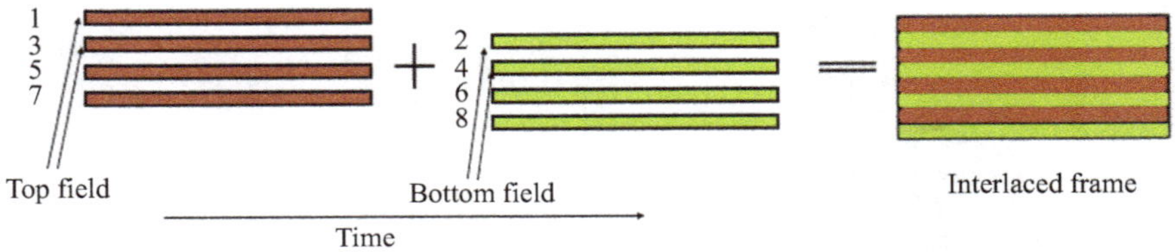

Figure 10.7 Top field and bottom field

and odd lines are grouped in the so-called top and bottom fields, respectively (Figure 10.7). Together the two fields form a frame, and are alternately displayed in two passes: the first pass displays the first and all odd numbered lines, from the top left corner to the bottom right corner, the second pass displays the second and all even numbered lines, filling in the gaps in the first scan.

Instead, inside the noninterlaced material, called progressive, a frame holds a single picture.

10.4 Representation of color

10.4.1 Fundamentals of radiometry and photometry

To describe the intrinsic phenomenon related to the perception of light, by human beings and display devices, two sciences need to be taken into account: radiometry and photometry.

The former provides a measure of the electromagnetic radiation in terms of energy, the latter refers to the sensitivity of the human eye, i.e., limited to visible radiations in the range of 380–780 nm. Given a radiometric measurement, the corresponding photometric value is obtained by weighting it according to the response of the visual system characterized by a non-uniform sensitivity at different wavelengths.

Table 10.1 includes the main radiometric and photometric measurements defined in the following:

- Radiant flux: It is the *radiant energy* emitted, reflected, transmitted or received, per unit time (Figure 10.8(a)).
- Radiant intensity: It is defined as the directed angular density of radiation from a source (Figure 10.8(b)).
- Irradiance: This is a measure of radiant flux incident an object's surface (radiant flux per unit area) (Figure 10.8(c)).
- Radiance: It is defined as the measure of the total radiant intensity per unit projected area (Figure 10.8(d)).

Table 10.1 Radiometric and photometric units and symbols

Radiometric units			Photometric units		
Quantity	**Symbol**	**Metric units**	**Quantity**	**Symbol**	**Metric units**
Radiant flux	Pe	W	Luminous flux	ϕ	lm
Radiant intensity	Ie	W/sr	Luminous intensity	I_v	cd (lm/sr)
Irradiance	Ee	W/m^2	Illuminance	E_v	lx (lm/m^2)
Radiance	Le	W/sr m^2	Luminance	L_v	cd/m^2

Figure 10.8 (a) Radiant flux, (b) radiant intensity, (c) irradiance and (d) radiance [1]

The luminance represents the quantity of light that reaches the eye from a certain direction and is the most relevant photometric unit for HDR (high dynamic range). The SI (international system of units) unit for luminance is candela per square metre (cd/m^2). A non-SI term for the same unit is the nit (nt).

10.4.2 Human-visual system

The eye is the organ responsible for the perception of visual stimuli. It focuses the light on the photoreceptive cells of the retina, through a lens called crystalline, and by adjusting the pupil diameter with respect to the intensity of the incoming light radiation.

The retina is made of about 130 million light-sensitive cells, called photoreceptors, which convert light stimuli into electrical impulses through the photo-transduction process. These signals are transmitted, via the optic nerve, to the central nervous system to be interpreted and processed in order to allow the perception of the visual image.

There are two types of photoreceptors, the cones and the rods, differentiated by their shape [10], which exhibit a different behavior with respect to the light radiation. The cones, about 5 million and located mainly in the center of the retina known as the fovea, are dedicated to diurnal or photopic vision, which occurs when the luminance level varies from 0.01 to 10^6 cd/m^2. They are also responsible for the perception of color, visual acuity and the ability to capture rapid changes. In contrast, the rods, concentrated on the sides of the retina, are more numerous (about 125 million) and intervene in nocturnal or scotopic vision, i.e., in low luminance conditions (from 10^{-6} to 10 cd/m^2), being extremely sensitive to light. The rods, by the way, do not allow to discriminate colors and provide a poor vision of details.

These photoreceptors can operate simultaneously in low-light conditions but not completely in dark, when the luminance varies from 0.01 to 10 cd/m^2 (mesopic or crepuscular vision).

The ability to perceive colors depends on the presence of three types of cones, indicated as cones S (short), M (medium) and L (long), that have different sensitivity to wavelengths. Sensitivity peaks are, respectively, close to 440, 545 and 565 nm [11].

These cones are sometimes called blue, green and red cones: a weighted combination of these three responses allows the brain to distinguish colors (trichromatic vision).

The colors perceived by the human brain are the results of the combination of three components: hue, saturation, brightness:

- Hue is the attribute linked to the perception of the light source (or the object reflecting the light) as red, yellow, blue, etc. It is determined by the dominant wavelength emitted by the source.
- Saturation refers to the intensity of color, from pure color (100%) to gray (0%), at the same brightness level.
- Brightness is the attribute that defines the amount of white or black present in the perceived color.

10.4.3 Color space CIE 1931 XYZ

In 1931, the Commission Internationale de l'Eclairage (CIE) issued the color matching system to define the tristimulus values X, Y, Z where Y represents the luminance of the perceived light signal, Z is almost equal to blue stimulation, or the S-cone response, and X is a mix (a linear combination) of cone response curves chosen to be nonnegative.

Given a three-dimensional (3D) space having X, Y and Z as axes, each color is identified by a point, whose coordinates are the XYZ tristimulus values.

The set of all these points constitutes a 3D solid, called the XYZ tristimulus space, which represents all the colors perceivable by the human eye.

The representation of color through the triplet (X, Y, Z) in 3D space is complex, and to make the interpretation easier, it is useful to separate the chromaticity information from the luminance one. The CIE has therefore introduced the chromatic coordinates x, y derived from the tristimulus values through the normalization [3]:

$$x = \frac{X}{X + Y + Z}$$
$$y = \frac{Y}{X + Y + Z}$$
$$z = \frac{Z}{X + Y + Z} = 1 - x - y$$

x and y correspond to the two coordinates of the chromaticity diagram (or CIE diagram), illustrated in Figure 10.9, obtained by projecting the XYZ tristimulus space on the equation plane $X + Y + Z = 1$.

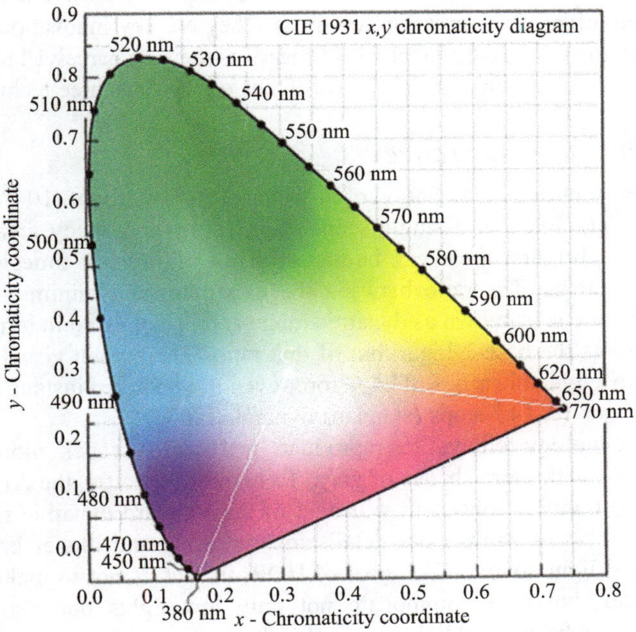

Figure 10.9 CIE 1931 xy chromaticity diagram

This diagram allows the evaluation of the hue and saturation and has the shape of a bell: on the curved perimeter, called the spectral place, all the hue values are represented with the saturation at its maximum. The saturation then decreases moving from the perimeter to the center (zero saturation), where the white point is placed (gray scale). Choosing two points on the diagram, all the possible colors formed by mixing these two chromaticity's, lie on the line that connects them.

10.5 State of the art of audio–video technologies

The technological evolution of the television and the broadcasting/streaming production chain translates into a better and more engaging viewing experience for the end user. In the recent years, device manufacturers and broadcasters have focused their efforts on raising the technical quality of pictures by increasing their resolution to the point of making it equivalent to that of the cinema (UHD) or beyond (8K).

Parallel to the increase in spatial resolution, a set of new parameters were added to improve the overall perceived quality of the "moving pictures." To this end, four additional technologies were developed:

- HDR (high dynamic range): Greater dynamic range of luminosity
- WCG (wide color gamut): Wider color resolution (colorimetric palette)
- HFR (high frame rate): Better temporal resolution (number of pictures per second)
- Next-generation audio: Immersive audio and 3D-Virtual Microphone System.

These innovations require the use of new technologies in receivers and the consequent redefinition of transmission resources. They are key innovations in modern tech TVs that aim to reproduce real-world scenes as they are perceived by the human eye and ear. The following paragraphs will focus on the first three technologies.

10.5.1 High dynamic range

Real world presents a wide variation of luminance, from starlight (10^{-6} nt) to direct sunlight (10^{9} nt). Even in a single scene (e.g., sunny and shady areas) the light intensity of the brighter areas can be thousands or millions of times greater than that of darker areas. The ratio between the maximum and minimum luminance value within a scene is known as dynamic range and is expressed in terms of *f*-stop, which represents the base-2 logarithm of that ratio. The human visual system can perceive up to a dynamic range of 24 *f*-stops, even if, in a given instant, this value is reduced to about 13–14 *f*-stops (standard dynamic range).

HDR technology allows to reproduce real-world scenes more faithfully through images with more shades of gray. To be more specific, the APL (Average Picture Level) remains consistent with that of the standard dynamic range (SDR) images while images show more details in low lights and higher brightness on reflections and light sources. The goal of HDR, in fact, is not to make the whole image brighter, but to accommodate not only low lights but also, and most importantly, very bright highlights (e.g.: the chromatic reflections, sunsets, clouds illuminated by the sun, fire, etc.) within the picture, without the risk of blowing them out. These highlights correspond to luminance levels greater than the white reference defined as a diffused white with Lambertian reflection. There are two types of highlights: specular reflections, useful for the perception of depth and three-dimensionality of objects, and the emission sources [10].

HDR aims to overcome the limits of the SDR to offer an experience closer to the human perception of luminance. In terms of *f*-stop, the definitions adopted by the MPEG Committee [11] are:

- standard dynamic range (SDR) $\leq$ 10 *f*-stops,
- enhanced dynamic range (EDR) > 10 *f*-stops and <16 *f*-stops and
- HDR $\geq$ 16 *f*-stops.

The increased dynamic range improves the micro-contrast and exploits a wider range of colors (defined by the BT.2020 color space) to allow a more natural saturation of bright colors.

On the broadcasters' side, HDR has an impact on the production and distribution workflow, from the shooting, to the transmission and reproduction on the TV sets.

10.5.2 Wide color gamut

HDR is used together with BT.2020 color space (Figure 10.10(a)), standardized in the UHD television (UHDTV) specifications [9]. This color space is larger than BT.709 for HDTV [10]. Specifically, the BT.2020 covers 63.3% of the chromaticity visible to the human eye, compared to 33.5% of the BT.709.

The combination of WCG and HDR makes the colors more saturated and brighter, resulting in a more natural and 3D effect in the image reproduction. As shown in Figure 10.10(b), the color volume of the HDR is considerably larger than SDR with the BT.709 color space. It is worth mentioning that the SDR technology can also support BT.2020 color range.

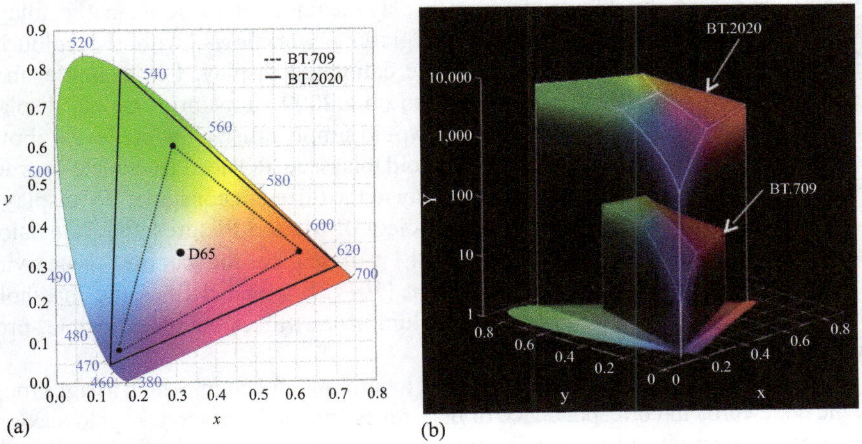

Figure 10.10 (a) BT.2020 and BT.709 in CIE 1931 [11] and (b) HDR with BT.2020 versus SDR with BT.709 [11]

10.5.3 Transfer functions: EOTF and OETF

HDR signals require new EOTF (electro-optical transfer function) and OETF (opto-electronic transfer function) to accommodate more shades of gray and the highlights. The linear signal captured by the camera is characterized by a wide dynamic range and, consequently, its encoding requires a large number of bits to avoid visible artifacts. On the other hand, the human-visual system exhibits a nonlinear behavior being more sensitive to differences between dark tones than bright tones. For this reason, the luminance information is mapped into a perceptive domain by means of the opto-electronic transfer function (OETF), typically implemented in the camera. In the reception side, the signal is retransformed into the linear domain by applying the EOTF function, implemented in the display. However, EOTF is not generally defined as the inverse of OETF: the combination of the two provides the so-called OOTF (opto-optic transfer function), used to match the difference between lighting conditions during the shooting and at the user's home [11].

ITU-R BT.2100 has standardized two transfer functions: the perceptual quantizer (PQ) and the hybrid log-gamma (HLG). It is important to note that also SDR uses transfer functions; in particular, the EOTF, called gamma, is a power function with an exponent of 2.4, defined for luminances ranging from 0.1 to 100 nt. Its formulation is reported in ITU Recommendation BT.1886. The OETF, called gamma correction, is a power function with an exponent equal to 0.45 and is specified in the BT.709 recommendation. The combination of EOTF and OETF defines the system gamma or OOTF, equal to 1.2 for SDR.

10.5.4 Perceptual quantizer

The PQ is the EOTF developed by Dolby and optimized for absolute luminance values from 0.005 to 10,000 nt (up to 24 *f*-stops of dynamic range). It is defined as a display-referred EOTF: the pictures being manipulated are transformed into the dynamic range of the TV production display being used to perform the image manipulations. This is the usual colour grading workflows, with the colourist grading the images to look correct on the calibrated display. For example, in a display-referred TV system, content graded on a 2000 cd/m^2 professional display and shown on a 1000 cd/m^2 consumer TV would simply crush all white levels above 1000 cd/m^2, if no adjustment is done. To avoid this issue, PQ needs metadata in order to adjust the signal at the professional display to the different consumer TV displays.

This function, standardized by the Society of Motion Pictures and Television Engineers (SMPTE) as SMPTE ST.2084, is based on the contrast sensitivity function (CSF) model proposed by Barten [10–12], which defines the threshold above which the difference between two luminance values becomes visible, producing visual artifacts such as banding.

The inverse of PQ, used for coding, is a logarithmic function, which assigns most of the codewords in correspondence of dark tones and half tones to take into account the greater sensitivity of the human-visual system at low values luminance and to minimize the banding. To avoid any visible artifacts, PQ requires at least 12 bits for encoding (below Barten's curve). However, numerous studies reveal that 10 bits are sufficient to avoid effects of posterization in the highlights (posterization occurs when a region of an image with a continuous gradation of tone is replaced with several regions of fewer tones, resulting in an abrupt change from one tone to another) and noise at low levels.

PQ formulation is

$$Y = 10,000\left(\frac{\max\left[\left(E'^{1/m_2} - c_1\right), 0\right]}{c_2 - c_3 E'^{1/m_2}}\right)^{1/m_1}$$

with

$$m_1 = \frac{2,610}{4,096} \times \frac{1}{4} = 0.1593017578125$$

$$m_2 = \frac{2,523}{4,096} \times 128 = 78.84375$$

$$c_1 = c_3 - c_2 + 1 = \frac{3,424}{4,096} = 0.8359375$$

$$c_2 = \frac{2,413}{4,096} \times 32 = 18.8515625$$

$$c_3 = \frac{2,392}{4,096} \times 32 = 18.6875$$

where Y represents the luminance and E' the luminance in the PQ domain, normalized between 0 and 1. The use of PQ in transmission requires the use of metadata to display correctly the signal on displays different from that used to master the content.

PQ is not backward compatible, which means that it is not possible to use the same stream to display the content on HDR and SDR displays at the same time.

10.5.5 Hybrid log-gamma

HLG is the opto-electronic transfer function developed jointly by BBC, the British public service broadcaster, and NHK, the Japanese public broadcaster, in order to provide a backward compatible signal on SDR displays that support the BT.2020 color space [4].

This function is obtained by extending the BT.1886 above 100 nt, with a logarithmic curve for the highlights according to the perception of the luminance by the human-visual system. Unlike PQ, HLG is scene related because the signal represents the light captured by the camera, a relative signal, normalized with respect to the reference white. This implies that the peak luminance of the HLG OETF coincides with the maximum luminance of the playback display and, therefore, the codeword allocation changes automatically according to the reproduction display without any need of additional metadata.

The formulation of HLG is

$$E' = \begin{cases} \sqrt{3E} & 1 \leq E \leq 1/12 \\ a \cdot \ln(E - b) + c & 1 \leq E \end{cases}$$

with

$$a = 0.17883277$$
$$b = 0.28466892$$
$$c = 0.55991073$$

where E represents the signal proportional to the scene light normalized in the range [0–1], and E' is the resulting nonlinear signal in the range [0–1].

10.5.6 Other HDR formats

Other HDR formats implemented in some TV sets available in the market are

- HDR10
  It was announced on August 2015 by the Consumer Technology Association and uses the BT.2020 color space, 10-bit color depth and the SMPTE ST.2084 (PQ) transfer function and the static metadata defined in ST.2086, encoded as supplemental enhancement information (SEI) messages within the video stream. HDR10 is supported by a wide range of companies, including monitor and TV manufacturers such as Dell, LG, Samsung, Sony and Vizio. UHD Blu-ray standard is based on HDR10.
- HDR10+
  It was announced on the April 2017, by Samsung and Amazon Video. HDR10+ uses dynamic metadata (scene or frame based) defined in SMPTE ST.2094-40 to improve the reproduction of the HDR signal regardless the reproduction display at home's user. It is backward compatible with TVs and other consumer devices incorporating HDR10 decoders.
- Dolby Vision
  Defined by Dolby Laboratories and currently used by providers like Netflix and Amazon Prime. It is based on PQ with a bit-depth of 12 bits and SMPTE ST.2094-10 dynamic metadata, in addition to static metadata. It is supported by LG and Vizio TV sets.
- SL-HDR (Single Layer)
  Proposed by Technicolor and Philips, adopted by LG, it uses dynamic metadata to adapt content to the properties of an HDR TV's display capabilities and convert HDR10 content on discs to SDR display.

10.5.7 High frame rate

The frame rate is the rate at which "moving pictures" are displayed and it is expressed as fps. In progressive videos, each picture in a video represents a frame and this implies that for a video captured and played at, e.g., 24 fps, 24 different still pictures are shown every second. That "speed" determines the perception of fluidity of movement by the human brain.

The frame rate greatly influences the style and visual experience of a video. Different frame rates produce different viewing experiences. Choosing a frame rate often involves a choice between variables such as the degree of video realism, the use of techniques such as slow motion or motion blur effects.

For example, movies at the cinema are usually displayed at 24 fps, since this frame rate is similar to the visual perception of the human eye and creates a very "cinematic" aspect. Videos with a lot of movement, such as the recording of a sports event or a video game, often have a higher frame rate, because a higher frame rate keeps the movement fluid and the details sharp.

In the following are some of the most common frame rates:

- 24 fps—This is the standard for movies at the cinema and has been determined to be the minimum speed required to capture video while maintaining realistic motion. Even if a movie is shot at a higher frame rate, it is often produced and displayed at 24 fps.
- 25 fps—This has been the standard for television since early times, and it is still widely used.
- 50+ fps—Anything over 25 fps is usually reserved for recording scenes with a lot of movement, such as video games, athletics or anything shown in slow motion. Sports are often recorded at an HFR, so they can be slowed down to show replays, while maintaining a clear and crisp video.

The reference standard is the ITU-R BT.2020 recommendation (Parameter Values for UHDTV System for Production and International Programme Exchange) which provides support for the HFR format up to 100/120 fps with HEVC encoder.

The research is still ongoing, and for the deployment of future HFR services, it will be essential a coding scalable over time with compatibility to current video services based on 50 fps, which is why the international body DVB (Digital Video Broadcasting) is working on the specification TS101154 [13] that determines how the HFR is mapped in the bitstream for compatibility with 50 Hz receivers (DVB-UHD phase 2). Shortly, the approach is to use a dual packet identifier: the bitstream will be decoded by phase 1 UHD receivers at half frame rate, while phase 2 UHD receivers at full frame rate (e.g., 100 Hz).

Research in the coming years will have to clarify the trade-off between the increase in bandwidth (due to the higher amount of fps, about 20% more considering an HD resolution) and the real benefit in the perception of the signal by the end user.

10.6 Hybrid video-coding scheme

From H261, hybrid video-encoding scheme has been the basic structure for all video-encoding standards; the term "hybrid" refers to the fact that temporal predictions between images in the video sequence and the techniques for encoding the prediction error are used together.

The diagram of the hybrid video-encoding scheme is shown in Figure 10.11. It is based on the classic differential pulse-code modulation (DPCM) loop: images are split into blocks and sent to the encoder. A prediction signal is generated and subtracted from the input signal to generate the prediction error. A transform is then applied to reduce the signal energy followed by a quantization process to remove all the irrelevant information before encoding the signal into a bit sequence. The parameters necessary to reconstruct the signal at the decoder side are also included in the bitstream.

Assuming an error-free transmission, encoders and decoders are aligned because the encoder includes the structure of prediction and reconstruction of the decoder signal (Figure 10.11). The decoded images are used for the temporal

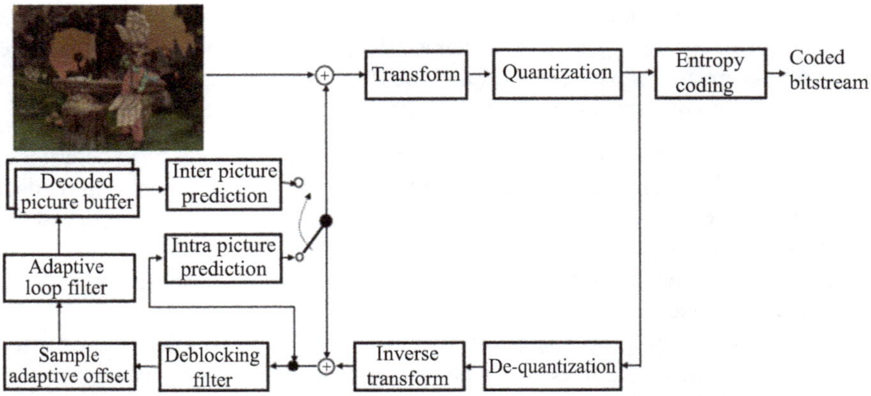

Figure 10.11 Hybrid video-coding scheme

prediction, called inter-prediction, of the current image. The encoder can also exploit neighboring blocks previously coded within the same image (intra-prediction). The first image of the video sequence is encoded as intra because there are no previous images, while for the subsequent ones, the encoder will take a decision based on an appropriate metric (e.g., rate-distortion criterion).

The image is segmented into nonoverlapping blocks and these blocks are processed one after the other; for the construction of the final prediction signal, the transformed and quantized prediction errors are reconstructed and added to the available prediction. This signal is processed by a loop filter (e.g., a deblocking filter) to be further improved.

The complete reconstructed image is then stored in the decoded image buffer and made available for temporal predictions.

For inter-prediction, there is a motion estimation phase during which the best prediction for the current block is searched in the decoded image buffer.

For the intra-prediction, the samples from neighboring blocks reconstructed in the same current image can be used for predictions.

In the diagram in Figure 10.11, there are no blocks to control the encoder. The implementation of the encoder requires a control engine to perform the decision about the best prediction modes, filter parameters and quantization parameters. This information is sent to the decoder via the bitstream.

The components of the hybrid video encoder are described in the following; the focus will be on the luma component but the behavior is similar for the chroma components.

10.6.1 Picture partitioning

Each image is partitioned into nonoverlapping blocks. In AVC, the basic structure of the block is called macroblock and formed by 16×16 samples. In HEVC, there is a generalization of the macroblock concept and the introduction of the coding tree block. It can be divided into several coding blocks with the quadtree mechanism.

The quadtree algorithm divides the block into four blocks of equal size and then the compliance of each block with respect to some homogeneity criteria is verified. If a block meets these criteria, it is not further divided. Otherwise, it is subdivided into four other blocks and the criterion is applied recursively (Figure 10.12).

10.6.2 Intra-prediction

The intra-prediction is used to remove the correlation within the same picture by calculating prediction values through extrapolation from already coded pixels for effective delta coding. The concept behind the intra-prediction is that the texture of a region is similar to the texture in the local neighborhood and can therefore be predicted from there.

For intra-prediction, samples from the reconstructed neighborhood of the prediction block under consideration are used to form the prediction signal. The direct neighbor samples are used for prediction, i.e., the samples from the line above the current block and the samples from the last column of the reconstructed block, close to the left of the current block. The values of the neighboring samples are combined to form directional or planar predictions (Figure 10.13).

The more directions are available for use, the more efficient signalling is needed in order to balance the prediction benefit over the signalling cost.

Intra-prediction is applied when no pictures for inter-prediction are available or when inter-prediction is more expensive in terms of a specific cost metric.

Figure 10.12 Picture partitioning example

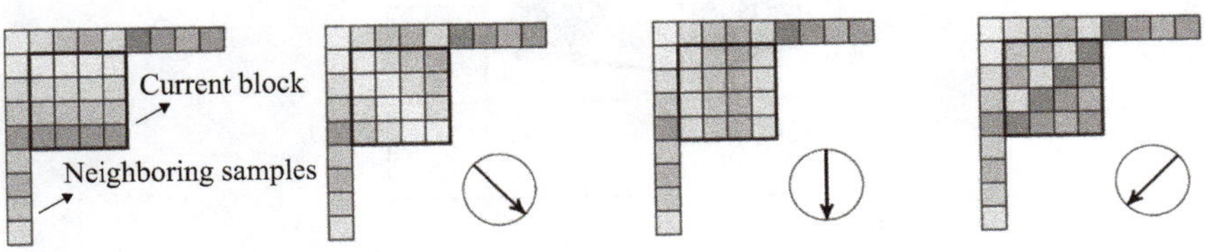

Figure 10.13 Intra-prediction example

10.6.3 Inter-prediction

The goal of this stage is to exploit the temporal redundancy between transmitted frames to compress video data. A predicted frame is built from previously encoded past or future frames and this predicted frame is subtracted from the current one. Thus, the better the prediction, the lower the energy of the residual frame.

The inter-prediction, at block level, is performed by selecting an available reference picture from the decoded picture buffer and indicating a displacement relative to the location of the current prediction block in the current picture.

Due to the fact that the frame rate in the video standard is 25 fps (e.g., in Europe), only small differences in the scene content are observed, and these differences are due to the motion. If the prediction is based on motion, the scene can be efficiently represented by a motion vector and a prediction error.

The motion vectors determine how the picture regions should be moved from the reference picture to the current picture to form the prediction.

The prediction error contains the part of the scene content that has not been described by the applied motion model.

The implicit assumption is the planar motion of rigid objects in the scene content (Figure 10.14). This very simple motion model cannot represent more

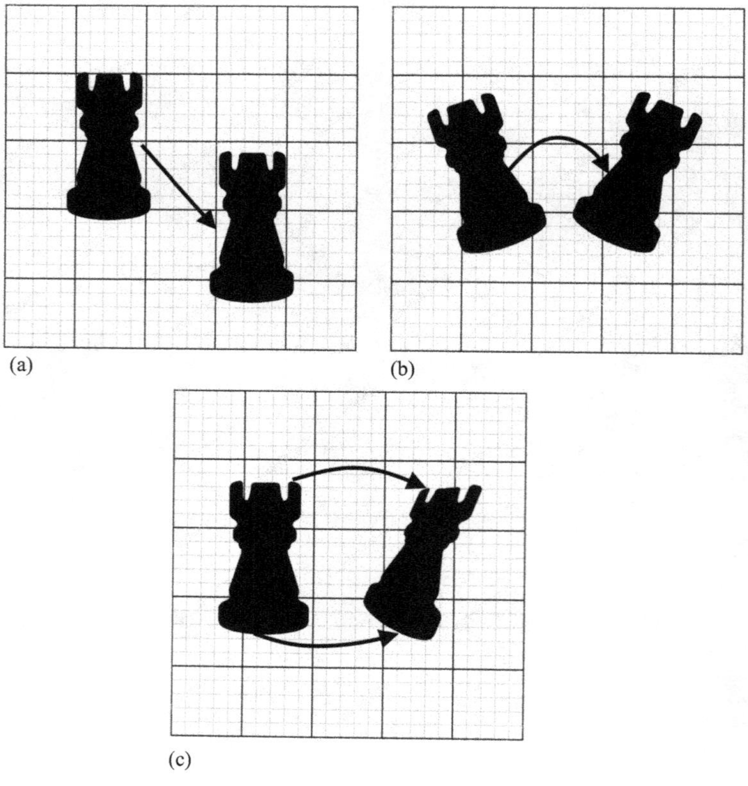

Figure 10.14 Motion models for objects between pictures: (a), (b) rigid transformation and (c) nonrigid transformation

complex motion like movement in the direction of the camera or rotation or motion of nonrigid objects.

10.7 Evolution of video services and video-compression technologies and standardization activities in MPEG

With the increasing number of multimedia applications and video consumption, video compression and content quality evaluation have become an interesting and challenging topic. Recently, a new coding tool has been developed under the JVET, called versatile video coding (VVC), with the main goal to provide a high bit rate saving compared to the HEVC standard. Results have shown that the VVC codec enables, subjectively, a quality enhancement up to 40%–50%. Objective quality improvement is about 30%. This coding gain comes at the expense of additional complexity (estimated to $\times 10$) at both encoder and decoder sides with respect to HEVC [8].

The emerging video applications, such as virtual reality (VR360) combined with HDR, offer an immersive and more natural viewing experience.

However, these new services require both higher quality and resolution (4K, 8K) to satisfy the quality of service required by the end users. Moreover, the usage of video services increases exponentially with the proliferation of mobile devices and higher usage of video applications: social networks, live video streaming, TV, network video games, E-learning, etc. To meet with these new challenges, the JVET at the 119th MPEG meeting concluded on July 2017 has concluded that evidence exists of a potential to build a new standard that could significantly outperform HEVC, either for conventional video material or for other domains such as HDR/WCG or 360° video.

In April 2018, there was the evaluation of the call for proposal and the work is still ongoing and will end in 2020.

10.7.1 Description of the VVC coding tools

The new coding scheme is called VVC and can be seen as an extension of the HEVC coding structure.

Compared to HEVC, the following new coding features are included in VVC:

1. Block structure
   - (i) Multi-type tree (MTT) partitioning with binary split or ternary split; larger coding tree units (CTUs): the software supports blocks with size 256×256, the default is 128×128.
   - (ii) The VVC has further enriched the flexibility of the partitioning through the introduction of binary trees and ternary trees which further divide the square blocks into rectangles. The residual signal is computed on a rectangular division, and a subsequent 2D rectangular transform is applied on the same area.
2. Intra-prediction
   - (i) 65 intra-prediction directions with improved intra mode coding (+2 mode: DC and planar).

(ii) 4-tap interpolation filter for intra-prediction.
(iii) Boundary filter applied to other directions in addition to horizontal and vertical ones.
(iv) Cross-component linear model (CCLM) prediction.
(v) Position-dependent intra-prediction combination (PDPC).
(vi) Reference sample adaptive filtering (RSAF).

3. Inter-prediction
   (i) Sub-PU level motion vector prediction.
   (ii) Adaptive motion vector resolution (AMVR).
   (iii) 1/16 pel motion vector storage accuracy.
   (iv) Overlapped block motion compensation (OBMC).
   (v) Local illumination compensation (LIC).
   (vi) Affine motion prediction.
   (vii) Pattern-matched motion vector derivation (PMMVD).
   (viii) Bidirectional optical flow (BIO).
   (ix) Decoder-side motion vector refinement (DMVR).
4. Transform
   (i) Large block-size transforms with high-frequency zeroing.
   (ii) Adaptive multiple transform (AMT).
   (iii) Mode-dependent non-separable secondary transforms (NSST).
   (iv) Signal-dependent transform (SDT).
5. Loop filter
   (i) Bilateral filter.
   (ii) Adaptive loop filter (ALF).
   (iii) Content-adaptive clipping.
6. Enhanced context-adaptive binary arithmetic coding (CABAC) design
   (i) Context model selection for transform coefficient levels.
   (ii) Multi-hypothesis probability estimation.
   (iii) Initialization for context models.

The main tools can be categorized in the following categories: (1) frame partitioning (MMT), (2) intra-prediction improvement (CCLM, PDPC), (3) inter-prediction improvement (PMMVD, BIO, AMVR, OBMC, Affine) and (4) transform improvement (AMT, NSST) and in-loop filters (ALF).

10.7.1.1 Frame partitioning

The video frame is split into square blocks in the HEVC standard (the highest level is denoted as CTU), using quadtrees, to adapt the local characteristics of the residual signal. This residual signal is either predicted from pixels inside current image (intra coding) or from others (inter coding).

The VVC has further enriched the flexibility of the partitioning through the introduction of binary trees and ternary trees which further divide the square blocks into rectangles. The method is called multi-type tree (MTT). The residual signal is computed on a rectangular division, and a subsequent 2D rectangular transform is applied on the same area. Moreover, VVC enables higher block sizes up to 256×256 instead of maximum block of 64×64 in HEVC.

10.7.1.2 Intra-prediction

The basic prediction mechanism of intra-prediction is kept unchanged (most of the prediction tools of VVC inherit from the HEVC standard), although the number of intra modes is increased from 35 in HEVC to 67 in VVC. A linear mode is operated to predict chroma components from luma. PDPC enables an additional prediction scheme, depending on the prediction mode. It combines filtered and non-filtered reference samples. PDPC parameter is signalled in the bitstream with a flag because it represents an additional coding choice. When PDPC is not used for a block, an RSAF can adaptively be performed to provide a smoother prediction when necessary.

10.7.1.3 Inter-prediction

PMMVD consists of an additional merge mode, enriching the HEVC merges, derived at the decoder side and therefore, it does not imply additional signalling. BIO refines the motion estimation using a pixel-wise motion compensation in the case of bi-prediction (B picture).

Temporal motion vector precision is also supplemented by more advanced mechanism and by increasing the resolution of the reference vectors.

The motion vector difference between the actual motion vector and its estimation can be adaptively coded using quarter-pel or integer-pel precision. This solution is called AMVR, where the resolution is signalled at the coding unit (CU) level.

OBMC has been also introduced in order to avoid blocking artifacts during motion compensation.

Finally, motion compensation is improved through the consideration of affine deformation which augments the traditional translational-only movements as considered in previous standards.

10.7.1.4 Transform improvements

Different types of DCT and discrete sine transform (DST) can be used for the case of intra coding, indicated by explicit signalling.

HEVC introduced an additional transform kernel to the well-known DCT-II: the DST-VII was determined as more efficient for small intra blocks (4×4). The usage of different trigonometric transforms has been extended in the VVC through the AMT tool. A set of kernels belonging to the DCT (I, II, V and VIII) and the DST (I and VII) families are combined to provide vertical and horizontal transform sets.

Intra-prediction modes exhibit oblique patterns, as a consequence, non-separable transforms were introduced to more efficiently handle such residuals. Those transforms are implemented as secondary transforms (hence the name NSST), where the lower frequency coefficients output from the primary transform at the AMT stage are processed with an additional transform stage.

10.7.1.5 Loop filter

Finally, ALF with block-based filter adaption are introduced in the VVC. The filter is operated in the prediction loop, and it consists of a set of explicitly signalled filters and performs a local classification of picture content to decide which is the best filter from the set. It has been introduced also the bilateral filter, having the property of denoising flat areas while preserving the edges.

10.7.1.6 Entropy coding

A new design of the CABAC is used, performing context model selection for dedicated transform coefficients levels, using a multi-hypothesis probability estimation and an improved initialization for context models.

10.8 The codec battle

The market today is dominated by the AVC/H.264 codec (ISO/IEC 14496-10, ITU-T H.264) released in 2004 due to its broad decoder support and well-functioning licensing scheme. HEVC/H.265 (ISO/IEC 23008-2, ITU-T H.265), published in 2013, was supposed to supersede its predecessor AVC/H.264, as it provides improved compression performance in the range of 50% bit-rate reduction for equal perceptual video quality, but it has stuttered under an unclear royalty picture; it is still difficult to evaluate the total cost of licensing standard-essential HEVC patents because, beside the three patent pools (MPEG LA, HEVC Advance, Velos Media), there are several other IP (intellectual property) holders not members of any of them.

For content owners and broadcasters looking to adopt HEVC, two patent pools offered by MPEG LA and HEVC Advance have been offering licensing terms around HEVC patents for some time. A third pool entered the market in 2016: Velos Media launched a new licensing platform that includes patents from Ericsson, Panasonic, Qualcomm Incorporated, Sharp and Sony. Technicolor licenses their HEVC patents independently.

This situation has undermined a widespread adoption of HEVC/H.265 that is now being challenged by new competing technologies promising better performances and/or royalty-free usage.

In particular, it is worth mentioning:

- AV1: It is an open royalty-free video coding developed by the Alliance for Open Media (AOMedia) [14] and released in March 2018 along with a reference software-based encoder and decoder. Governing members of AOMedia are *Amazon*, *Apple*, *ARM*, *Cisco*, *Facebook*, *Google*, *IBM*, *Intel*, *Microsoft*, *Mozilla*, *Netflix*, *Nvidia*. Many of the members already have their own work on next-generation codecs:
  - Cisco has *Thor*.
  - Mozilla has been working on *Daala*.
  - Google on VP9 and VP10.

  They joined their effort and merged their codecs to get a final codec with an improvement on HEVC. AV1 follows the typical hybrid block-based approach also used in the standards from the MPEG/ITU family.

Reasonable conclusions are [15]
- AV1 and HEVC are roughly comparable.
- AV1 encoder is significantly more complex than HEVC, so its use is likely to be restricted to non-real-time applications.
- Further work is needed to improve run time encoding and processor resource requirements to allow practical software implementations.

- XVC: It is a new codec released in September 2017. According to the information available at [16]. "The codec is a modern and flexible software-defined video compression format which includes a reference encoder and a reference decoder. The xvc codec is primarily based on compression tools that originate from MPEG standards such as AVC/H.264 and HEVC, but it also includes technology that goes beyond those standards, enabling unprecedented compression performance and it is endowed with the unique feature of "one-stop shop" license covering all rights to use the software including patent rights"; the codec has constructed to make easy the identification of the built-in tools and processing steps in order to disable them when there is a potential risk of infringing third-party patents.

The answer of MPEG community to this challenging codec battle was to exploit the standard outside the audio–video filed, investigating the compression of DNA sample reads coming out of high speed sequencing machines with the establishing of the MPEG-G (ISO/IEC 23092 genomic information representation). The aim was to provide critical solutions to industries that do not possess or cannot develop the necessary expertise and, even if they could, the result would be a technologically isolated solution. Investigating "neural network compression", motivated by the advanced stage of application of this technology especially in mobile devices. Investigating technologies like Point Cloud.

Those "non-media" compression activities show that the body of MPEG technology can indeed be exploited outside of its native "audio-video compression" field. Moreover, MPEG is considering to address many other application domains where huge amounts of data are stored, processed and transmitted, and trying to offer them data-compression technologies for effective storage and systems technologies for effective sharing, transmission and processing.

The video coding performance improves around 50% every 10 years under the cost of increased computational complexity and memory usage. Nowadays it encountered great challenges to further significantly improve the coding efficiency and to deal efficiently with novel sophisticated and intelligent media applications such as face/body recognition, object tracking, image retrieval, etc. So there exist strong requirements to explore new video coding directions and frameworks as potential candidates for future video coding schemes, especially considering the outstanding development of machine learning technologies and neural networks.

References

[1] "Generic Coding of Moving Pictures and Associated Audio Information—Part 2: Video", ISO/IEC 13818-2 MPEG-2 and ITU-T Recommendation H.262, 1994.

[2] "Video Coding for Low Bit Rate Communication", ITU-T Recommendation H.263, version 1, Nov. 1995; version 2, Jan. 1998; version 3, Nov. 2000.
[3] "Coding of Audio-Visual Objects—Part 2: Visual", ISO/IEC 14496-2 MPEG-4 visual, Apr. 1999.
[4] G. Bjøntegaard, "H.26L Test Model Long Term Number 1 (TML-1) draft 2", August 1999.
[5] "Advanced Video Coding (AVC)", ITU-T Recommendation H.264 and ISO/IEC 14496-10, May 2003.
[6] "High Efficient Video Coding (HEVC)", ITU-T Rec. H.265 and ISO/IEC 23008-2, Apr. 2013.
[7] C. Reader, Patent landscape for royalty-free video coding, Proc. SPIE 9971, Applications of Digital Image Processing XXXIX, 99711B, Sep. 2016.
[8] B. Bross, "Versatile Video Coding (Draft 1)", JVET-J1001, San Diego, US, 10–20 Apr. 2018
[9] "Parameter values for ultra-high definition television systems for production and international programme exchange". ITU-R Rec, BT.2020-0. http://www.itu.int/rec/R-REC-BT.2020/en (2012)
[10] P. D. S. P. G. W. K. M. E. Reinhard, and W. Heidrich, "High dynamic range imaging: acquisition, display, and image-based lighting". Morgan Kaufmann, 2010
[11] P. N. J. S. Ronan Boitard, and Mahsa T. Pouzarad, "Demystifying High-Dynamic- Range Technology: A new evolution in digital media", Consumer Electronics Magazine, IEEE, vol. 4, no. 4, pp. 72–86, October 2015
[12] O. B. Rolf R. Hainich, "Displays: Fundamentals and Applications". A K Peters/CRC Press, 2017
[13] ETSI TS 101 154 V2.4.1 (2018-02) https://www.etsi.org/deliver/etsi_ts/101100_101199/101154/02.04.01_60/ts_101154v020401p.pdf
[14] https://www.v-net.tv/2019/03/25/new-decade-new-codec/
[15] Jerome Vieron and Ian Trow, AV1: Implementation, Performance and Applications
[16] https://xvc.io/

Index